Manual of Indian Mangroves

Author

Dr. Kumudranjan Naskar (*born* 27-01-1949), National Fellow & Principal Scientist, ICAR was Awarded as 'National Fellow' *w.e.f.* 24-01-1995 and posted at Central Inland Fisheries Research Institute, Kolkata. Dr. Naskar earned his M.Sc. Degree (Botany) during 1972 and Ph.D. Degree in 1985 from the University of Calcutta. Since his joining at CIFRI, Dr. Naskar investigated on the Economic Botanical aspects related to fisheries potentialities on inland water bodies and estuaries and roles played by aquatic plants and mangroves on fish and fisheries.

After 1903 (D. Prain's Publication), the 1st Book of Indian Mangroves was published among Indian Scientist by Dr. Naskar and his teacher, Dr. D.N. Guha Bakshi During 1987 and the second book was published on the 'Aquatic and Semi-aquatic Plants of the Lower Ganga Delta', during 1990. Till date, Dr. Naskar has published total 7 books and along with his teacher and students have published about 120 scientific papers in National and International journals, proceeding, etc. He has published about 80 popular scientific articles both in Hindi and Bengali to popularize and highlight the importance of mangroves and aquatic plants in rural ecology and economy, especially to enhance the fisheries and agriculture production on a sustainable manner.

Since February, 1993 to April, 1996 Dr. Naskar had coAnducted as the Principal Investigator of the Ministry of Environment and Forest, sponsored project entitled 'Potentialities of the tidal mangrove forests of the Sundarbans with special reference to the estuarine fish and fisheries development' and since 1998-2000 he had worked as the Principal Investigator of a Environment Development, W.B. sponsored project entitled "Development of Mangrove Ecological Park at Jharkhali, Sundarbans'.

Dr. Naskar has also guided and presently guiding total 10 students leading to Ph.D. Degree of different Universities and 5 students for M.Sc. Dissertation. Besides these, Dr. Naskar has also developed a 'Mangrove Ecological Park' at Jharkhali and a 'Mangrove Ecological Museum' in the Laboratory of the National Fellow at CIFRI in collaboration with the Calcutta Wildlife Society, Director, CIFRI and active support of the Environment Department, W.B., Minister of Sundarbans Affairs Department, Minister of the Fisheries Department, W.B. and many other scientist, forest officials, administrators of State and Central Institutes/ Organisations etc. With their help and support, Dr. Naskar has also arranged Seminar on Sundarbans Mangroves during 8-9 November, 1996 and since 1997, the 3rd June of the last 7 years were also celebrated as the 'Sundarbans Day' to highlight the importance and impact of the mangroves on the sustainable use. Several researchers, students of university, college and school, ministers, planners, administrators, scientists, teachers have visited the mangrove ecological park and museum, since last 8 years and with the continuous effort and activities of Dr. Naskar and time to time the delivering of his lectures on different seminars, symposia, scientific and group discussion Dr. Naskar highlighted the role and importance of mangrove ecosystem and its impact on estuarine fisheries.

Manual of Indian Mangroves

Kumudranjan Naskar

2026

Daya Publishing House®

A Division of

Astral International Pvt. Ltd.

New Delhi – 110 002

First Published 2004
Reprinted, 2026
ISBN: 9788170353034 (Hardbound)

Published by : **Daya Publishing House®**
A Division of
Astral International Pvt. Ltd.
– ISO 9001:2015 Certified Company –
4736/23, Ansari Road, Darya Ganj
New Delhi-110 002
Ph. 011-43549197, 23278134
E-mail: info@astralint.com
Website: www.astralint.com

Digitally Printed at : **Replika Press Pvt. Ltd.**

This copy is for sale in Indian Sub-Continent.

Dedicated to

My most respected
Dr. Dwijendra Narayan Guha Bakshi (Dwijen Da), My teacher &
Dr. Gaur Gopal Maiti (Gaur Da), my friend
– the main guiding forces for my involvement in Mangrove Research

ACKNOWLEDGEMENT

I must acknowledge the Indian Council of Agricultural Research, New Delhi for providing the necessary support to carry out research work in the field of mangrove ecological investigation in Sundarbans; without this logistic support it was difficult for me to collect all these information presented in this monograph.

Words are too adequate to express my gratefulness to my teachers, students, friends, my colleagues and my family members, whose real guidance, untired help, time to time assistance and support given me the inspiration and impetuous to involve in the mangrove research. Indian Council of Agricultural Research, New Delhi has also supported me for the research work in the field of Mangrove Ecological Investigations by awarding me as the 'National Fellow' and the Director, Central Inland Fisheries Research Institute, Barrackpore also provided time to time the necessary support for this research work both in the field and the laboratories, for these I am especially thankful to them.

Under the above circumstances I would also like to express my heartiest gratefulness:

– to my respected teacher Dr. Dwijendra Narayan Guha Bakshi, a renowned/well known *'man of heart'* and the scientist in the field of plant science research in West Bengal and Dr. Gaur Gopal Maiti, who really inspired me in the field of Plant Science Research and their untired, unfailing help, guidance and suggestion in course of this field and laboratory investigation;

– to my students Dr. Rathindranath Mandal, Scientist, CIFA, Bhubaneswar, Orissa, Dr. Neera Sen, Research Associate, National Fellow Scheme, CIFRI, Mrs. Dolanchampa Sarkar, SRF Sri Asok Ghosh, Junior Research Fellow, National Fellow Scheme, Mrs. Manjistha Das Gupta, Senior Research Fellow, National Fellow Scheme, Sri Amal Sarkar, Lecturer Botany, West Bengal College Service, Mr. Subrat Mukherjee, IFS & CF & Ex Deputy Field Director, Sundarbans Tiger Reserve and Sri P. Sanyal, Additional Chief Conservator of Forest, Forest Department, West Bengal; Sri Asijit Ganguli, EMRC, Kolkata, Sri Ratan Das, the most important and real Assistance of my research work in the field and in the laboratory; without their untired help and assistance my research work in Sundarbans may not continue for long;

– to my other Scientist friends & colleagues, *viz.*, Dr. Ajoy Kumar Ghosh, Retd. Principal Scientist & Ex Director of CIFRI, Barrackpore.

– to the other Staff Members of the National Fellow Scheme and the Staff Members of the CIFRI Station, Salt Lake, Kolkata, *viz.*, Dr. P. K. Pandit, Ex- Principal Scientist, Sri H.C. Karmakar, Principal Scientist and the Officer-in-Charge, CIFRI Station, Kolkata, Sri Amoy Kumar Barui, Mrs. Suvia Bhattacharyya, Sri Debasis Saha, Sri Joydeb Patra, Sri Abani Bhushan Patra, Sri B. B. Das, Sri T. J. Chatterjee, Mrs. Sukla Das, Mrs. Giti Majumdar, Sri P. Sen, Sri Tapan Mukherjee, Sri Samir Kumar Das, Sri Girindra Nath Mridha, Boat-Man, Sri Nihar Mandal, Smt. Khuku Mandal, Sri Swapan Mandal, Sri Sasanka Baidya, Research Scholar, Sri Achinta Bhunia, Research Scholar, and Sri Sonatan Mandal, Boat Man, Sundarbans and others for their co-operation, sincere help and assistance in the laboratory and in the field, as well.

– to my wife, Smt. Sipra Naskar, my boudi, Smt Biva Naskar and my elder sister Smt. Usha Rani Sarkar and Dinataran Sarkar, my brother - in - law (Jamai Babu), whose help and support also able me to engage myself in the research work of mangrove ecological studies at Sundarbans, Andaman, Goa, Bombay and other parts of the country.

– to Mr. Anil Mittal, Daya Publishing House, Delhi-110035, for his keen personal interest for publication of this Manual.

– to all my teachers, friends and officials of the different departments, colleges, universities and others whose help, their suggestion and advice also gave me impetuous for writing this book.

– last but not least, I would be failing in my duties, if I do not mention the yeomen services rendered by my both sons Sri Soumya Naskar, B.E. & Sri Saikat Naskar, M.Sc. and Dr. Neera Sen, Ph.D. my student for helping me in preparation of this manuscript; Soumya help me a lot by designing the complete manuscript in computer and scanning the photographs and designing the cover page, as well.

Kumudranjan Naskar

PREFACE

Mangrove plants and the mangrove ecosystem or the mangal in totality are the most important in the perspectives of its economic potentialities, as mangroves provide ideal habitats for economic species of fish, shrimps, prawns, crabs and good number of threatened, rare and endemic but most important species of aquatic and terrestrial fauna. All these coastal or estuarine vegetation along with its dense growth and wide spreading habitats play physical support and protective roles in these fragile coastal or estuarine zones in the tropical and sub-tropical countries in the Globe in general and in the Indian mangals in particular. Bio-mass from the mangrove plants continuously supply nutrients to coastal, estuarine and offshore aquatic environment, resulted healthy environment and also raised or maintained productivity of the important water phases. Besides these, mangrove timber and mangrove wood have immense value in carpentary work and as fuel wood to the rural people of Sundarbans.

For raising the productivity and playing protective roles, the worldwide threatened of these most fragile mangrove ecosystem have drawn attention to the scientists in particular and planners, policy makers, nature lovers and conservationists in general. During the recent times, mangrove research, its status identification, investigation of its important roles are most fascinating topic not in India alone, but throughout the Globe.

All these needs for identification of these mangrove plants, mangrove dwelling fauna and the problems and prospects of these mangrove areas, as well. It is also felt most urgent and necessary to investigate worldwide about the mangroves, its importance in these fragile ecosystems in totality and their uniqueness.

Considering all these impacts, ecological importances and other important roles of mangroves and mangrove ecosystem it is most urgent to focus in its true perspectives and to prepare an important tool for identification of these mangrove species and its surrounding physical and biological environment, which are most urgent and rather important.

The mangroves and mangrove associated plant species are dealt here critically and the mangrove habitat or mangrove inhabiting fauna are also mentioned and listed to focus their abundances, importance, existence and holistic relationship, in brief.

With this aims and objectives, all these available mangroves and mangrove associated flora of the Sundarbans, Andaman Islands, Bitarkanika, Goa, Mumbai and other parts of India are presented alongwith their habitat photographs and close view of flora and fauna; as such, this monograph must be the useful tool for identification of mangrove plants in the field itself.

If this monograph can help the readers, researchers, students and teachers for the mangroves identification and identification of the problems and necessary conservation needs that will be the best award and satisfaction to the author.

Suggestions towards improvement of this publication will be thankfully acknowledge.

Kumudranjan Naskar

CONTENTS

Chapter	Subjects	Pages
1	**Introduction**	**1**
2	**Indian Mangals**	**17**
	Other dominant Indian mangrove areas and mangrove species	18
	i) Mangroves of the Andaman & Nicobar Islands of the Bay of Bengal	19
	ii) Mangroves of Bhitarkanika of the Mahanadi delta and other Mangals of the Orissa coast	19
	iii) Mangroves of Coringa, Kankinada Bay, the Krishna and the Godavari estuaries of the Andhra Pradesh coast	20
	iv) Mangroves of Muthupet, Pichavaram and Chhatram area of the Cauveri delta in the Tamil Nadu coast	20
	v) Mangroves of the Cochin estuary of the Kerala coast	21
	vi) Mangroves of Coondapur and Malpe Bay of the Karnataka coast	21
	vii) Mangroves of the Mandovi, the Zuary and other estuaries of the Goa	21
	viii) Mangroves of Ratnagiri and other Estuaries of the Maharashtra State	22
	ix) Mangroves of the Gulfs of Cambay and Kachchha of the Gujarat State	22
	x) Coral reef Mangroves of Lakshadweep Atols & Minicoy Islands of the Atlantic Ocean	22
3	**Indian Sundarbans**	**29**
	Roles Played by these Mangroves in Indian Sundarbans	37
4	**Mangrove Zones and Succession**	**38**
	The Terms and Definitions of Mangroves and the Mangal	40
	Uniqueness of these Sundarbans Mangals	41
	The Unique Sundarbans Mangals for its highest bio-diversity	42
5	**Floral Succession in the Mangal**	**43**
6	**Sundarbans - The Marshy Tiger-land of the World**	**47**
	Common and higher groups of fauna of the Sundarbans	51
	Mangal support coastal and estuarine fisheries	65
	Tropic level on the Sundarbans mangrove ecosystem	66
	Tidal amplitude/fluctuations of the Sundarbans mangal	67
7	**Flora of the Indian Mangals**	**70**

1. Rhizophoraceae- 73, 2. Avicenniaceae - 88, 3. Sonneratiaceae- 93, 4. Combretaceae -98,
5. Arecaceae - 100, 6. Meliaceae - 104, 7. Euphorbiaceae-109, 8. Myrsinaceae- 111,
9. Aegialitidaceae-113, 10. Sterculiaceae- 115, 11. Rubiaceae -118, 12. Acanthaceae-121,
13. Apocynaceae- 125, 14.Asclepiadaceae-126, 15.Bignoniaceae-131, 16. Malvaceae- 132,
17.Tiliaceae- 136, 18. Rutaceae - 137, 19.Poaceae - 138, 20. Verbenaceae -141,
21. Tamaricaceae- 143, 22. Solanaceae - 145, 23. Sapotaceae- 146, 24. Sapindaceae-147,
25. Amaryllidaceae-148, 26.Araceae - 150, 27. Aizoaceae- 151, 28. Barringtoniaceae-152
29. Boraginaceae-153, 30.Fabaceae - 154, 31.Caesalpiniaceae-159, 32.Chenopodiaceae-161,
33. Clusiaceae- 163, 34.Convolvulaceae-164, 35.Cyperaceae- 165, 36.Loranthaceae - 166,
37. Opuntiaceae- 168, 38. Pandanaceae - 169, 39. Pteridaceae - 170, 40. Ruppiaceae - 171

8 **Possibilities and Conservation Needs** 172
Sundarbans Mangals – the Largest Single Mangrove Chunk of the Globe 172
Sundarbans Mangals acts as the Sink for Metropolitan Pollutants 172
Protection vs. Loss of Bio-diversity 173
Major Problem in Indian Sundarbans 174
Large Scale Deforestation of Forest Land 175
Large Scale Netting Out of Shrimp Seeds vs. Damage of the Other Fish Juveniles 175
Siltation on River-bed Cause Frequent Flood 175
Transport Facilities is not much Developed 175
Uncontrol Population Growth 175
Agriculture not much Developed 176
Enormous Pressure on the Natural Forest and Inevitable Destruction 176
Effect of Developmental Activities without Environmental Impact Assessment 176
Renewal of Natural Wealth and Prospect 176
Present Scenario of the Sundarbans Mangals and Few Comments 177
9 **Bibliography of the Mangroves and Mangrove Ecosystem of India** 179
10 **Index to the Mangroves and Mangrove Ecosystem** 211

INDEX TO THE TABLES

Table No.	Subject	Page
1	Showing the Different World Mangrove Zones (Both Old World and New World) - Based on Different Publications	10
2	Distribution of Mangroves and Mangrove Associated Plants in India and other Important World Mangrove Zones	11
3	Showing the Indian Mangroves (Modified after Govt. of India Publication, 1987)	17
4	Showing declining trend of Mangrove Areas in Indian Territory (Modified after Jagtap, *et al.*,1993)	17
5	Geographical Areas and Nature of the Landscape in Indian Sundarbans	30
6	Total Areas, Cultivated Areas and Population in Different Mangrove Reclaimed Areas of 24 Parganas Districts (S & N)	31
7	Population Growth in Indian Sundarbans Mangrove Reclaimed Zone, India (Modified after Das, 1996 & Kanjilal, 2000)	32
8	Different Forest Blocks under Sundarbans Tiger Reserve (STR), West Bengal	34
9	Areas Under Forest Blocks in the Sundarbans Reserve Forest, under 24 Parganas South(Outside STR)	34
10	The National and the International Programmes and Research Projects in Indian Sundarbans	35
11	Showing the Different Regions and the Project Areas in the Indian Sundarbans	35
12	Natural Calamities, *viz.*, Cyclones, Floods, Earth Quakes and Famines reported from Sundarbans	36
13	Different Groups & Numbers of Flora and Fauna Reported from the Indian Sundarbans Mangal	42
14	Common Mangroves and Mangrove Associated Plant Species of the Indian Sundarbans	44
15	Rare, Threatened and Endangered Mangroves, Mangrove Associates in the Indian Sundarbans Mangals	46
16	Different Animal and Faunal Habitats in the Indian Sundarbans	46
17	Tiger Population in the Indian Sundarbans (Informations Taken from STR Reports)	49
18	Tiger Straying Incidents in Indian Sundarbans (Source : STR Reports)	49
19	Common and Higher Group of Fauna of the Indian Sundarbans	55
	(**A.** Common Mammals, **B.** Common Reptiles, **C.** Common Amphibians, **D.** Avi-Fauna /Birds, **E.** Fish species, **F.** Echinoderms, **G.** Shrimp, Prawns & Crabs & **H.** Mollusca & Others)	
20	The Threatened, Endangered and Extinct Fauna of the Sundarbans Mangal	64
21	Tidal Amplitudes in the Indian Sundarbans	67
22	Key Characters of the Dominant Indian Mangroves and Mangrove Associated Families	71
23	Differences among these four Genera of the Family - Mangrove Rhizophoraceae	73
24	Differences between the two species of *Rhizophora* L. in Sundarbans	74
25	Difference between the four species of the Genus- *Bruguiera* Lamk. of the Indian Sundarbans	74
26	Difference between the two species of *Ceriops* Arnold	74
27	Different characters of viviparous germinated hypocotyles of Mangrove - Rhizophoraceae	75

Table No.	Subject	Page
28	Key to the Genera under the family - Mangrove Rhizophoraceae (Tribe - Rhizophorae)	76
29	Key to the species under the Genus - *Rhizophora* L.(Species distributed in Indian Sundarbans)	76
30	Key to the species under the Genus - *Bruguiera* Lamk.	77
31	Key to the species under the Genus - *Ceriops* Arnold	77
32	Comparison between the four Indian species of *Rhizophora* L.	77
33	Other New World Mangrove species under the Family - Mangrove Rhizophoraceae	87
34	The distinct characteristic features between three species of *Avicennia* L. Sundarbans	88
35	Key to the species under the Genus- *Avicennia* L.	89
36	Key characters of the species under *Sonneratia* Linn.f. (Based on P.B. Tomlinson, 1986)	93
37	The distinct characteristic features between the members of the Genus - *Sonneratia* Linn.f.	94
38	Key characters between two species of *Lumnitzera* Willd. from the Indian Mangals	98
39	Difference between the two mangrove habitat palm species of the Indian Mangals	100
40	Key characters among the two mangrove genera under the family - Meliaceae	104
41	Key characters among the three species under *Xylocarpus* Koen.	104
42	Differences between the two species of *Xylocarpus* Koen.	105
43	Key characters between the two common Indian species of *Heritiera* Aiton.	115
44	Key characters of Indian species under the Genus - *Acanthus* L.	121
45	The distinct characteristic features of three species of *Acanthus* L.	121
46	Key to the genera under the Family - Asclepiadaceae in Sundarbans Mangal	126
47	Key to the Species under the Genus- *Sarcolobus* R. Brown	127
48	Mangrove genera under the Family - Malvaceae	132
49	Key to the Mangrove Associated or Back Mangrove species under the Genus - *Hibiscus* L.	132
50	Key to the Mangrove Associated or Back Mangrove Species under the Genus - *Thespesia* Soland	132
51	Key to the Species under the family - Poaceae, collected from the mangrove habitats	138
52	Key to the Genera under the family Verbenaceae of the inter-tidal habitats in Sundarbans	141
53	Key to the species under the Genus - *Clerodendron* L. from the inter-tidal habitat in Indian Sundarbans	141
54	Key to the species under the Genus - *Tamarix* L.	143
55	Key to the species under the Genus - *Solanum* L.	145
56	Key to the Species under the Genus - *Barring tonia* L.	152
57	Key to the different genera under the family - Fabaceae	154
58	Key to the species under the Genus - *Derris* Lour.	155
59	Key to the species under the Genus - *Caesalpinia* L.	159
60	Key to the species under the Genus - *Suaeda* Forsk. ex Scop.	161
61	Key to some common genera under the family - Cyperaceae of the Sundarbans Mangals	165
62	Key to genera under the Family - Loranthaceae	166

INDEX TO PHOTO PLATES

Photo Plate Number	Subjects of the Photo Plates	Page
1.	Mangroves of Sundarbans, showing the river bank vegetation of *Rhizophora mucronata* Lamk.	1
2.	River-bank Mangroves in Indian Sundarbans, showing *Nypa fruticans* Wurmb and other mangroves.	2
3.	Aerial Root Systems of Different Mangrove species, *Avicennia* spp. in Indian Sundarbans	3
4.	Aerial Root Systems of Different Mangrove Species, in the Indian Sundarbans	4
5.	Canopy Structures and Root Systems of Different True Mangrove Species	5
6.	Aerial Root Systems of the Mangroves, *Avicennia* spp. in Indian Sundarbans	6
7.	Viviparous germinated hypocotyles of the Mangrove-Rhizophoraceae in Indian Sundarbans	7
8.	Crypto-viviparous seed Germination of *Avicennia* spp. in Indian Sundarbans	8
9.	Algal habitats and different species of brackish-water Algae of Sundarbans	9
10.	Natural Habitats of the Mangroves in the Inter-tidal zones in Sundarbans and Andaman Island	15
11.	Exploitation of Mangrove wood and barraging of flowing river for Human settlement at Sundarbans	16
12.	Mangroves of Port Blair. A. *Rhizophora* sp.,B & C.*Acanthus ebracteatus, Avicennia* sp. & *Nypa fruticans*	23
13.	Mangroves of The Andaman Islands, Port Blair	24
14.	Mangrove Areas at Mumbai with the Distinguished Mangrove Researchers (During 1990)	25
15.	Mangroves of the Subarnarekha mouth & Bitarkanika, Mahanadi Delta of Orissa	26
16.	Mangroves of the Mandovi & the Zuari Estuaries of Goa (During 1994)	27
17.	Exploitation of Mangrove Wood for Timber and Fuel Wood	28
18.	Extensive Stilt roots of *Rhizophora mucronata* Lamk. & other mangroves in Sundarbans, India	34
19.	Mangrove succession on newly silted up land- *Porteresia coarctata* followed by *Avicennia* sp.	35
20.	River bank *Rhizophora* sp. plant, showing tide water carrying silt mark on the leaves	36
21.	A. Succession of *P. coarctata* & *S. apetala*. B. Secondary succession, *Bruguiera. parviflora*	38
22.	Salt-crust on the surface soil of mangrove cleared inter-tidal habitat in Indian Sundarbans	40
23.	Fishing at Sundarbans by Cast Nets	41
24.	A. Dense & tall mangrove Forest, B. Brackish water mixed *Heritiera* forest, C. Deforested mangrove swamp, D. Palm swamp, dominated by *Phoenix* sp. E.Brackish water *Nypa* sp. & *Heritiera* sp. forest, F. *Heritiera* fruits and G. Tall *Heritiera fomes* tree with buttress roots.	43
25.	Tiger released from cage at Indian Sundarbans (Source - STR)	49
26.	Swimming Tiger in Sundarbans River (Source : Sundarbans Tiger Reserve)	49
27.	Tiger- human conflict is the age old incidents in the Sundarbans area	50
28.	Tiger in the Sundarbans Tiger Reserve. A. Tiger is being released from the cage by the forest staff. B. Released Tiger run away out of fear. C.Tiger crossing the vast river. D. Tiger after release from the cage	52
29.	Wild Animals of Sundarbans. A. Spotted Deer, *Cervis axis*.B.Rhesus monkey, *Macaca mulatta*. C. Jungle Fowl, *Gallus gallus*. D. Wild Boar, *Sus scrofa*. E. Olive Ridley, *Lepidochelys olivacea*. F. Estuarine Crocodile- *Crocodilus porosus*	53

Photo Plate Number	Subjects of the Photo Plates	Page
30.	Few Other Fauna of the Indian Sundarbans. A. Sea gull, B. Eggs of Marine Turtle. C. Snake Bird. D. Bones/ jaw of *Rhinoceros.* E. Snake. F. Local Bird. G.Water monitor, Varanus salvator	54
31.	Aquatic Food Chain – Detritus Food Chain of Estuarine Fish and Shrimp Species in the Sundarbans Mangal	65
32.	Mangrove Litter Collection by litter traps for bio-mass estimation	66
33.	A. Dense River Bank Mangrove Forest . B. Tidal inundated *Rhizophora* sp. forest, Sundarbans	67
34.	Fishermen along with their poor fish catch	68
35.	Large scale exploitation of shrimp seeds and damage of other fish, prawn and aquatic fauna (juveniles) is alarming at Sundarbans	69
36.	A. Plantation of *Avicennia* spp. at river flat. B. Natural succession of *Porteresia* sp. followed by *Avicennia* spp. C. *Rhizophora mucronata* Lamk. in its natural habitats at Sundarbans Mangal.	70
37.	*Rhizophora mucronata* Lamk. A. Full grown *R. mucronata* Lamk. tree in the inter-tidal river flat land at Sundarbans, India. B. Stilt roots and viviparous germinated hypocotyles hang from the mother plant. C. Full grown hypocotyles in hand, Mayabandar, Andaman. D. Twigs with two full grown hypocotyles of *R. mucronata* Lamk. at Sundarbans. E. Flowering Twig of *R. mucronata* Lamk, Sundarbans.	78
38.	*Rhizophora apiculata* Blume. A. Full grown tree of *R. apiculata* Blume at Sundarbans. B. View of branches with flowers and viviparaous germinated hypocotyles. C. Close view of branch with flowers. D. Branch with the initiation of viviparous germination hypocotyles. E. Branch with full grown hypocotyles in hand.	79
39.	*Bruguiera gymnorhiza* (L.) Lamk. A. Full grown tree of *Bruguiera gymnorhiza* (L.) Lamk at Sundarbans. B. *B. gymnorhiza* (L.) Lamk. within tidal water. C. View of broom like fused stilt roots. D. Close view of branch with flowers. F. Branch with viviparaous germinated hypocotyles. G. Newly germinated seedling Plant.	80
40.	*Bruguiera sexangula* (Lour.) Poir. A. Full grown tree of *Bruguiera sexangula* (Lour.) Poir. at Sundarbans. B. Close view of branch with flowers. C. Close view of branches with viviparous germinated fruits. D. Close view of branch with flowers. F Branch with viviparaous *germinated hypocotyles.* G. Close view of branch with persistent twisted calyx tube.	81
41.	*Bruguiera cylindrica* (L.) Blume. A. Close view of branches with flowers of *Bruguiera cylindrica* (L.) Blume at Sundarbans. B. Close view of branch with viviparous germinated hypocotyles. C. Close view of branches with viviparous germinated hypocotyles. D. Close view of branches with viviparous germinated hypocotyles.E.Seedlings of *Bruguiera cylindrica* (L.) Blume	82
42.	*Bruguiera parviflora* W. & A. ex Griff. A. Long view of *Bruguiera parviflora* W. & A. ex Griff. at Sundarbans. B. Close view of branch with flowers. C. Close view of branches with viviparous germinated fruits. D. Close view of seedling plant	83
43.	*Ceriops decandra* (Griff.) Ding Hou. A. Full grown plant at Sundarbans. B. Close view of branch with flowers. C. Close view of branches with viviparous germinated fruits. D. Viviparous germinated hypocotyles. E.Viviparous germinated hypocotyles hang in the branches. F. Viviparous germinated hypocotyles.	84
44.	*Ceriops tagal* (Perr.) Robinson. A. Full grown small tree at Sundarbans. B. Close view of branch with flowers. C.Long view of tree on the river bank. D. Viviparous germinated hypocotyles.	85
45.	*Kandelia candel* (l.) Druce. A. Full grown small tree at Sundarbans. B. Close view of branch with flowers. C. Long view of flowers. D. Viviparous germinated hypocotyles in hand. E. Close view of viviparous germinated hypocotyles in hand. F. Viviparous germinated hypocotyles in hand	86
46.	*A.& B. Rhizophora mucronata* Lamk. with dominant stilt roots in Port Blair, Andaman(Bay Islands)	87

Photo Plate Number	Subjects of the Photo Plates	Page
47.	(A & B). *Avicennia officinalis* L. - with pneumatophores and stilt roots, in Sundarbans Mangal	88
48.	*Avicennia officinalis* L. A. Full grown tree on the river side - with stilt roots. B. Stilt roots developed on the trunk.C. Flowring branch of *A. officinalis* L. D. Close view of the flowering branch. D. Mature fruits of *A. officinalis* L. E. Crypto-viviparous germinated seeds of *Avicennia officinalis* L.	90
49.	*Avicennia alba* Blume. A. Full grown tree - with stilt roots and pneumatophores. B. Stilt roots developed on the trunk base.C. Branch with fruits & Close view of fruits.D. Trunk base with pneumatophores. E & F. Mature fruits with close view	91
50.	*Avicennia marina* (Forsk.) Vierh. A. Plant in the natural habitat. B.Twig with Flower buds. C.Pneumatophores. D. Branch with fruits. E. Germinated seeds with tender roots. F. Seedling plants in the earthen pots in the laboratory for growth study	92
51.	A.& C. *Sonneratia alba* J.Smith & B. S. *caseolaris* (L.) Engler,collected from Goa, Mandovi estuary	94
52.	*Sonneratia apetala* Buch.– Ham. A. Tall full grown tree along with *Nypa fruticans* and *Avicennia* sp. on the river flat silted up soil. B. Tall full- grown tree on the river side silty soil. C. Flowering branch, close view. D. Branch with round fruits, E. Branch with fruits	95
53.	*Sonneratia griffithii* Kurz. A. Tall full grown tree along the river flat silted up soil. B. Close view of branch with fruits. C. Flower buds in branch, close view. D. Branch with flower buds. E. Seeds of *S. griffithii.* F. Close view of fruits with persistent calyx teeth. G. Branch with mature flat fruit.	96
54.	*Sonneratia caseolaris* (L.) Engler. A. Tree in natural habitat and close view of fruit. B. Full grown tree on the river flat and with pneumatophores. C. Close view of branch with fruits. D. Flower and flower buds in branch, close view. E. Branch with flower buds and fruits. F. Close view of fruits with persistent calyx teeth. G. Seedling plant	97
55.	*Lumnitzera racemosa* Wild. Branch with close view of the reddish fruits. Mohanpur, Maya Bandar-Mayabandar, Andaman	98
56.	*Lumnitzera racemosa* Willd. A. Close view of the plat on the river bank at Sundarbans & B. Inset Photograph of Herbarium specimen for *Lumnitzera racemosa* Willd.	99
57.	*Nypa fruticans* (Thunb.) Wurmb. A. Natural habitat in Sundarbans & B. Close View in Sundarbans	100
58.	*Nypa fruticans* (Thunb.) Wurmb. A. Natural habitat in the river slope land at Sundarbans. B. Natural habitat with dicotomous branching of stoloniferous rhizomataous stem. C. Planted in fresh water zone above the tidal limit. D.*Nypa fruticans* (Thunb.) Wurmb. in above tidal zoner at Sundarbans - with inflorescence stalk	101
59.	*Nypa fruticans* (Thunb.) Wurmb. A. Natural habitat of *N. fruticans* at Port Blair B. Seedling plants in the river falt land at Sundarbans.C.*Nypa fruticans* & *Cocos nucifera* grow side by side at Sipi Ghat, Port Blair((2001) D.Fruits at mature female inflorescence. E. Single fruit. F. Two halves of fruit, showing the cotyledon	102
60.	*Phoenix paludosa* Roxb. A. Natural habitat of *P. paludosa* at Sundarbans. B. Natural habitat at Sundarbans showing the pneumatothods and C. Natural habitat at Sundarbans showing the fruit stalk at spadix.	103
61.	*Xylocarpus* spp. at Sundarbans. A.Full grown tree of X. *granatum* B. Full grown tree of X. *mekongensis.* C. Hanging fruit of X. *granatum.* D. Two halves of the fruit of X.*granatum* with seeds. E. Single fruit of X. *mekongensis* with leaves	105
62.	*Xylocarpus granatum* Koenig. A.Tall tree in the natural mangrove forest, Sundarbans. B. Tree with fruits. C. Close view of twig with flowers. D. Branch with inflorescence and flowers. E. Close view of fruit, X. *granatum* Koen.	106

Photo Plate Number	Subjects of the Photo Plates	Page
63.	*Xylocarpus mekongensis* Pierre. A.Tree with yellow leaves in the natural mangrove forest at Sundarbans. B. Tree with aerial breathing roots. C.Yellow leaves before sheading. D. Flowering branch. E. Mature fruits in hands of *X. mekongensis*	107
64.	*Aglaia cucullata* (Roxb.) Pellegrin. A. Small tree in the natural habitat at Sundarbans. B. Close view of branch. C. Potted seedling at the Laboratory. D. Newly germinated seedling in the laboratory.	108
65.	*Excoecaria agallocha* L. A. Full grown tree on the inter-tidal flat land. B. Horizontal roots of *Excoecaria agallocha.* L. C. Natural habitat of the *Excoecaria agallocha* L. in Sundarbans within the inter-tidal zones	109
66.	*Excoecaria agallocha* L. A. Natural habitat on the tidal river side. B. Male plant with catkin inflorescence. C. Female plant with tri-lobed fruits. D. Female plant with tri-lobed fruits. E. Close magnified view of fruits, *Excoecaria agallocha* L.	110
67.	*Aegiceras corniculatum* (L.) Blanco. A. Natural habitat on the tidal river side/ river falt land. B. Branches with mature fruits. C. White scented flowers on the umbel inflorescence - honey-bees collecting nectar from flowers	111
68.	*Aegiceras corniculatam* (L.)Blanco.A. Natural habitat on the tidal river falt land. B.Branches with mature fruits. C. Branches with white fragrant flowers.D. Close view of branches with white flowers. E. Fruits on branch.	112
69.	*Aegialitis rotundifolia* Roxb.A.Plants on the forest floor. B.Floating fruit gathered on the sandy soil. C. Natural seedling plants on the forest floor. D. Close view of the fruiting branches. E. Close view of the fruiting branches.	113
70.	*Aegialitis rotundifolia* Roxb. A. Shrub in the natural habitat. B. Branch with flowers and fruits. C. Close view of the flowering branch. D. Close view of the branch with fruits.	114
71.	*Heritiera fomes* Buch. Ham. A. Tree in the natural habitat at Sundarbans. B. Tree showing the buttress. C. Branch with flowers. D. Mature fruits on branch. E. Germinated seed of H. fomes, F. Collection of specimen in Sundarbans.	116
72.	*Heritiera littoralis* Dryand. A. Full grown up tree at Sipi Ghat Farm, Port Blair. B. Seedling plant of *H. littoralis.* at Wandoor Bay, Andaman. C. Branch with mature fruits. D. Showing single fruit. E. Two fruits hang from the branch.	117
73.	*Scyphiphora hydrophyllacea* Gaertn. f. , View of the dense foliages with flowers and fruits	118
74.	*Scyphiphora hydrohyllacea* Geartn. f. A. Shrub in the natural habitat at Sundarbans. B.Flowering branch. C. Close view of the flowering branch of *Scyphiphora hydrohyllacea* Geartn. f.	119
75.	*Hydrophylax maritima* Linn. f. - flowering twig in hand.	120
76.	A. *Acanthus ilicifolius* L. *B. A. volubilis* Wall. and C. *A. ebracteatus* Vahl	121
77.	*Acanthus ilicifolius* L. A. Plants in the Natural habitat in Sundarbans. B.Twig with flowers and fruits. C. Twig with flowers. D. Flowering plant in the natural habitat. E. Inflorescence with flowers & fruits. F. Inflorescence with fruits	122
78.	*Acanthus volubilis* Wall. Flowering twig in hand, collected from Indian Sundarbans	123
79.	*Acanthus ebracteatus* Vahl. A. Plants in the natural habitat at Sipighat, Port Blair, Andaman. B. Flowering branches, close view. C. Flowering branches, close view. D. Flowering branches, close view.	124
80.	*Cerbera manghas* L. A. Flowering branch of *Cerbera manghas* L. B. Two mature fruits hang from the branch. C. Seedling plant in the earthen pot at laboratory condition.	125
81.	*Sarcolobus carinatus* Wall. in the natural habitat at Sundarbans	126

Photo Plate Number	Subjects of the Photo Plates	Page
82.	*Pentatropis capensis* (Linn.) Bullock - twinning twigs on the *Phoenix* sp. hedges in natural condition.	127
83.	*Sarcolobus carinatus* Wall. A. Flowering twig. B. Twig with single flower - close view. C. Natural habitat in Sundarbans Mangal. D. Mature winged fruits - close view.	128
84.	*Sarcolobus globosus* Wall. - A. Twig with flowers. B. Twig with fruits.	129
85.	*Finlasonia obovata* Wall. - Plant in hand, with flowers and fruits	130
86.	*Hoya parasitica* Wall.- Parasite on the Xylocarpus mekongensis tree at Sundarbans	130
87.	*Dolichandrone spathaceae* (Linn.f.) K. Schum. A. Showing the canopy structure. B. Branch with the mature fruits in the branch. C. Branch with the mature fruits in the branch.	131
88.	*Hibiscus tiliaceous* Linn. A. Long view of the branches. B. Close view of the branch with flowers	132
89.	*Thespesia populnea* (Linn.) Soland ex Correa - flowering branch	133
90.	*Hibiscus tiliaceous* Linn. A. Branch with flower buds. B. Branch with flower buds.	134
91.	*Hibiscus tortuosus* Roxb. - Single branch with flowers	135
92.	*Brownlowia lanceolata* (Linn.) Kosterm.A, B & C. Branches with flowers and wrinkle fruits	136
93.	*Atalantia correa* Roem. A. Branch with flowers. B. Branch with fruit. C. Plant in the natural habitat. D. Three branches. E. Branch of *Atalantia correa* Roem.	137
94.	*Porteresia coarctata* (Roxb.) Takeoka Natural habitat in the river flat land with inflorescence	138
95.	*Porteresia coarctata* (Roxb.) Takeoka A. Vegetative growth in the river falt at Sundarbans B. Enlarged view of single leaf with serrated margin. C. Single stalk of panicle inflorescence. D. Single stalk of panicle inflorescence, close view. E. Grass with mature inflorescence. F. Germinated seeds of *Porteresia coarctata* (Roxb.) Takeoka.	139
96.	*Myriostachya wigtiana* (Nees ex Steud) Hook. f. A. Single inflorescence and foliage. B. Tuft of the Grass, *Myriostachya wigtiana* (Nees ex Steud) Hook. f.	140
97.	*Clerodendron neriifolium* var. *macrocarpum* L. A. Plant in the natural habitat at the mangrove forest of Sundarbans Tiger Reserve. B. Close view of the branches. C. Close view of the fruits of *Clerodendron neriifolium* var. *macrocarpum*	141
98.	*Clerodendron inerme* (Linn.) Gaertn. A. Plant in the natural habitat. B. Close view of the flowers on twig. C. Flowers on the branches. D. Foliages and flowers in natural habitat.	142
99.	*Tamarix troupii* Hole. A. & B. Branches with white cat-kin inflorescence.	144
100.	*Solanum trilobatum* Linn. A. Twig with fruits. B. Natural habitat in the thickets of the river bank. C.Close view of flowers	145
101.	*Manilkara hexandra* (Roxb.) Dub. - Close view of the branch with foliage and flower stock.	146
102.	*Dodonaea viscosa* (Linn). Jack. A. & B. Twigs/ branches with flowers and fruits	147
103.	*Crinum defixum* Ker-Gawler- mixed with *Cryptocoryne ciliata* (Roxb.) Fischy. ex Wydler. A. Dense growth on the tidal river-flat . B. On the tidal river-flat also showing the flowers, at Ghusighata areas of Sundarbans, India.	148
104.	*Crinum defixum* Kar - Gawler- A. Natural habitat along with close view of single plant. B. Close view of flowering plant. C. Natural habitat in the Kulti River bed, Sundarbans.	149
105.	*Cryptocoryne ciliata* (Roxb.) Fischy. Ex Wydler- in the natural habitat of inter-tidal river slope.	150
106.	*Sesuvium portulacastrum* Linn. A. Natural habitat in dry coastal area - on river bed. B. Close view of foliages and fruits, C. Close view of a flower and twigs	151
107.	A. & B. *Heliotropium curasavicum* Linn. A. Natural habitat. B. Close view of flowering twig.	153
108.	*Cynometra ramiflora* Linn. A. Flowering branch. B. Flowering branch. C. Close view of branch.	154

Photo Plate Number	Subjects of the Photo Plates	Page
109.	*Dalbergia spinosa* Roxb. A. Plant in the Natural habitat. B. Branches showing the flowers and pod.	155
110.	*Derris trifoliata* Lour. A. Twining shrub in the natural habitat. B.Twig with flowers C. Twig with flowers.	156
111.	*Derris scandens* Benth. A. Plant in the natural habitat. B. Close view of twig with fruits (Pods). C. Branch with fruits (Pods).	157
112.	*Derris indica* (Lamk.) Bennet- A. Branch with pods. B. Flowering branch. C. Branch with pods(fruits)	158
113.	*Caesalpinia crista* Linn. A. Branch with fruits. B. Branch with flowers.	159
114.	*Caesalpinia bonduc* (Linn.) Roxb. A. Spreading branches with Inflorescence. B. Branch with fruits. C. Branch with Inflorescence.	160
115.	A. & B. Twigs of *Salicornia brachiata* Roxb.	161
116.	A. & B. Natural habitats of *Suaeda maritima* Dumort.	161
117.	*Suaeda nudiflora* Roxb. A. Natural habitat in the inter-tidal river flat land. B. New growth on river flat. C. Twig in hand. D. Close view of the twig. E. Branched twig.	162
118.	*Calophyllum innophyllum* Linn. A. & B. Branches with dark green leaves but without flower & fruits.	163
119.	*Ipomoea pes-caprae* (Linn.) R. Br. - Natural habitat in the sand dune along with flowers.	164
120.	*Cyperus* sp. in the marshy areas at Goa.	165
121.	*Viscum orientale* Willd. A. Natural habitat on *Excoecaria* sp. and B. Collected branch in hand.	166
122.	*Dendrophthoe falcata* (Linn. f.) Ettingsh, flowering branches.	167
123.	*Opuntia dillenii* (Ker- Gawler) Haw. A. Natural habitat in the sandy coastal area of Sundarbans. B. Close view of single phylloclade with yellow flower. C. Close view of the phylloclades, showing the spines after leaf modification.	168
124.	*Pandanus odoratissimus* Soland ex Parkinson - Collected from Wandoor Bay, Andaman.	169
125.	*Acrostichum sp.* A. *Acrostichum aureum* Linn. - grow in the mangrove degraded area at Sundarbans. B. *Acrostichum aureum* Linn. - Collected from the Indian Sundarbans.	170
126.	A. Luxuriant growth of *Ruppia maritima* Linn. in brackishwater fisheries. B. Collected specimen in hand, close view. C. Submerged condition in brackishwater fisheries at Sundarbans. D. Collection of *Ruppia maritima* Linn.- samples for studies	171
127.	Tide water current in the mangrove forest of Indian Sundarbans, West Bengal.	172
128.	Sewage discharge point at Ghusighata, Kulti	173
129.	Aerial view of the forest and river, Sundarbans	173
130.	Barraging of flowing river at Jharkhali	174
131.	Forest Guard in STR	176
132.	Present scenerio of the Sundarbans Mangals. A, B & C.. Clearing of mangrove forest area for Brackishwater Fisheries Development at Jharkhali, Sundarbans. D. Degradation of coastal forest areas of the Sundarbans	177
133.	Different Research Activities and Investigations in the Sundarbans Mangal	178

INDEX TO MAPS, PIE CHART, DIAGRAMS, FLOW CHARTS & GRAPH

Map	Subject	Page
1	The World Distribution of the Mangroves	2
2	Showing the Major Indian Mangrove Zones	17
3	Undivided Sundarbans of Bangladesh & India (World's Largest Single Mangrove Chunk)	29
4	Area and Position of Indian Sundarbans and Sundarbans Tiger Reserve STR	30
5	Indian Sundarbans, covering 9,630 km^{a} areas of Mangroves and Mangrove Reclaimed 24 Parganas South and 24 Parganas North Districts of West Bengal	32
6	Satellite Map of the Sundarbans Under the District 24-Parganas South	33
Pie Chart - 1	Showing the Different Mangrove Zones in India along with their Percentages	18
Diagram - 1	Showing the Roles Played by the Mangrove Ecosystem of Sundarbans, India	37
Diagram - 2	Inundation Classes of Different Mangroves in the Ganga Delta Regions, Sundarbans in India	40
Flow Chart-1	Tropic Level in Indian Sundarbans	66
Flow Chart-2	Compartmental Model of the Food Chains Operating in the Sundarbans Mangal	68
Sketches-1	Different Hypocotyles of the Species under Mangrove - Rhizophoraceae	75
Graph-1	Tidal Amplitude in the Mangals of Indian Sundarbans During Different Tidal Phases	67

1
INTRODUCTION

Sundarbans, in the coastal West Bengal (India) and Bangladesh coast is well known for having its interesting and extensive mangrove flora and mangrove dwelling fauna. Adaptation and succession of these mangrove plants and mangrove dwelling fauna in the deltaic islands and estuarine mouths of the rivers, Ganga-Brahmaputra is unique and important. These mangrove plants grow in intertidal deltaic lands having higher percentage of salts, *viz.*, Sodium Chloride, Magnesium Chloride and Magnesium Sulphate. These salts are present in soil substrates and water phases. As such, mangrove plants and mangrove dwelling fauna have developed saline resistance mechanism. Normally, these plants form aerophores or air-breathing roots; these are known as pneumatophores, blind root suckers, plank roots, root knees and pneumatothods (in *Phoenix paludosa* Roxb.) along with conventional, characteristic taproots, adventitious fibre roots, *etc.* These aerial roots perform effective root zone respiration in this anoxic, water logged, compact intertidal saline soil and also provide mechanical support to these plants. Besides these air breathing roots, supporting stilt roots or prop roots, root buttresses,knee roots, plank roots are characteristic features of these major mangrove species (Photo Plate- 1, 2, 3, 4, 5 & 6). These dominant mangrove species mostly have unique viviparous, crypto-viviparous or pseudo- viviparous germination mechanism, by which these plants can get a better chance of germination in these inter-tidal silted up saline soil (Photo Plate - 7 & 8).

Other than these mangroves or halophytic plants few other plants can normally grow in these intertidal coastal, estuarine and deltaic regions. Most of these mangrove species can survive and grow in these saline soil and frequent salt water inundated zones, where other non-halophytic and non-mangrove plants cannot survive or grow for absence of their selective salt screening or salt excretion mechanism. Normal saline water absorption is impeded for these non-halophytic and non-mangrove plant species. Warming (1909) categorized these mangroves as helophilous types; these helophilous halophytes again are of following categories, *viz.*, salty swampy type, salty desert type and littoral swamp forest type; the last one is *'true-mangroves'*.

As such, these mangroves and the good number of other halophytic plants are able to protect these coastal deltaic lands from frequent soil erosion, protect the coastal zone from cyclones, sea surges and high tidal thrust, as buffer. These coastal mangroves and halophytic plants may also supply considerable volume of bio-mass and provide nutrients to these coastal soil and water and enrich the soil and water phases considerably and raises the productivity of this interesting intertidal ecosystem. For having these characters, Mac Nae (1968), Naskar & Guha Bakshi (1987), Naskar & Mandal (1999) and Guha Bakshi, Sanyal & Naskar (1999) referred mangrove forests as *'highest productive zone'* in comparison to the other ecosystems in tropical zones of the Globe.

Photo Plate - 1. Mangroves of Sundarbans, showing the river bank vegetation of *Rhizophora mucronata* Lamk.

Photo Plate - 2. River-bank Mangroves in Indian Sundarbans, showing *Nypa fruticans* Wurmb.and other mangroves.

For having higher salt concentration in soil and water phases in this coastal physiologically dry soil, osmotic pressure values of mangrove plants are generally high. Foliages of these major mangrove species are thick, succulent, evergreen, leathery textured with wax coating; cuticles are thick and well-developed; palisade tissues are present on their lamina.

Mangrove plants have special salt tolerance mechanisms, *viz., salt-extrusion, salt-exclusion* and *salt-accumulation*. By these unique salt tolerance abilities, mangrove species can adapt best in saline habitat, within the inter-tidal zones at coastal areas. *'Mangroves are in broad sense the tropical & sub-tropical coastal inter-tidal vegetation and also the ecosystem'.*

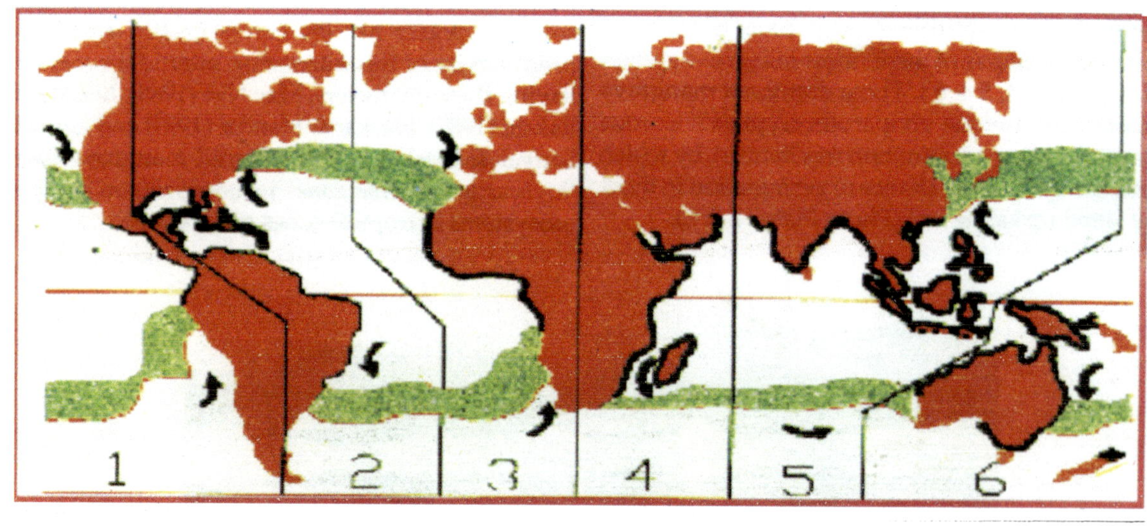

W.America E.America W.Africa E.Africa Indo-Malaysia Australasia

ATLANTIC EAST PACIFIC REGION (New World)

INDO - WEST PACIFIC REGION (Old World)

Map.1.The World Distribution of the Mangroves
(Mangrove Areas Shown by Thick - Black Dark Line & Arrows) -Modified after Duke, 1992.

A. Stilt roots of *Avicennia officinalis* L.

B. Stilt roots & pneumatophores *A. officinalis* L.

C. Aerial stilt roots of *A. officinalis* L.

D. Aerial stilt roots *of A. alba* Blume

E. Aerial stilt- roots *of A. marina* Vierh

F. Pneumatophores of *A. marina* Vierh

G. Pneumatophores of *A. alba* Blume

H. Pneumatophores of *A. alba* Blume

I. Pneumatophores of *A. marina* Vierh

Photo Plate - 3. Aerial Root Systems of Different Mangrove species, *Avicennia* spp. in the Indian Sundarbans.

A. Buttress root of *Heritiera littoralis* Dryand, Andaman Island

B. Broom like stilt roots of *Ceriops tagal* Robin, Sundarbans

C. Broom like stilt roots of *Ceriops tagal* Robin, Sundarbans.

D. Blind root-suckers of *Heritiera fomes* Buch-Ham, Sundarbans

E. Stilt roots of *Bruguiera* sp. & Buttress of *Xylocarpus* sp., Sundarbans

F. Pneumatothods of *P. paludosa* Roxb, Sundarbans

G. Pnematophores of *Sonneratia* sp., Indian Sundarbans.

Photo Plate - 4. Aerial Root Systems of Different Mangrove Species, in the Indian Sundarbans

A. *Bruguiera parviflora* W & A. ex Griff. trees in the Indian Sundarbans

B. Buttress roots and root knees of *Xylocarpus* sp.

C. Broom like stilt- roots on trunk base of *Bruguiera* sp.

D. Broom like stilt root on trunk base of *Bruguiera* sp.

Photo Plate - 5. Canopy Structures and Root Systems of Different True Mangrove Species

A. Pneumatophores of *Avicennia marina* Vierh.

C. Stilt roots & pneumatophores of *Avicennia alba* Blume

B. Stilt roots and pneumatophores of *Avicennia alba* Blume

D. Stilt roots of *Avicennia officinalis* L.

E.. Stilt roots and pneumatophores of *Avicennia marina* Vierh.

F. Stilt roots and pneumatophores of *Avicennia officinalis* L.

Photo Plate - 6. Aerial Root Systems of the Mangroves, *Avicennia* spp. in Indian Sundarbans

A. *Rhizophora mucronata* Lamk; B. *Rhizophora apiculata* Blume; C. *Bruguiera gymnorhiza* Lamk; D. *Bruguiera sexangula* Poir; E. *Bruguiera cylindrica* Blume; F. *Bruguiera parviflora* W. & A ex Griffth

G. *Ceriops decandra* Ding Hou

H. *Ceriops tagal* Robin

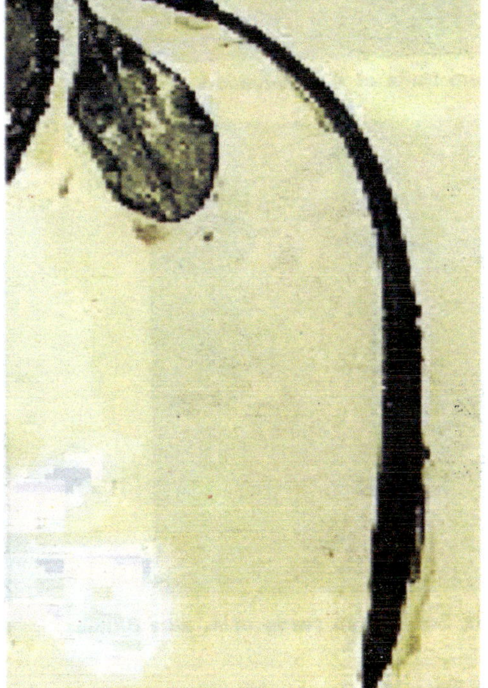

I. *Kandelia candel* Druce

Photo Plate - 7. Viviparous germinated hypocotyles of the Mangrove-Rhizophoraceae in Indian Sundarbans.

A. Crypto-viviparous Germinated seeds of *A. alba* Blume

B.Crypto-viviparous Germinated seeds of *A. marina* Vierh

C. Mature fruits of *A. officinalis* L.

D. Crypto-viviparous Germinated seedling of *A. marina* Vierh.

E. Crypto-viviparous Germinated seeds of *A. alba* Blume

F. Crypto-viviparous Germinated seeds of *A. officinalis* L.

Photo Plate - 8. Crypto-viviparous seed Germination of *Avicennia* spp. in Indian Sundarbans

Deltas, sheltered bays, intertidal coastal areas and muddy estuarine mouths are the ideal habitats for these mangrove plants, where large groups of marine and terrestrial fauna are also abundant. As such, mangrove zones may be considered as *'sea-land interface* zones'. The dominant mangrove zones of the world are spread over in the Old World and New World tropics and sub-tropics (Map-1, The World Mangrove Zones). The World's Mangrove zones and the World's Mangroves and associated species are shown in Table 1 & Table 2.

Several mangrove researchers had referred these intertidal zones and the overall intertidal ecosystem as *'mangroves'*. Lear & Turner (1977) also suggested that the term *'mangroves'* may be referred to express the 'coastal ecosystem in a holistic manner, including its common habitat or inhabitant flora and fauna'.

However, Mac Nae (1968) had defined the community of these mangroves as 'Mangal'. Mac Nae (1968) had stated this total ecosystem as *'mangal'* and the intertidal plant assemblage as *'mangroves'*. But, till date different mangrove researchers had defined these terms mangroves differently; for that the number of world's mangrove species is variable from one to other. For different types of definition the world-renowned scientist V. J. Chapman (1976) had mentioned the total world mangrove species as 90, but UNESCO (1984) had mentioned that the number of world mangrove species is only 65. Saenger, *et al.* (1983) in IUCN Working Plan had reported 60 mangrove species exclusively from the tidal zones and recognized these as true mangroves. Mepham & Mepham (1984) had also raised several questions and had mentioned 900 plant species to define their exact status in the list of world's mangroves. On the other hand, Tomlinson (1986) had stated total 40 mangrove species from the Old World and only 8 mangrove species from the New World. Furthermore, Tomlinson (1985) had also categorised these world's mangroves as (i) *major elements of mangroves*, (ii) *minor elements of mangroves* and (iii) *back mangroves* or *mangroves associates.*

Naskar (1983) had mentioned that these mangroves are of different types, *viz.*, true mangroves, mangrove associates and obligatory mangroves; this classification is based on their habits, habitats and distribution in different saline zones. Besides these, several algal species are also present in these mangals.

A. Inter-tidal forest with *Avicennia* spp. & *Sonneratia* sp.

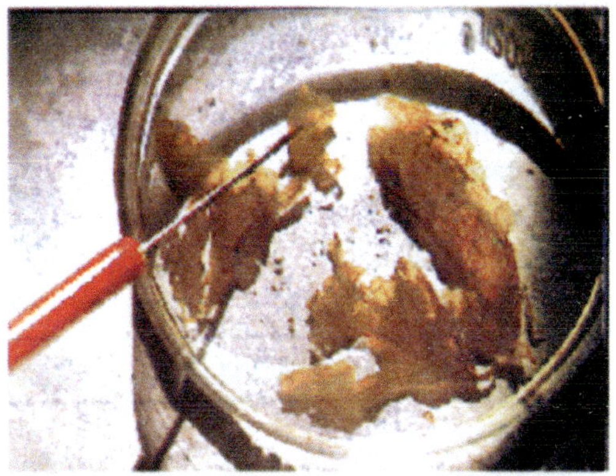

B. *Ulva lactuca* - Green Algae, in inter-tidal river bed

C. Microscopic view of *Catenella* sp.- Red Algae

D. Green Algae- *Boodleopsis sundarbanensis*

Photo Plate- 9. Algal habitats and different species of brackish-water Algae of Sundarbans

Table 1 : Showing the Different World Mangrove Zones (Both Old World and New World) –Based on Different Publications

Sl. Nos.	Country / Zones of the Mangroves	Area Under Mangals (km²)	Latitudes	Longitude
1.	Indonesia	36,000	6°N - 10°S	95° E - 150° E
2.	Australia	11,617	11°S - 35°S	115° E - 155° E
3.	Bangladesh	10,000	21°N - 22°30 'N	89° 5' E - 92° 1' E
4.	India	6,419	7°N - 23°N	69° E - 89° 5 ' E
5.	Thailand	6,020	4°N - 15°N	99° E - 10° 3'E
6.	Mayanmar	5,200	15°N - 20°N	93° E - 97° E
7.	Papua New Guines	4,116	0- 7°30 'S	132 °E - 150° E
8.	Philippines	4020	7 °N - 18 °N	119 ° 30' E-126°E
9.	Sabah	3,653	5° N - 7° 30' N	115° E - 119° E
10.	Malagasy Republic	3,207	12° 39' S - 25° 30'S	40°30' E - 50°20' E
11.	South Vietnam	2,864	9° N - 21° 30'N'	105° E - 108° E
12.	Pakistan	2,495	24°N - 25°N	62° E - 68° E
13.	Sawara	1,500	3°S - 5°S	113° E - 114° E
14.	Peninsular Malaysia	1,500	1° 30 N - 7° N'	100 °30' E- 140° E
15.	Mozambique	850	17 °S - 26° S	32° E - 40° E
16.	Kenya	590	2° S - 4° S	39° E - 41° E
17.	Kampuchea	500	4° N - 15° N	99° E - 103° E
18.	Tanzania	500	5° S - 10° S	38° E - 40° E
19.	New Zealand	198	31° S - 34° S	173° E - 178° E
20.	Fiji	197	16° S - 20° S	178° E - 180° E
21.	Sri Lanka	32	10° N - 10° 30 ' N	79° 30' E - 81°30'E
22.	Singapore	28	6° N - 10° N	80° E - 85° E
23.	Hong Kong	65	22° 30 ' N - 23° N	113° 30' E - 114° E
24.	South Africa	Negligible	32° S - 14° S	17° E - 30° E
25.	Other Pacific Islands	100	35° N - 35° S	120° E - 120° W
	Total Areas in Old World =	1,02,631 km²		
1.	Brazil	15,000	0° S - 30° S	41° W - 52 ° W
2.	Cameroon	2,000	2° N - 5° N	8° E - 10° E
3.	Columbia	4,400	1°30' N - 12° N	72° W - 74° W
4.	Ecuador	2,159	1° 20' N - 3° N	79°30' W - 81° W
5.	Gabon	2,500	0° 30' N - 1° N	6° 30' E - 10° E
6.	Mexico	6,600	15° N - 31° N'	87° W - 114° W
7.	Nigeria	9,730	5° N - 5° 30' N	5° E - 9° E
8.	Panama	4,860	7° N - 9° N	78° W - 82° W
9.	Senegal - Gambia	5,000	12° N - 17° N	16° W - 17° W
10.	U. S. A.	2,050	65° N - 26° N	60° W - 120° W
11.	Venezuela	6,730	9° N - 12° N	60° W - 72° W
	Total Areas in New World =	71,029 km²		
	Total World Mangrove Zones	1,73,660 km²		

Besides these above world mangrove areas , few mention may be made for the West African Mangrove areas, *viz.*, Liberia, Angola, Zaire, Gambia and Ivory coast, *etc.;* in these areas mangrove patches are present scatteredly.

Table 2 : Distribution of Mangroves and Mangrove Associated Plants in India and other Important World Mangrove Zones

Family Name & Species	Habit and Habitat	Sundarbans, Hugli Delta, W. Bengal	Andaman & Nicobar Islands	Mahanadi Delta, Orissa	Godavari & Krishna Delta, A.P.	Cauvery Delta, T. N.	West Coast of India
1. Rhizophoraceae (The Mangrove Rhizophoraceae has 4 genera and under these 17 species in the world mangals.							
1. *Rhizophora mucronata* Lamk.	Tree	+++	+++	+++	+++	+++	+++
2. *R. apiculata* Blume	Tree	++	++	++	++	++	++
3. R. *stylosa* Griff.	Tree	-	++	+?	-	-	-
4. *R. X lamarckii* Montr.	Tree	-	+	-	-	-	-
5. *R mangle* L.	Tree	The New world mangrove species					
6. *R samoensis* (Hochr) Salv.	Tree	The New world mangrove species					
7. *R racemosa* Meyer	Tree	The New world mangrove species					
8. *R X selala* (Salv.) Tom.	Tree	The New world mangrove species					
9. *Bruguiera gymnorhiza* (L.) Lamk.	Tree	+++	+++	+++	+++	+++	+++
10. *B. sexangula* (Lour.) Poir.	Tree	++	++	+	-	-	-
11. *B. cylindrica* (L.) Bl.	Tree	++	++	+	+	+	++
12. *B. parviflora* W. & A. ex Griff.	Tree	++	++	+	+	-	+
13. *B. exaristata* Ding Hou	Tree	Old world species but not reported from India					
14. *B. hainesii* Rogers	Tree	Old world species but not reported from India					
15. *Ceriops decandra* ((Griff.) Ding Hou	Shrub	+++	+++	+++	++	++	+
16. *C. tagal* (Perr.) Robinson	Tree	+++	+++	+++	+	+	+
17. *Kandelia candel* (L.) Druce	Tree	++	++	++	+	+	++
2. Avicenniaceae (This family has 1 genus, *Avicennia*, under this 11 species are reported from the mangals of the Globe)							
18. *Avicennia alba* Blume	Tree	++++	+++	+++	+++	+++	++
19. *A. officinalis* L	Tree	++++	+++	+++	+++	+++	++
20. *A. marina* (Forsk.) Vierh	Tree	++++	++++	+++	+++	+++	+++
21. *A. lanata* Ridley	Tree	Old world species but not reported from India					
22. *A. eucalyptifolia* (Zipp. ex Miq.) Molden.	Tree	Old world species but not reported from India					
23. *A. schaueriana* Stapf.& Lechman ex Molden.	Tree	New World Species					
24. *A. germinans* (L.) Stearn	Tree	New World Species					
25. *A. africana* P. Beauv.	Tree	New World Species					
26. *A. bicolor* Standley	Tree	New World Species					
27. *A. balanophora* Stapf.& Molden. ex Molden.	Tree	New World Species					
28. *A. marina* var. *acutissima* Stapf & Molden.	Tree	A variety of *A. marina* grow in the West coast, India					
3. Sonneratiaceae (This family has 1 genus - *Sonneratia;* under this 5 species are reported from Old World mangals)							
29. *Sonneratia caseolaris* (L.) Engler	Tree	+	+	+	-	-	++
30. *S. griffithii* Kurz.	Tree	++	++	++	-	-	-
31. *S. apetala* Buch. - Ham.	Tree	++++	++	+++	++	+	+
32. *S. alba J. Smith.*	Tree	?	++	+	-	-	++
33. *S. ovata* Backer	Tree	Old world species but not reported from India					
4. Combretaceae (This family has 4 genera; under these there are 6 species reported from the different World mangals)							
34. *Lumnitzera racemosa* Willd.	Tree	+++	+++	+++	++	++	+
35. *L. littorea* (Jack.) Voigt.	Tree	-	++	-	-	-	-
36. *L. rosea* (Gaud.) Presl.	Tree	According to Tomlinson (1986) a putative sterile hybrid					
37. *Laguncularia racemosa* (L.) Gaertn. f.	Tree	New world mangrove species					

Contd...

Family Name & Species	Habit and Habitat	Sundarbans, Hugli Delta, W. Bengal	Andaman & Nicobar Islands	Mahanadi Delta, Orissa	Godavari & Krishna Delta, A.P.	Cauvery Delta, T. N.	West Coast of India
38. *Terminalia catappa* L.	Tree	Not Mangrooves but occasionally in back mangrove area					
39. *Conocarpus erectus* L.	Tree	Mangrove associated in the Tropical America					
5. Arecaceae (This family has 5 genera; under these 5 species are reported from the different world mangals)							
40. *Nypa fruticans* (Thunb.) Wurmb.	Palm	++	+++	-	-	-	-
41. *Phoenix paludosa* Roxb.	Palm	+++	+++	++	-	-	-
42. *P. reclinata*	Palm	occasional in the back mangrove thickets					
43. *Calamus erinaceus* (Becc.) Dansfield	Palm	climbing palm on the sandy back mangrove area					
44. *Oncosperma tigillarium* (Jack.) Ridley	Palm	tall many stemmed palm on back mangrove thicket					
Tomlinson (1986) also mentioned the other palm species , *i.e.*, *Euterpe* sp., *Manicaria* sp., *Mauritia* sp. and *Raphia* sp. from tropical swamp communities of Africa and America.							
6. Meliaceae (This family has 2 genera and under these there are 4 mangroves and/or mangrove associated species in the Old World)							
45. *Xylocarpus granatum* Koenig.	Tree	+++	+++	+++	+	-	+
46. *X. mekongensis* Pierre	Tree	+++	+++	+++	+++	?	+
47. *X. moluccensis* (Lamk.) Roem.	Tree	-	++	++	-	+	-
48. *Aglaia cucullata* (Roxb.) Pellegrin	Tree	+	-	+	-	-	-
7. Euphorbiaceae(There are 3 genera and 5 species reported from mangroves and back-mangrove communities).							
49. *Excoecaria agallocha* L.	Tree	+++	+++	+++	+++	+++	+++
50. *Excoecaria indica* (Willd.) Muell.	Tree	+	+	-	-	-	-
51. *Excoecaria dallachyana* (Baill.) Benth.	Tree	Queensland and New South Wales					
52. *Hippomane mancinella* L.	Tree	As back mangrove in the New World tropics					
53. *Glochidion littorale* Bl.	Tree	As back mangrove in the New World tropics					
8. Myrsinaceae (This is the mangrove associated family and 3 species under 2 genera are reported so far from mangals).							
54. *Aegiceras corniculatum* (L.) Blanco	Tree	+++	+++	+++	++	++	++
55. *A. floridum* Roem. & Schult.	Tree	An old world mangrove associate, but not in India					
56. *Ardisia elliptica* Thunb.	Tree	An old world mangrove associate, rare in India					
57. *Myrsine umbellulatum* A. DC.	Tree	An old world mangrove associate, but not in India					
9. Aegialitidaceae (This family has two mangrove associated species - out of these one is reported from India)							
58. *Aegialitis rotundifolia* Roxb.	Shrub	+++	+++	++	-	-	-
59. *Aegialitis annulata* R. Brown)	Shrub	An old world mangrove associate, not in India					
10. Sterculiaceae (This family has three mangrove associated species- out of these two are reported from Indian mangals							
60. *Heritiera fomes* Buch. - Ham.	Tree	+++	-	+++	-	-	-
61. *H. littoralis* Dryand	Tree	-	+++	++	-	-	-
62. *H. globosa* Kosterman	Tree	An old world species, but not in Indian mangal					
11. Rubiaceae (This family has two mangrove associated species - out of these one is reported from India)							
63. *Scyphiphora hydrophyllacea* Gaertn. f.	Shrub	+++	+++	+++	-	-	-
64. *Rustia occidentalis* (Benth.) Hemsl.	Shurb	A new world mangrove associated species					
12. Acanthaceae (This family has three mangrove associated species - all these three are reported from Indian Mangals)							
65. *Acanthus ilicifolius* L.	Shrub	+++	++	+++	++	++	++
66. *A. ebracteatus* Vahl	Shrub	-	+++	-	-	-	-
67. *A. volubilis* Wall.	Shrub	+	++	-	-	-	-

Contd...

Family Name & Species	Habit and Habitat	Sundarbans, Hugli Delta, W. Bengal	Andaman & Nicobar Islands	Mahanadi Delta, Orissa	Godavari & Krishna Delta, A.P.	Cauvery Delta, T. N.	West Coast of India
13. Apocynaceae (This has 5 mangrove associated species - out of these 2 are reported from Indian Mangals)							
68. *Cerbera odollam* Gaertn. f.	Tree	+	+	+	+	+	-
69. *C. manghas* L.	Tree	+	+	-	-	-	-
70. *C. floribunda* Schum. - Initially reported from New Guinea; as an associate of *Rhizophora* sp.							
71. *Rhabdadenia biflora* (Jacq.) Muell.- in the back mangrove communities in Caribbean region							
14. Asclepiadaceae(Six species were collected in Sundarbans; either as mangrove associate or as parasite)							
72. *Sarcolobus globosus* Wall.	As mangrove associate on river flat land in Sundarbans						
73. *Sarcolobus carinatus* Wall.	As mangrove associate on river flat land in Sundarbans						
74. Pentatropis capensis (Linn.f.) Bullock	Throughout in the mangrove thickets, above tidal zone						
75. *Finlaysonia obovata* Wall.	- Climbing herb on the mangrove tree in Sundarbans & Andaman Islands						
76. *Hoya parasitica* Wall.	As parasitic on *Xylocarpus* sp. tree of Sundarbans						
77. *Tylophora tenuis* Bl.	Occasional shrub on the Sundarbans area, above tidal zones						
15. Bignoniaceae (Five species are collected and reported from the back mangal communities, throughout the tropics)							
78. *Dolichandrone spathacea* (Linn.f.) Schum.	As back mangal, throughout the coastal areas in India						
79. *Amphitecna latifolia* (Mill.) A. Gentry	Mangrove associate in Pacific coast of T. America.						
80. *Anemopaegma chrysoleucuma* (H.B.K.) Sand.	Mangrove associate in northern S. America						
81. *Phryganocydia phellosperma* (Hemsl.) Gentry	Restricted in Pacific side of Colombia in T. American mangroves						
82. *Tabebuia palustris* Hemsl.	A mangrove associate, distributed in Pacific region.						
16. Malvaceae (Seven species are collected and reported from the World mangrove zones)							
83. *Hibiscus tiliaceus* L..	Occasional above the frequent tidal inundated zones throughout in Indian mangal						
84. *H. tortuosus* Roxb.	Occasional alone the above tidal zones in the mangrove reclaimed areas						
85. *Thespesia populnea* (L.) Soland. ex Corr.	Distributed throughout the coastal areas in India						
86. *T. populneoides* (Roxb.) Kostel	Distributed throughout the coastal areas in India						
87. *T. lampus* (Cav.) Dalz. & Gibs.	Distributed throughout the coastal areas in India						
88. *Pavonia rhizophorae* Killip.	Distributed in the Tropical America						
89. *P. spicata* Cav.	In Mangals, salt marshes and coastal areas of Tropical America						
17. Poaceae (Four species are common in the inter-tidal river flats of the coastal areas and mangrove habitats).							
90. *Porteresia coarctata* (Roxb.) Takeoka	Common in newly silted up river flat land as pioneer species						
91. *Myriostachya wightiana* (Nees ex.Steud) Hk. f.	Occasional in the inter tidal river flat lands of Sundarbans						
92. *Aeluropus lagopoides* (L.) Trin. ex Thw.	On sandy coastal areas at Sundarbans and elsewhere.						
93. *Phragmites kakra* (Retz.) Trin. ex Steud.	On the silted up river flats at Sundarbans and elsewhere						
18. Rutaceae (Only one species is collected & reported from the mangrove habitats in the Sundarbans)							
94. *Atalantia correa* Roem =*Merope angulata*	Collected from the mangrove habitats in the Sundarbans						
19. Tiliaceae (Only one species collected & reported from Indian Sundarbans mangrove forests)							
95. *Brownlowia lanceolata* (L.) Kosterman	Occasional in the inter-tidal river flats and ridge forests						
96. *B. tersa* (L.) Kosterman	Occasional in the inter-tidal river flats and ridge forests						
97. *B. argentata* Kurz	Occasional in the inter-tidal river flats and ridge forests						
20. Verbenaceae (Three species are collected as the mangrove associated species in the Indian Sundarbans)							
98. *Clerodendron inerme* (L.) Gaertn. f.	Common in the inter-tidal coastal/mangrove areas.						
99. *C. neriifolium* var *macrocarpa*	Occasional in the inter-tidal coastal/mangrove areas.						
100. *Premna corymbosa* Rottb. & Willd.	Occasional in the inter-tidal coastal/mangrove areas.						

Contd...

Family Name & Species	Habit and Habitat	Sundarbans, Hugli Delta, W. Bengal	Andaman & Nicobar Islands	Mahanadi Delta, Orissa	Godavari & Krishna Delta, A.P.	Cauvery Delta, T. N.	West Coast of India
21. Tamaricaceae (There are four species distributed in dry habitats of the coastal regions throughout tropics)							
101. *Tamarix* aphylla (L.) Kantze	In the dry sandy coastal area, as back mangrove species or as associate.						
102. *T. dioica* Roxb.	In the dry sandy coastal area, as back mangrove species or as associate.						
103. *T. gallica* L.	In the dry sandy coastal area, as back mangrove species or as associate.						
104. *T. troupii* Hole	In the dry sandy coastal area, as back mangrove species or as associate.						
22. Solanaceae(Two species are available in the coastal sandy soil as back mangroves in Sundarbans)							
105. *Solanum trilobatum* L.	Shrub,on mangrove thickets or back-mangrove areas						
106. *Solanum surattensis* Burm. f.	Shrub of the coastal sandy beaches						
23. Sapotaceae (One species is reported from the back mangrove communities of the Old World tropics)							
107. *Pouteria obovata* (R.Br,) Baehni	A back-mangrove species in Old World Tropics, including India						
24. Sapindaceae (One species is reported from the back mangrove communities of the Old World tropics)							
108. *Allophyllus cobbe* (L.) Bl.	A mangrove associate in India, Sri Lanka, Malesia, New Guinea						
109. *Dodonaea viscosa* (L.) Jack.	Occasionally found in the sandy beach area in S.T.R.						
25. Amaryllidaceae (One species is reported from the back mangrove communities of the Old World tropics)							
110. *Crinum defixum* Ker- Gawler	A mangrove associate in the inner estuary						
26. Araceae (This family has one species in the river flat moist land of Old World Tropics)							
111. *Cryptocoryne ciliata* (Roxb.) Fisch. et Wydler	Erect herb of the inter-tidal river flats at Sundarbans.						
27. Aizoaceae (This family has one species in the river flat moist land of the Old World).							
112. *Aerva lanata* (L.) Juss. ex Schulte.	Occasionally grow on the river bank						
28. Barringtoniaceae (This family has two species in the back mangrove communities of the Old World)							
113. *Barringtonia acutangula* (L.) Gaertn. f.	This Species grow in the river bank, moist land.						
114. *B. racemosa* Roxb.	This Species grow in the river bank, moist land.						
29. Boraginaceae(This family has only one herbaceous species in the river bank beyond normal high tide zone)							
115. *Heliotropium curassavicum* L.	Almost common throughout the coastal area on the naked river slopes.						
30. Caesalpiniaceae(This family has few species in back mangrove zones as mangrove associated species).							
116. *Caesalpinia crista* L.	Almost throughout the coastal back mangrove areas						
117. *C. bonduc* (L.) Roxb.	Almost throughout the coastal back mangrove areas						
31. Fabaceae (This family has few species in the back mangrove zone, as mangrove associated species).							
118. *Cynometra ramiflora* L.	Occasionally in Sundarbans mangal and throughout coastal areas						
119. Dalbergia spinosa Roxb.	Occasionally in Sundarbans mangal and throughout coastal areas						
120. *Derris trifoliata* Lour.	In the back-mangrove zones at Sundarbans and throughout						
121. *D. scandens* Benth.	In the back-mangrove zones at Sundarbans and throughout						
122. *D. indica* (Lamk.) Bennet	In the back-mangrove zones at Sundarbans and throughout						
123. *Intsia bijuga* (Colebr.) Kuntze	In the back-mangrove zones at Sundarbans and throughout						
124. *Mucuna gigantea* (Willd.) DC.	In the back-mangrove zones at Sundarbans and throughout						
32. Chenopodiaceae (This family has three species in the back mangrove zones).							
125. *Suaeda maritima* Dumort.	In back-mangrove zones at Sundarbans and throughout as halophytic herb.						
126. *S. nudiflora* Roxb.	In back-mangrove zones at Sundarbans and throughout as halophytic herb.						
127. *S. monoica* Forsk. ex Gmel.	In back-mangrove zones at South India as halophytic herb.						
33. Clusiaceae (This family has 1 species in the back mangrove zone as associated species).							
128. *Calophyllum inophyllum* L.	Typical in beach communities, in Old World; also in Sundarbans						

Contd...

Family Name & Species	Habit and Habitat	Sundarbans, Hugli Delta, W. Bengal	Andaman & Nicobar Islands	Mahanadi Delta, Orissa	Godavari & Krishna Delta, A.P.	Cauvery Delta, T. N.	West Coast of India
34. Convolvulaceae (This family has few herbaceous species in the sandy coast as back mangrove species).							
129. *Ipomoea pes-caprae* (L.) R. Brown	Throughout the coastal sandy soil as sand binder herb						
35. Cyperaceae (This family has few species in the back mangrove zone as halophytic sedge species).							
130. *Scirpus littoralis* Schrad.	Throughout the marshy places and river sides in Sundarbans						
36. Flacourtiaceae (This family has few species in the back mangrove zone as mangrove associated species).							
131. *Scolopia macrophylla* (Wight & Arnold) Clos.	Indochina, Malay Peninsula, Sumatra, Borneo to Java.						
37. Lauraceae-The family has few parasitic species, *viz., Cassytha filiformis* L. found as parasite in coastal Sundarbans							
38. Loranthaceae (The family has few parasitic and / or epiphytic species in the coastal tree species)							
132. *Dendrophthoe falcata* (Linn.f.) Etting.	Epiphytic herb, occasionally in mangrove trees in Sundarbans						
133. *Viscum orientale* Willd.	Epiphytic herb, occasionally in mangrove trees in Sundarbans						
134. *Viscum monoicum* Roxb.	Epiphytic herb, occasionally in mangrove trees in Sundarbans						
135. *Macrosolen cochinchinensis* (Lour.) van Teig	Epiphytic herb, occasionally in mangrove trees in Sundarbans						
39. Pellicieraceae (This family has one mangrove associated species in the New World zone)							
136. *Pelliciera rhizophoreae* Triana and Planchon	Pacific coast of the northern part in S. America also Atlantic coast						
40. Celastraceae (This family has few back mangrove species in the Old World zone)							
137. *Cassine viburnifolia* (Juss.) Ding Hou	Wet coastal communities, in mangrove swamps, Andaman Is.						
41. Opuntiaceae (This family has 1 back mangrove species in the coastal sandy area of Old World zone)							
138. *Opuntia dillenii* (Ker. Gawler) Haw.	This species is common on the sandy dry coastal area						
42. Pandanaceae (This family has few back mangrove species in the coastal area of Old World zone)							
139. *Pandanus odoratissimus* Soland. ex Park.	On the above tidal zone in Sundarbans and Andaman						
43. Pteridaceae (This family has 3 back mangrove species in the coastal area of Old World zone)							
140. *Acrostichum aureum* L.	Common in back mangrove zones throughout in India						
141. *A. speciosum* Willd.	Common in back mangrove zones throughout in India						
142. *A. danaeifolium* Langsdorff et Fischer	Common in the island in America- New World Species						
44. Ruppiaceae (This family has 1 species in the brackish water at the coastal areas of Old World zone)							
143. *Ruppia maritima* L.	Submerged, delicate, much branched herb, in Sundarbans						
45. Anacardiaceae (This family has 1 back mangrove species in the coastal area of Old World zone)							
144. *Gluta velutina* Blume	Old world tropical mangrove areas.						
46. Asteraceae (This family has few mangrove associated species in the coastal area of Old World zone)							
145. *Tuberostylis axillaris* Blake	Mangrove associate and recorded from tidal forest of Colombia.						
146. *T. rhizophorae* Steetz.	Epiphytic on mangrove trees at Colombia, Central America						
47. Batidaceae (This family has 1 back mangrove species in the coastal area of New World zone)							
147. *Batis maritima* L.	Halophytic, adjacent to mangal community.						
48. Cuscutaceae -Species, *Cuscuta reflexa* Roxb. is common parasitic herbaceous species at *Excoecaria* sp..							
49. Asteraceae - Several species under this family are frequent in mangrove areas beyound tidal inundation.							
50. Ebenaceae- Few species are reported under this family from the back mangrove communities of the Old World Tropics.							

A. Tide inundated mangrove forest at Indian Sundarbans

B. River bank - row of *Rhizophora* sp. tree at Andaman Is.

Photo Plate - 10. Natural Habitats of the Mangroves in the Inter-tidal zones in Sundarbans and Andaman Island.

A. Mangrove wood (*Ceriops* sp.) at Ghusighata

B. Exploitation of *Ceriops* stalk at Sundarbans

C. Mangrove wood loading at truck for transport to city

D. Mangrove wood - seized from poachers, Sundarbans

E. Clearing of Mangrove Area by the local people

F. Barraging of Flowing river by the local people

G. Exploitation of Mangrove wood by the rural people

H. Exploitation of fuel wood at Sundarbans areas

Photo Plate - 11. Exploitation of Mangrove wood and barraging of flowing river for Human settlement at Sundarbans

2

INDIAN MANGALS

In India, mangrove habitats are of the following six types (Table- 3 & Map - 2); which have very specific and characteristic features. These are –

- **The Estuarine and/or Deltaic Mangroves** - abundant on the river mouths of the Ganga (Hugli River), the Mahanadi, the Krishna, the Godavari, the Cauveri, the Mandovi and the Zuari. Mostly, these dominant deltaic or estuarine mangroves exist on the East Coast of the Indian Territory, while the Mandovi and the Zuari river mouths are of West Coast;
- **The Island Mangroves** of the Andaman & Nicobar Islands are scatteredly distributed on the Bay Islands, and is considered as the second dominant and best mangrove zone in Indian territory.
- **Sea-shores or Coastal Mangroves** of Karnataka, Goa, Maharastra, Gujarat States in W. Coast;
- **Mangroves of Marshy Back Water** of Kerala coastal areas;
- **Mangroves on the Gulfs**, *i.e.*, Gulf of Cambay, Gulf of Kuchchh and
- **The Mangroves of the Coral Reef,** ***i.e.*,** Lakshadweep Atolls Mangroves of Arabian Sea.

Table 3 : Showing the Indian Mangrooves
(Modified after Govt. of India Pub., 1987)

S.No.	Delta / Estuary / Coasts etc.	Areas Km^2
1.	Ganga Delta/Sundarbans in West Bengal	4267
2.	Andaman & Nicobar Islands of the Bay Islands	1152
3.	Mahanadi Delta / Bhitarkanika of Orissa State	120
4.	Krishna & Godavari Delta and Coringa Mangroves	184
5.	Cauvery Delta/Pichavanam, Muthupet and Chhatram	26
6.	Kerala Coast/Cochin estuary	Negligible
7.	Karnataka Coast/Coondapur, etc.	60
8.	Goa Coast/Mandovi, Zuari, etc.	20
9.	Maharashtra Coast/Ratnagiri	330
10.	Gujarat/Gulfs of Cambay & Kuchchh	260
11.	Lakshadweep & Minicoy Islands	Negligible
	Total	**6419**

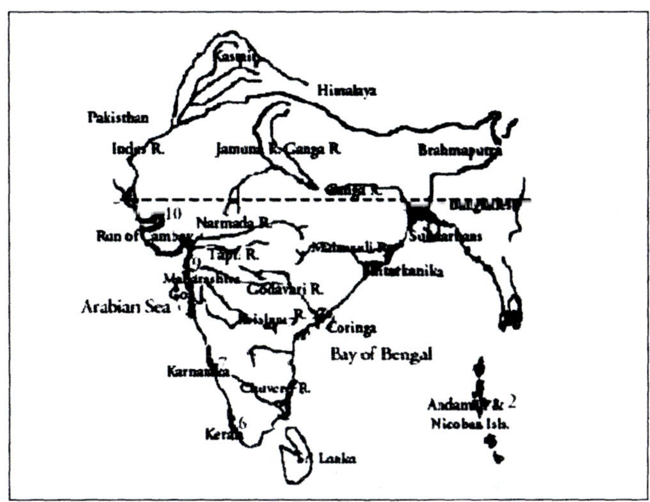

Map-2. Showing The Major Indian Mangrove Zones (Green shade)

Table 4 : Showing Declining Trend of Mangrove Areas in Indian Territory (Modified after Jagtap, *et. al.*, 1993)

Regions	During 1975	During 1990-91	Reduction (%)
East Coast	2210 km^2	1635 km^2	28%
West Coast	1150 km^2	650 km^2	44%
A. & N. Is	1190 km^2	780 km^2	32%
Total	**4550 km^2**	**3065 km^2**	**33%**

Among these, the estuarine mouths and deltaic zones of the East Coast of India have about 70% of the total Indian mangals. Dominant mangrove areas (4267 km^2) and major mangrove ecosystem (66%) along with about 90% of the total Indian mangrove species were collected, identified and reported from the Ganga delta of Indian Sundarbans and about 18% mangrove areas (1152 km^2) and 90% mangrove species are distributed in the Andaman & Nicobar Islands. But, most of these Indian mangrove ecosystem is in very much degraded and threatened condition due to anthropogenic pressure (Table - 4).

Other Dominant Indian Mangrove Areas and Mangrove Species

Besides Sundarbans mangal, the other dominant mangrove areas within the Indian territory are in (i) Mangroves of Andaman & Nicobar Islands of the Bay of Bengal, (ii) Mangroves of Bhitarkanika of the Mahanadi delta of Orissa coast, (iii) Mangroves of Coringa and Kankinada Bay of Andhra Pradesh - in the Krishna and Godavari estuaries, (iv) Mangroves of Muthupet, Pichavaram and Chhatram areas of the Cauveri delta in Tamil Nadu coast, (v) Mangroves of Cochin estuary at Kerala coast, (vi) Mangroves of Coondapur and Malpe Bay at the Karnataka coast, (vii) Mangroves of Mandovi, Zuari and other estuaries of Goa, (viii) Mangroves of Ratnagiri and other estuaries of Maharashtra coast, (ix) Mangroves of the Gulf of Cambay and Mangroves of the Gulf of Kuchchh of Gujarat and (x) Mangroves of Coral reef of Lakshadweep Atolls & Minicoy Islands of Arabian Sea. Naskar and Mandal (1999) had collected and studied the information regarding the total Indian mangrove areas and the mangrove species diversity in India. Details of informations about the Indian mangrove areas and mangrove species composition are also highlighted by Naskar & Guha Bakshi (1987), Naskar & Mandal (1999) in the books, entitled (i) *'Mangrove Swamps of Indian Sundarbans- An Ecological Perspectives'* & (ii) *'Ecology and Biodiversity of Indian Mangroves and Its Global Status'*. Indian mangrove areas are highlighted in the Pie Chart-1. All these Indian and World mangrove species and mangrove areas are shown in Table- 2 & Table - 3.

ESTIMATION OF MANGROVES AND MANGROVE AREA IN THE INDIAN SUB-CONTINENT

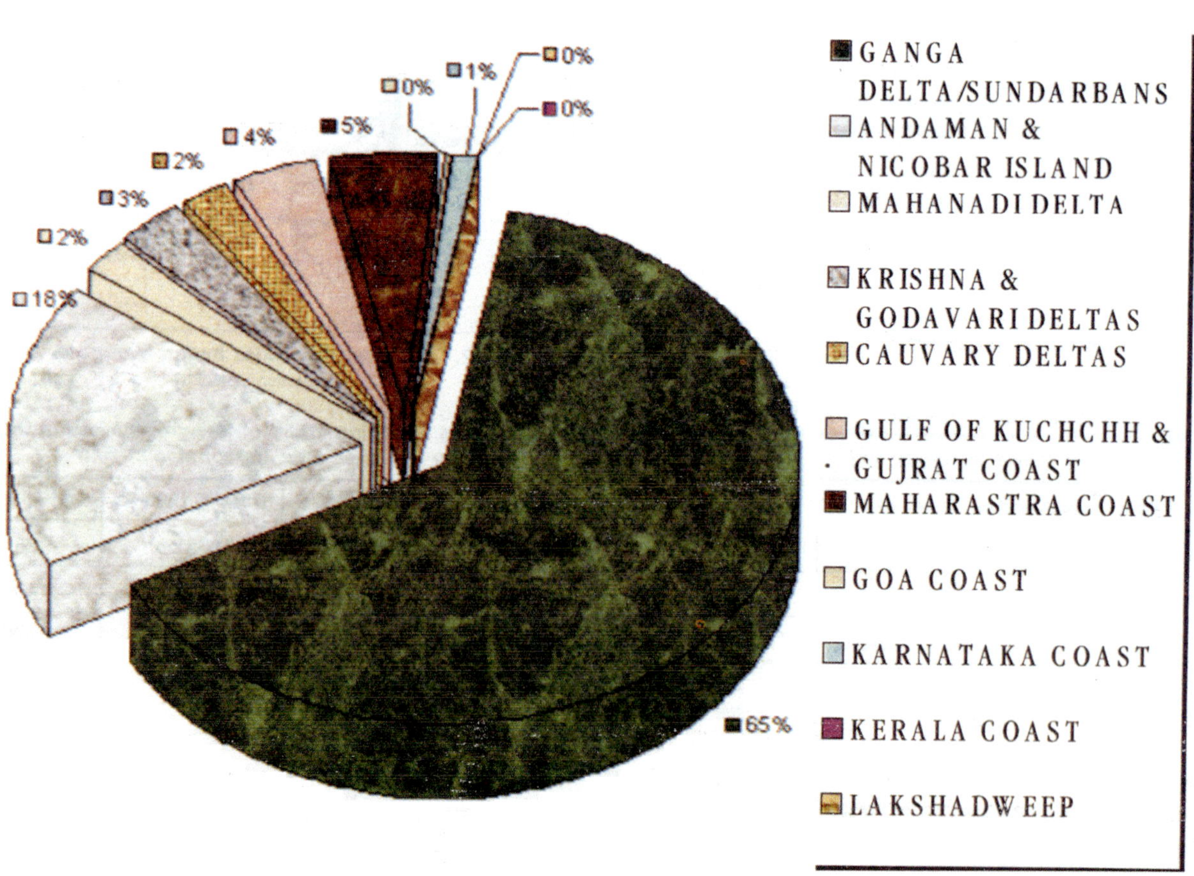

Pie Chart - 1. Showing the different mangrove zones in India along with their percentage

(i) Mangroves of the Andaman & Nicobar Islands of the Bay of Bengal

The Andaman Group of Islands are situated within the latitude 10° 30′ N - 14° 0′ N and longitude 92° E - 94° E; while the Nicobar Group of Islands are spread over within latitude 6° 45′ N - 9° 15′ N and longitude between 92° E and 94° E. In these Andaman Groups, there are about 278 major islands (*viz.*, Andaman, Middle Andaman, South Andaman and Little Andaman); while in the Nicobar Groups there are about 30 major islands. These mangrove species are mostly distributed in narrow patches on the sheltered coastal areas of different islands. Here the tidal inundation exist but the wave action is minimum. Different mangrove workers had estimated the mangrove areas and mangrove species in Andaman & Nicobar Islands differently. According to the estimate of Dagar, *et al.* (1991) the dominant mangrove flora in these Andaman & Nicobar Islands are 34 species; these are *Rhizophora apiculata, R. mucronata, R. lamarkii, R. stylosa, Bruguiera cylindrica, B. gymnorhiza, B. parviflora, B. sexangula, Ceriops tagal, C. decandra, Kandelia candel, Avicennia marina, A. officinalis, Sonneratia alba, S. apetala, S. caseolaris, S. griffithii, Heritiera littoralis, Xylocarpus granatum, X. mekongensis, X. molluccensis, Aegiceras corniculatum, Aegialitis rotundifolia, Scyphiphora hydrophyllacea, Lumnitzera littorea, L. racemosa, Excoecaria agallocha, Cynometra iripa, C. ramiflora, Nypa fruticans, Phoenix paludosa, Acanthus ilicifolius, A. ebracteatus, A. volubilis* (Naskar & Mandal, 1999 and Debnath & Naskar, 2000).

These mangrove flora and the mangrove zones of the Andaman & Nicobar Islands are least disturbed and best preserved in comparison to the others mangrove areas within the Indian Territories (Naskar & Mandal, 1999).

(ii) Mangroves of Bhitarkanika of the Mahanadi delta and other Mangals of the Orissa coast

The Orissa coastal mangal and Bhitarkanika mangrove areas are situated within the latitude 19°N - 22°N and longitude between 85°E and 87° E. The river system, *viz.*, Brahmani and Baitarani form the deltaic mangrove zones in Orissa; besides these, Sanyal, *et al.* (1999) had reported about 20.0 km^2 mangrove areas from the mouth of the river Subarnarekha. The Bhitarkanika mangrove in the Orissa costal areas is the 3rd important mangrove area in India, *i.e.*, after Sundarbans and Andaman & Nicobar Islands mangal (Naskar & Mandal, 1999). Mangrove areas of the Orissa coast are both dense and sparse, which are estimated through Satellite imagery as 108.5 km^2 and 84.4 km^2, respectively (Anonymous, 1992). In Government of India publication (1987), Indian Mangrove Status Report (1987) has mentioned 120 km^2 mangroves areas present in the Mahanadi and 30 km^2 mangroves in the other coastal zones/deltas in Orissa State. Jagtap, *et al* (1993) have reported the total mangrove area in the Orissa coast as 210 km^2 and the total mangrove species as 36 numbers. But, Blasco (1975) has reported only 50 km^2 mangal present in the Orissa state; these mangroves are distributed in three zones, *viz.*, (I) Mangroves of the Mahanadi Delta, (ii) Mangroves of the Brahmani - Baitarini Delta and (iii) Mangroves of the Balasore coast. Naskar & Mandal (1999) have reported the following mangroves and mangrove associated species from the Orissa coast based on the reports of other workers and their own studies, *viz.*, *Rhizophora apiculata, R. stylosa, Bruguiera gymnorhiza, B. sexangula, B. parviflora, B. cylindrica, C.decandra, C. tagal, K. candel, A. alba, A. marina, A. officinalis, A. marina* (var. *acutissima*), *S. apetala, S. alba, S. caseolaris, S. griffithi, X. granatum, X. molluccensis, X. mekongensis, A. cucullata, A. corniculatum, H. fomes, H. littoralis, H. kanikensis, L. racemosa, E. agallocha, S. hydrophyllaceae, D. trifoliata, D. heterophylla, A. correa, C. ramiflora, A. ilicifolius, S. nudiflora, S. maritima, S. monoica, S. brachiata, S. portulacastrum, P. coarctata, M. wightiana, N. fruticans, P. paludosa, C. manghas, D. spinosa, S. globosus,S. carinatus, D. spathaceae, H. curassavicum, C. bonduc, C. crista, H. tiliaceous, T. populnea, T. populneoides, T. dioca, T. troupii, Brownlowia tersa, Clerodendrum inerme, Scirpus littoralis, Tylophora tenuis, Finlaysonia obovata,, Barringtonia racemosa, Stenoclaena palustris, Acrostichum aureum, Intsia bijuga, Mucuna gigantea, Ipomoea tuba, Cynometra iripa, Pluchera indica, Salvadora persica, Salacia prinoides.* Bhitarkanika mangal is also declared as an important sanctuary, with several endangered species of flora and fauna and also declared as the '*Wildlife sanctuary*' for having wide spectrum of '*Genetic diversity*'. This Bhitarkanika mangal is the important natural breeding place for estuarine crocodiles, *Crocodilus porosus* and several species of threatened sea turtles, *viz.*, *Lepidochelys olivacea, Batagur baska* and the king crab, *i.e.*, *Carcinoscorpius rotundicanda* and *Tachypleus gigas.* Several dolphin species, *viz.*, *Platinista gangetica, Neomeris phocaenoides, Sotalia plumba* are also reported from this Bhitarkanika mangal (Naskar & Mandal, 1999).

Haines (1921) reported the following species of mangroves and mangrove associates from Orissa coastal areas, *viz.*, *Kandelia candel* (Syn. *K. rheedii*), *Salvodora persica*, *Heritiera fomes*, *Aegialitis rotundifolia*, *Aegiceras corniculatum* (Syn = *A. majus*), *Phoenix paludosa*, *Acrostichum aureum*, *Rhizophora mucronata*, *Rhizophora apiculata* (syn = *R. conjugata*), *Bruguiera gymnorhiza*, *Bruguiera cylindrica* (syn. = *B. caryophylloides*), *Ceriops decandra* (syn. = *C. roxburghiana*), *Excoecaria agallocha*. Haines (1921) also reported three species of salt worts, *viz.*, *Suaeda nudiflora*, *Suaeda maritima* and *Suaeda monoica*. Haines (1935) also reported *Nypa fruitcans* from the Orissa coast. Blasco (1975) reported *Kandelia candel*, *Sonneratia caseolaris*, *Scyphiphora hydrophyllacea* from the Mahanadi delta. Haines (1921). reported plenty growth of *Phoenix paludosa* from the river banks and sea shores of Orissa, which is now scarce.

(iii) Mangroves of Coringa, Kankinada Bay, the Krishna and the Godavari Estuaries of Andhra Pradesh coast

In Andhra Pradesh coastal areas the mangroves are situated within the latitude between 13° 30′ N and 19° 00′N and longitude between 80° 05′ E and 85°05′ E . The major estuaries, *viz.*, the Godavari and the Krishna form the deltas in the coastal areas of Andhra Pradesh. These estuaries and deltas comprises the bays, tidal creeks, extensive tidal mudflats, pits, bars, mangrove swamps, marshes, ridges and swale areas and coastal alluvium plains (Anonymous, 1992). Few pockets of mangroves are present on the inter-tidal flat land at the mouth of the Sarada river and near Visakhapatnam. These mangrove areas are also distributed in several minor islands, mouth of the estuaries, tidal flats, on both sides of the tidal creeks and the river mouths. Dense mangrove areas are reported as 263.3 km^2 on the Nizampatam Bay, Machilipatnam Bay, on the western side of the Krishna delta. Sparse mangroves of an area about 88.7 km^2 is also reported on the land-ward side of the high and inter-tidal mud flats in the eastern side of the Krishna delta; scattered mangroves are also reported from an area of about 4.7 km^2 area (Anonymous, 1992). On the junction of the Coringa river at Bay near Kakinada, several patches of mangroves are reported (Naskar and Guha Bakshi, 1987). According to the recent estimate, the mangrove areas in the Andhra Pradesh is 322.7 km^2 (1992); while Sidhu (1963) has reported 1,3304 ha mangrove areas from the mouth of the Godavari and 5,120 ha mangrove areas from the mouth area of the Krishna river. Blasco (1975) has reported only 100 km^2 mangrove area in Andhra Pradesh and the Mangrove Status Report, Government of India (1987) has reported about 200 km^2 mangrove areas of the Andhra Pradesh.

The mangrove species reported from the Andhra Pradesh are as follows, *viz.*, *Rhizophora mucronata*, *R. apiculata*, *Bruguiera gymnorhiza*, *Ceriops decandra*, *Avicennia alba*, *A. marina*, *A. officinalis*, *Sonneratia apetala*, *Lumnitzera racemosa*, *Excoecaria agallocha*, *Acanthus ilicifolius*, *Aegiceras corniculatum*, *Scyphiphora hydrophyllacea*, *Suaeda maritima*, *S. monoica*, *S. nudiflora*, *S. brachiata*, *Arthrocnomum fruticosum*, *Porteresia coarctata*, *Myriostachya wightiana*, *Xylocarpus moluccensis*, *Sarcolobus carinatus*, (Anonymous, 1992).

(iv) Mangroves of Muthupet, Pichavaram and Chhatram area of the Cauveri delta in Tamil Nadu coast

From the coastal zone of Tamil Nadu State only about 30 km^2 mangrove areas are reported (Anonymous, 1987); these mangroves are spread over in the river mouths at Pichavaram in Cuddalore, Vadaranyam, Point Calimere, Muthupet, Chhatram and Tuticorin. The Pichavaram mangrove area alone covers about 13.6 km^2 mangrove areas (Venkatesan, 1996). In Muthupet patches of mangrove areas had reported by the earlier workers (Anonymous, 1987). Though the Tuticorin and Pamban Island had dense mangrove forest few decades back, but large scale exploitation and cattle grazing had caused maximum damage and turned these mangrove areas to naked coastal sands. According to different workers, these mangrove areas at the Cauvery delta are variable largely. Tirumalai Rai (1959) had reported the effected forest area are only 10.08 km^2; but Sidhu (1963) had estimated as 26.50 km^2 mangrove areas from the Tamil Nadu State. Venkatesan (1966) has estimated as 148.97 km^2 mangrove area including Pichavaram- 13. 58 km^2, Muthupet -68.03 km^2, Chhatram - 53.47 km^2. Blasco (1975) has also estimated about 5-10 km^2 mangrove areas in Tamil Nadu coast line.

(v) Mangroves of the Cochin estuary of the Kerala coast

The Kerala coast has beautiful beaches, minor estuaries and lagoons. Cochin estuary, Kumargom, Dharmadon, Cheltuva, Nadavavu, Pappinisseri, Kunjimangalam, Chateri, Veli, *etc.* have small scattered mangrove patches. Only sporadic mangrove patches are present in this Kerala coast, covering about 45 ha (Anonymous, 1987). Ramchandran, *et al.* (1986) has reported the following mangrove areas from the Kerala coast, *viz.*, (a) Veli- estuary, (b) Quilon, (c) Kumarakon, (d) Kannamali (Cochin), (e) Chetwai, (f) Nadakkavu (Calicut), (g) Edakkad, (h) Pappinisseri, (i) Kungimangalam and (j) Chittari.

(vi) Mangroves of Coondapur and Malpe Bay of the Karnataka coast

The Karnataka coast line is extended about 320 km long from north to south, within the latitude 12° 45′N-15° 00′ N and longitude between 74° 00′E and 75° 00′E. In this coast line, the mangroves are fringing type, in the inter-tidal zones, covering about 50 km^2 area (Untawale & Wafar, 1986); based on the satellite imagery about 12.5 km^2 mangrove area is reported Anonymous (1992). According to the the Government of India Mangrove Status Report (1987) the mangrove areas in the Karnataka coast is about 60 km^2. Several rivers, after originating from the Western Ghats meet the Arabian sea and form wide and extensive estuaries, lagoons or backwaters and in these inter-tidal zones the mangrove vegetations are fringing type and in its back on the upland the sand dune vegetation grows. But in these coastal water marine algae and sea-weeds are plentiful (Rao, *et al.*, 1980). The soil in these coastal area is mostly clayee; water and soil are polyhaline to moderate saline.

The major mangrove areas in these Karnataka coast are Coondapur estuarine complex, along the river Sharavati at Karki fall, Haldipur, Dhareshur, Khonavar. Coondapur is the confluence zone of 3 rivers, *viz.*, Chakra, Kolluru and Haladi. Here about 307 ha inter-tidal lands have developed with good mangrove plants; these plants are *Avicennia officinalis, A. marina, Rhizophora mucronata, R. apiculata, Acanthus ilicifolius* (Untawale & Wafar, 1986). Next to Coondapur, the distinct mangrove patches are noticed at Malphe and Karwar (Untawale, 1980). Besides all these, the Kalinadi, the Bemdarnadi, the Ganga valli and the Agnashini are the other rivers, where the mangrove species grow in narow strip fringing type. Other estuaries in Karnataka are Sita Swarna, Mulki and Kopal, where minor mangrove habitats are existing. Fringing mangroves are also noticed in the Netravati and near Kumta. Untawale & Wafar (1986) had reported total of 14 mangroves, mangrove associated and sand dune flora under 9 genera and 7 families. These species are *Avicennia officinalis, A. marina, Rhizophora mucronata, R. apiculata, Kandelia candel, Sonneratia caseolaris, S. alba, Ceriops tagal, Aegiceras corniculatum, Excoecaria agallocha, Acanthus ilicifolius, Clerodendrum inerme, Kandelia candel, Sonneratia caseolaris, Ceriops tagal, Ipomoea pes-caprae* and *Spinifles littorales, etc.*

(vii) Mangroves of the Mandovi, the Zuari and other Estuaries of Goa

This coastal region is situated in the Central-Western Coast of India, facing the Arabian sea and extended north to south. The total length of the coast line at the Territory of Goa is approximately 120 km and within the latitude between 15°00′N -15°52′N and longitude between 73°38′E -74°44′ E. These inter-tidal zones of seven minor estuaries are mostly the flat lands and are formed with silty-sand and silty-clay, along with abundant organic matters. On both the sides of these minor rivers, the soils are mostly aluvial deposit and laterite soil with high percentage of Iron (Fe) and Manganese (Mn). The different phytozonation of this coastal region are influenced by salinity gradients. This coastal stretch is indented with seacliffs and notches (Ahmed, 1972).

In these coastal zones, the minor estuaries and rivers from north to south are (i) the Terakhol, in the border of Maharashtra and Goa, (ii) the Chapora, (iii) the Mandovi, (iv) the Zuari, (v) the Sal, (vi) the Talpona, and (vii) the Galgibag, in the border of Goa and Karnataka. Besides these, the connecting canal, Cumbargia link these two major estuaries Mandovi and Zuari and in its side fringing mangroves are also studied. These total mangrove areas in the Goa state is estimated about 20 km^2 and the dominant mangroves and the mangrove associated species are *Avicennia officinalis, A. marina, Rhizophora mucronata, R. apiculata, Bruguiera gymnorhiza, B. cylindrica, Kandelia candel, Sonneratia caseolaris, S. alba,*

Ceriops tagal, Aegiceras corniculatum, Excoecaria agallocha, Acanthus ilicifolius, Acrostichum aureum, Derris trifoliata, D. heterophylla, D. indica, Clerodendrum inerme, Kandelia candel, Sonneratia caseolaris, Ceriops tagal, Ipomoea pescaprae, Premna sp., Sesuvium portulacastrum, Salvadora persica, Dolichandrone spathacea, Hibiscus tiliaceous, Thespesia populnea, Aeluropus lagopoides and *Spinifles littorales;* the other species, *viz., Xylocarpus granatum, Bruguiera cylindrica, Sonneratia apetala* and *Cynometra ramiflora* are also reported from the Goa coastal mangals (Untale, 1985).

Government of India had declared the mangrove area of the Chorao Island as the 'Biosphere Reserve'. But from the critical field studies it was revealed that the mangrove areas in Goa, particularly at Agassi - Kortalin mangrove areas are damaged considerable for the construction of railway bridge on the Zuari river and development of shrimp farm near the Chorao island and also in the other mangrove areas; these mangrove areas in Goa are in very much degraded condition due to heavy human pressure and several developmental activities (Naskar & Mandal, 1999).

(viii) Mangroves of Ratnagiri and other Estuaries of the Maharashtra State

Untawale (1987) has stated that till the year 1670, the Bombay coast was covered with luxuriant mangroves in its seven islands. All these fringes and dense mangrove areas were cleared for human settlement, development of port and for rapid industrialisation. Now in and around Mumbai the following bays and beaches are noticed with scattered and fringing mangroves; these areas having the past sign of mangrove vegetation are Manoribay, Vikrolibay, Mahimbay, Backbay, Colaba beach, Malabar hill, Worli hill, Bandra beach, Manoli beach, Versova beach, Juhu beach, Girgaon beach, *etc.* Besides these, mangrove patches are also noted from Ratnagiri, Raigad, Thane, Mahim, Elephanta Island, Waghotan, Rajapur, Dharma, Vasai (Ulhas), Shastri and Vikroli, *etc.* From these fringing and scattered mangrove patches few mangrove species were also reported by the recent publication (Jagtap, *et al.*, 1993); these species are *Avicennia officinalis, A. alba, A. marina, A. marina* var. *acutissima, Rhizophora mucronata, Bruguiera gymnorhiza, Kandelia candel, Sonneratia apetala, S. caseolaris, S. alba, Ceriops tagal, Lumnitzera racemosa, Aegiceras corniculatum, Excoecaria agallocha, Acanthus ilicifolius, Acrostichum aureum, D. heterophylla, Clerodendrum inerme, C. soratum* and *Salvadora persica.*

Navalkar (1940, 1951, 1952, 1956, 1959, 1961 & 1974) and Navalkar & Bharucha (1949 & 1959) and few others had studied and reported details of the ecology, succession, distribution, floristic pattern and importance, chemical properties of the mangroves of the Maharashtra coast. But presently, most of these mangrove areas are in very much degraded or fully cleared for the rapid urbanisation activities.

(ix) Mangroves of the Gulfs of Cambay and Kuchchh of the Gujarat State

Gujarat coast line is extended within the longitude 68° E and 73°E and latitude between 20° 10'N and- 23° 10' N. These entire coastal zones are much variable as varied physiography, geomorphology, neotectonic movement of rivers, creeks and ocean, where continuous river discharge prevailed. Besides these, the nature of sand, silt, rock and overall climatic factors are also variable from zone to zone in the Gujarat coast. Based on these geographical situation, geomorphological position and variable nature of land or physiognomical condition, the Gujarat coastal zones may be divided under following 5 regions, *viz.*,(i) the Runn of Kuchchh, (ii) the Gulf of Kachchh, (iii) the Gulf of Cambay (Khambhat or Khambyat), (iv) the Saurashtra coast and (v) the South Gujarat coast. Out of these five regions, two Gulfs constitute the major mangrove zones of the Gujarat State. The substrate constituents, temperature, rainfall pattern, topography and the tidal ingress varies within these arid zones. Inspite of all these, mangals of these zones are more or less similar in their qualitative and quantitative aspects. In these regions sporadic mangroves are distributed and the major reported species are *Avicennia officinalis, A. marina, Rhizophora mucronata, Suaeda fruticosa, Aeluropus lagopoides, Cenchrus setigerus, C. ciliaris, Sporobolus marginatus, Cyperus rotundus, Salsola baryosama, Haloxylon salicornicum, Prosopis chilensis, Acacia nilotica, Tamarix dioica* and *Salvadora persica.* All these mangrove areas of the Gujarat coast are very much degraded condition due to fire wood exploitation and camel and wild ass grazing (Anonymous, 1987).

(x) Coral reef mangroves of Lakshadweep Atolls & Minicoy Islands at Atlantic Ocean

In the Arabian Sea, the coral islands (Atolls) were previously called Laccadive, Minicoy Islands and Amindivi Islands. All these islands are presently (since 1973) known as Lakshadweep. These Lakshadweep Islands remain within the Administrative control of the Union Territory of Lakshadweep. This group is of about 36 islands and is situated 200 - 470 km apart from the Kerala coast, within the latitude, between 10° 00' N and 12° 00'N and longitude between 71° 00′ E and 74° 00′ E. Furthermore, the Minicoy Islands are situated between the 8°18′N latitude and 73° 00′E longitude, which are the southern most islands and present about 180 km away from the Lakshadweep. All these coral Atolls have north-east and south-west orientations, with the low lying islands on the east and a broad well developed reef on the west. In its centre, 1-4 km broad lagoons exist. Steep slope on the wind wards (east) side and lagoons on the west shelf deposit are of marine origin and all these islands are in the process of formation. The only mangrove species, *Bruguiera parviflora* is reported from these Atolls (Untawale & Jagtap, 1984).

A. *Rhizophora* sp. juvenile plant with stilt roots, Andaman

B. Mangrove thickets at Sipighat, Port Blair, Andaman

C. *Acanthus ebracteatus*, *Nypa fruticans* and *Avicennia marina* at Sipighat, Port Blair

PhotoPlate-12. Mangroves of Port Blair. A.*Rhizophora* sp., B&C. *Acanthus ebracteatus, Nypa fruticans* & *Avicennia* sp.

A. Radhanagar beach, Havelock, Andaman-Mangrove Vegetation(2001)

B. Beach flora at Havelock, Andaman

C. Prop roots of *Rhizophora* sp. at Old Wandoor Bay, Andaman (1991)

D. Prop roots of *Rhizophora* sp. at Old Wandoor Bay, Andaman (1991)

E. Dense coastal flora, Middle Strait-Andaman

F. Inter-tidal coastal mangrove flora at Mohanpur, near Mayabandar

G. *Rhizophora* sp, tree at Baratang, Andaman

Photo Plate - 13. A, B, C, D, E, F and G -- Mangroves of the Andaman Islands, Port Blair.

A. Mangroves of the Elephanta Cave, Mumbai (1990)

B. Mangroves of the Mahim Creek, Mumbai (During 1990)

C. Discussion with Dr. M. Vannucci at Mahim (1990)

D. Tidal Creek & Mangroves at Vicroli, Mumbai (1990)

E. Discussion with Dr. K. Krishnamurthy, at Mahim

F. Discussion with Dr. A. G. Untawale, at Mahim (1990)

Photo Plate - 14. Mangrove Areas at Mumbai and author with the Distinguished Mangrove Researchers (During 1990)

A. River bank Mangroves of the Bhitarkanika, Orissa

B. Mangroves of the Bhitarkanika, Orissa

C. Mangroves of the Bhitarkanika, Orissa

D. Mangroves of the Bhitarkanika, Orissa

E. Mangroves and Tidal river of the Bhitarkanika, Orissa Coast

F. Tidal creek, Subarnarekha Mouth, Orissa

G. Mangroves and Tidal river of the Bhitarkanika, Orissa Coast

H. Shrimp seed collection, Balasore coast, Orissa

Photo Plate -15. Mangroves of the Bhitarkanika, Mahanadi Delta and the Subarnarekha mouth of Orissa

A. Konkan Raiway Bridge Construction at Zuari (Agassi-Kortalin) & Destruction of Mangrove Habitat, 1994

B. Konkan Railway-Bridge Construction at Zuari (Agassi-Kartalin) & Destruction of Mangrove Habitat, 1994

C. Mangrove forest of Chorao Island, Goa (1994)

D. Mangroves of Goa, *Rhizophora* sp. seedling (1994)

E. Mangrove forest of Chorao Island, Goa (1994)

F. Mangroves of Mandovi River side, Goa (1994)

G. Mangroves of Agassi-Kortalim, Zuari river, Goa (1994)

H. Mangroves on the river side at Mandovi, Goa (1994)

Photo Plate - 16. Mangroves of the Mandovi and the Zuari Estuaries of Goa (During 1994)

A. Mangrove wood at Ghusighata, Sundarbans. B. & C. Country boats carried mangrove wood from Sundarbans forest. D, E & F. Mangrove wood at Ghusighata. G. Exploitation of mangrove wood by local people. H. Mangrove wood at Ghusighata.

Plate Plate - 17. Exploitation of Mangrove Wood for Timber and Fuel Wood.

3

INDIAN SUNDARBANS

Sundarbans, the name is derived from the common and local name of the plant, *Sundari* (*Heritiera fomes* Buch.-Ham.). Others believe that this inter-tidal forest and vast water-ways looks beautiful or '*Sundar*'; so the name is derived from *Sundar* (beautiful) + ban(forest). Other views are that the forest facing 'Sumudra' or the Ocean, so it is Sundarbans; this area was the past Kingdom of 'King of Chandra Dweep' and from '*Chandradweep*' to Sundarbans was derived. However, this inter-tidal deltaic forest in the estuarine mouths of the two vast rivers, the Ganga and the Brahamaputra has formed the single largest deltaic land. Here the diverse mangrove flora and vast group of terrestrial, avian and aquatic fauna are present. This forest is also considered as the intermediate stage between the sea and land; which is important for migration of aquatic organisms, from sea to land.

Three centuries back forest and river areas in this undivided Sundarbans, of present day's India and Bangladesh, was more than 20,000 km² and restricted within the latitude 21°N - 22°31´N and longitude between 88°10´E and 92°15´E. This is extended from the mouth of the Hugli river to the Chattagram coast; which covered four coastal districts of Bangladesh and two coastal districts of West Bengal. Most of this mangrove forest area was reclaimed for human settlement, *i.e.*, renovation of agricultural fields, fish culture farms, salt pans and development of ports or harbours. Within three centuries, this mangrove forest zone including its surrounding rivers was reduced to half, *i.e.*, to about 10,000 km²; and in its back remained mangrove reclaimed areas (Map-3). The early settlers had exploited the mangrove forest indiscriminately. As a result, the virgin but unique mangrove forest in Sundarbans is now under threat and continuous degradation.

During partition of Bengal (1947) more than 66% of the total mangrove areas of Sundarbans had fallen within the jurisdiction of Bangladesh and only 34% mangrove forest remained within India, between the latitude 21°31´N and 22°30´N and longitude between 88°10´E and 89°51´E (Map-3 & Map-4). Till date, the total area of Sundarbans in India is only 9,630 km²; out of these the mangrove forest areas including six major estuarine rivers, other rivers, creeks and canals is about 4,266.66 km² (Table - 5). During the last two centuries more than 50% of these mangrove areas in Indian Sundarbans alone was reclaimed, renovated to human habitations, agricultural fields, brackishwater fisheries, roads, railways and for other purposes. All these reclaimed human habitat areas of the Indian Sundarbans are now under 19 rural blocks; 13 blocks under South 24 Parganas and 6 blocks under North 24 Parganas (Map - 5 and Table – 5 & 6). During the beginning of the 21st Century, the population in these 19 rural blocks of Sundarbans was about 42.0 lakh, which was only 4.87 lakh during 1901 in the undivided Sundarbans(Table-7). This mangrove forest is now losing its glamour and several plant and animal species have already become extinct or threatened. On the contrary, Government of India and the local State Government have taken up several steps towards its conservation, protection and management by developing or setting up several management policies and projects (Table- 8, 9, 10 & 11).

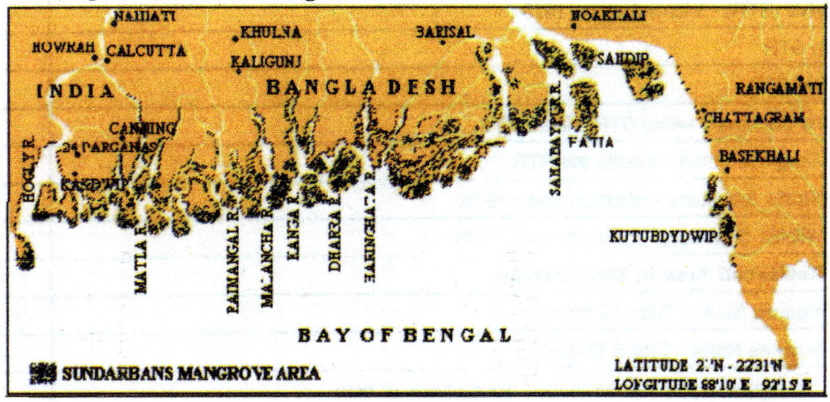

Map 3. Undivided Sundarbans of Bangladesh & India (World's Largest Single Mangrove Chunk)

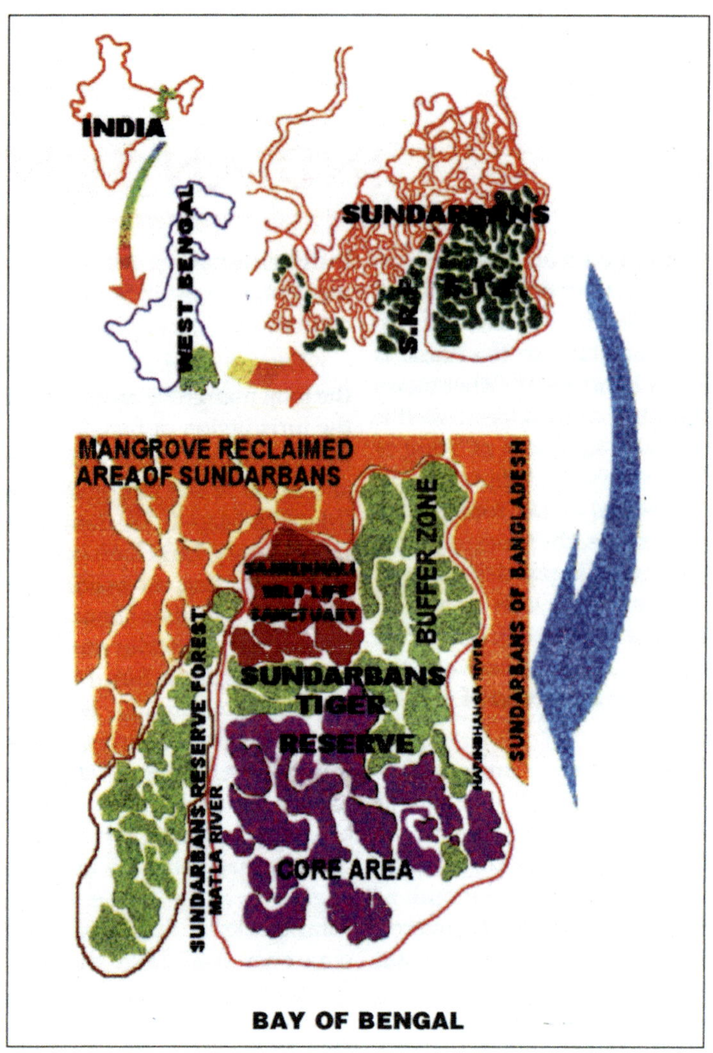

MAP 4. Area and position of Indian Sundarbans Mangrove Areas and Sundarbans Tiger Reserve (STR)

Table 5 : Geographical Areas and Nature of the Landscape in Indian Sundarbans

Total Areas in Indian Sundarbans, 24-Parganas South and 24-Parganas North Districts	**9630.00 km².**
A. Total Sundarbans Mangrove Areas (24-Parganas South & North)	4266.66 km².
(i) Sundarbans Tiger Reserve (STR) - initiated, since 1973	2585.10 km².
(a) Core Area within STR	1330.10 km².
(b) Buffer Area within STR	892.60 km².
(c) Sajnakhali Wildlife Sanctuary within STR, since 1976	362.40 km².
(ii) Sundarbans Reserve Mangrove Forest Outside the STR	1681.50 km².
(a) Lothian Island Wildlife Sanctuary - initiated, since 1976	38.00 km².
(b) Holyday Island Wildlife Sanctuary - initiated, since 1976	5.95 km².
B. Sundarbans Mangrove Reclaimed Area in West Bengal	5363.40 km².
(i) Under the District 24-Parganas South - Total 13 Blocks	3407.49 km².
(ii) Under the District 24-Parganas North - Total 6 Blocks	1106.77 km².
(iii) Rivers, Canals, Roads, Railways, etc. in the Reclaimed Sundarbans of both South & North 24-Parganas Districts	849.14 km². (approximately)

Table 6 : Total Areas, Cultivated Areas and Population in Different Mangrove Reclaimed Areas of 24-Parganas Districts (S & N).

Sl. No.	Blocks	Geographical Area (hectare)	Cultivated area (hectare)	Irrigable area (hectare)	Population (1991 Census)
A. Thirteen (13) Blocks of Sundarbans Under the District 24-Parganas (South)					
1.	Gosaba	33,725 ha	24,990 ha	4,530 ha	2,00,514
2.	Basanti	29,000 ha	23,359 ha	980 ha	2,26,974
3.	Canning-I	20,568 ha	14,252 ha	1,045 ha	2,06,100 (app.)
4.	Canning-II	22,528 ha	18,090 ha	2,240 ha	1,51,635
5.	Joynagar-I	12,711 ha	9,903 ha	1,740 ha	1,85,271
6.	Joynagar-II	17,518 ha	14,778 ha	1,295 ha	1,77,335
7.	Kultali	23,948 ha	18,826 ha	1,160 ha	1,56,450
8.	Mathurapur-I	14,838 ha	12,120 ha	1,860 ha	1,41,888
9.	Mathurapur-II	23,050 ha	18,425 ha	3,306 ha	1,72,982
10.	Patharpratima	46,950 ha	31,455 ha	2,865 ha	2,45,601
11.	Namkhana	22,723 ha	16,880 ha	1,720 ha	1,34,354
12.	Sagar	47,080 ha	15,544 ha	2,091 ha	1,54,202
13.	Kakdwip	26,110 ha	21,080 ha	1,670 ha	1,90,088
	Total	**3,40,749 ha +**	**2,39,702 ha +**	**26,502 ha +**	**23,43,394 ha +**
B. Six (06) Blocks of Sundarbans under the District 24-Parganas (North)					
14.	Minakhan	20,095 ha	11,590 ha	1,687 ha	1,37,361
15.	Haroa	15,675 ha	9,770 ha	4,456 ha	1,51,100
16.	Sandeshkhali-I	17,679 ha	12,000 ha	457 ha	1,20,539
17.	Sandeshkhali-II	19,500 ha	12,000 ha	446 ha	1,18,895
18.	Hasnabad	14,688 ha	11,500 ha	971 ha	1,51,115
19.	Hingalganj	23,040 ha	14,000 ha	522 ha	1,42,291
	Total	**1,10,677 ha =**	**70,860 ha =**	**8,539 ha =**	**8,21,301 =**
	Grand total	**4,514,26 sq. km.**	**3,105,62 sq. km.**	**350,41 sq. km.**	**31,64,695**

- Total Area Under the reclaimed Sundarbans Mangroves = 5363.40 km^2.
- Total Area Under the Reclaimed Agricultural Land in Sundarbans = 4,514.26 km^2
- Total Areas Under the Rivers, Canals, Roads, Railways, in Sundarbans = 849.14 km^2.

Population of 19 rural blocks of Sundarbans during 2001 census (ADB Report)

1. Basmati - 278543, 2. Canning-I - 244354, 3. Canning-II - 195967, 4. Gosaba - 222764, 5. Joynagas-I - 219009, 6. Joynagar-II - 209136, 7. Kakdwip - 239381, 8. Kultali - 187942, 9. Mathurabur-I - 164585, 10. Malhurapur-II - 198261, 11. Namkhana - 160630, 12. Patharpratima - 288259, 13. Sagar - 185630, 14. Haroa - 182499, 15. Hasnabad - 177470, 16. Hingalganj - 156568, 17. Minakhan - 168233, 18. Sandeshkhali-I - 140446, 19. Sandeshkhali-II - 136247.

Total Population in Sundarban = 37,55,934 (2001 Census)

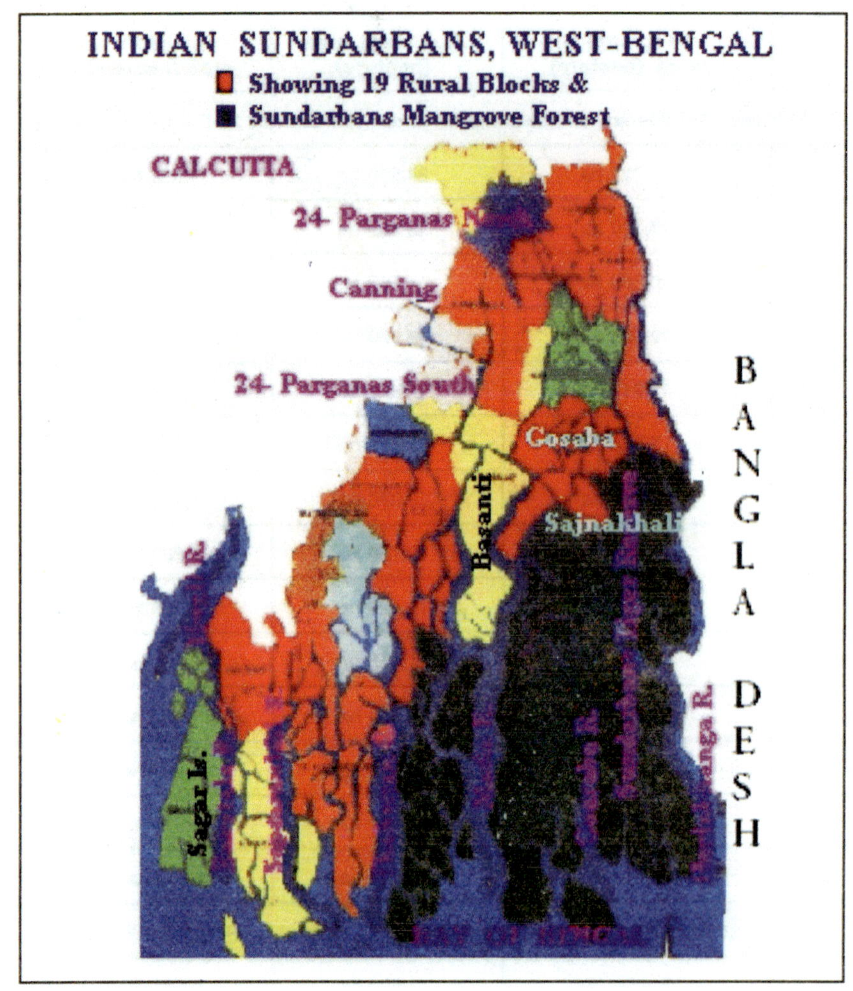

Map 5. Indian Sundarbans, covering 9,630 km² areas of Mangroves and mangrove reclaimed 24- Parganas South & North Districts of West Bengal

Table 7 : Population Growth in the Indian Sundarbans Mangrove Reclaimed Zone, India (Modified after Das, 1996 & Kanjilal, 2000)

Sl. No.	Census Report, since 1772	Total Population	Per cent of Rise (%)
1.	Census Report, since 1772	2,96,045	–
2.	Census Report, 1901	4,87,377	65%
3.	Census Report, 1931	7,54,421	55%
4.	Census Report, 1951	11,59,559	54%
5.	Census Report, 1961	14,46,282	–
6.	Census Report, 1971	About 18 lakh	–
7.	Census Report, 1981	About 27 lakh	–
8.	Census Report, 1991	31,54,890	176% after independence
9.	Estimated population during 2001	About 37.5 lakh	

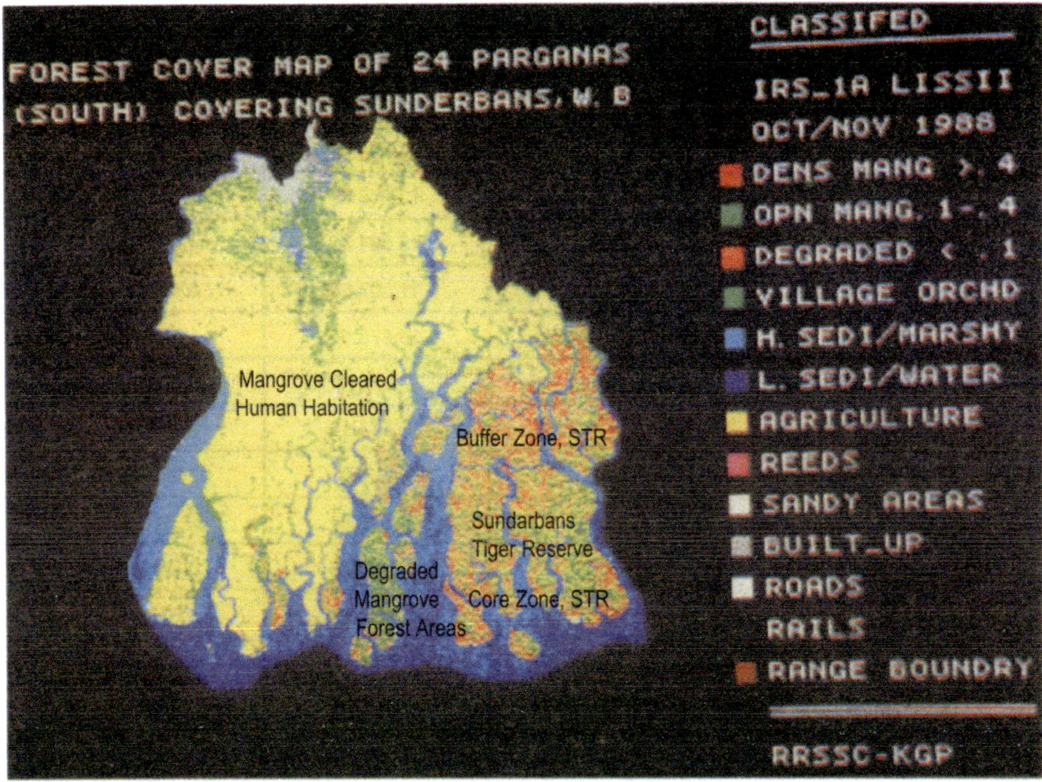

Map 6. Satellite map of the Sundarbans under the District 24 Parganas South

With the help of satellite map (Map - 6), remote census tools and techniques, the present status of the mangrove forest coverage within Indian part of the Sundarbans has clearly focussed the threatened and degrading condition. Now, the dense mangrove forest cover is restricted mainly in the south-eastern part, within the Sundarbans Tiger Reserve. The population pressure has also played dominant role for declining these mangrove forest coverage. Along with this, the tilting effect of the Ganga water flow towards eastern part at Bangladesh also plays a role for declining of this forest. Still now, good number of mangrove species diversity exists in the eastern part of the Indian Sundarbans, within the STR, covering the inter-tidal mangrove forest land areas of about 1,680.10 km^2. These mangrove forest land is also under threat from the wood poachers, wildlife hunters and also due to inadequate supply of up stream perennial fresh-water, blockage or diversion of the earlier ice-melting flow through the Great River Ganga in its northern limits. But, it is also noted that, these dense mangrove cover areas of Sundarbans play vital role in stabilisation and physical protection of these coastal areas from frequent natural calamities along with having their economic potentialities towards supply the nutrients to estuarine and off-shore species of fish, shrimp/prawn, crabs, *etc*. For having these natural resources the vast number of local people have engaged themselves in fishing activities in the estuaries/ rivers of the Sundarbans.

The nutrient status of these inland/ estuarine rivers, creeks and canals are self-maintained for the existence of these dense mangrove forest at Sundarbans. But, the present trend of declining nature and rapid growing threat on these forest area degradation have raised several alarming signals towards the natural phenomenon, like accretion on river-bed, diversion of the river flow and others. The recent geological events have also turned these mangrove areas of Sundarbans towards rapid rate of declining of the important gene pool. Several endemic species of mammals, reptiles, birds, fish, shrimp and other aquatic and forest fauna have also disappeared from these mangal along with some of the unique mangrove species. Only, mere conservation measures by adopting several national and international projects and declaring these vulnerable areas as the *'World's Heritage Site'* or declaring the Sundarbans mangal as the important *'Ramsar site'*, the effective preservation or conservation of these mangrove forest will not be possible. Problems and prospects need to be identified on an urgent basis with the experts, those who have actual knowledge and noble interest for real conservation and protection ideas. Several decades had already past, but still time is there to undertake unbiased and true conservation measures and also for undertaking proper measure for diversion of the forest dwellers from their exploitation activities of this unique but threatened forest in Sundarbans mangal.

Table 8 : Different Forest Blocks under Sundarbans Tiger Reserve (STR), West Bengal

Sl. No.	Forest Blocks Within the Sundarbans Tiger Reserve & Nos. of Compartments	Total Area Covered (ha)	Forest Land Area Within STR (ha)	Water Area Within STR (ha)	Remarks/ Comments
1.	Panchamukhani - 5 nos.	17,665.96	13,946.77	3,719.19	Buffer Zone
2.	Pirkhali - 7 nos.	18,576.12	13,684.52	4,891.60	Buffer Zone
3.	Matla - 4 nos.	17,629.94	11,507.64	6,122.30	Core Area
4.	Chamta - 8 nos.	22,068.69	16,875.99	5,192.70	Core Area
5.	Chotohardi - 3 nos.	17,566.81	9,260.75	8,306.06	Core Area
6.	Goasaba - 4 nos.	17,173.03	10,514.10	6,658.93	Core Area
7.	Gona - 3 nos.	13,903.46	8,562.64	5,340.82	Core Area
8.	Bagmara - 8 nos.	29,393.35	16,576.51	12,816.84	Core Area
9.	Mayadwip - 5 nos.	27,336.26	12,998.15	14,338.11	Core Area
10.	Arbesi - 5 nos.	15,042.69	9,663.83	5,378.86	Buffer Zone
11.	Jhilla - 5 nos.	12,313.80	8,750.71	3,573.09	Buffer Zone
12.	Khatuajhuri - 3 nos.	13,241.37	9,547.68	3,693.69	Buffer Zone
13.	Harinbari - 3 nos.	11,686.91	8,391.04	3,296.87	Core Area
14.	Netidhopani - 3 nos.	9,300.00	6,447.68	2,852.32	Core Area
15.	Chandkhali - 4 nos.	15,590.00	11,293.15	4,297.50	Core Area
	Total = 15 Blocks	**2,58,489.04 ha**	**1,68,011.16 ha**	**90,477.88 ha**	
	2,585.10 Km²	**1680.10 Km²**	**905.00 Km²**		

Photo Plate- 18. Extensive Stilt roots of *Rhizophora mucronata* Lamk. & other mangroves in Sundarbans, India

Table 9 : Areas under Forest Blocks in the Sundarbans Reserve Forest, under 24-Parganas South (outside STR)

Sl. No.	Name(s) of the Forest Blocks in Sundarbans, W.B., outside STR	Sundarbans Reserve Forests under the control of South 24-Parganas Forest Department, SBR
1	Herobhanga Block	South 24-Parganas Forest Department
2	Ajmalmari Block	South 24-Parganas Forest Department
3	Dhulibhasani Block	South 24-Parganas Forest Department
4	Chulkati Block	South 24-Parganas Forest Department
5	Thakuran Block	South 24-Parganas Forest Department
6	Saptamukhi Block	South 24-Parganas Forest Department
7	Lothian Block	South 24-Parganas Forest Department

Table 10 : The National and The International Programmes and Research Projects in Indian Sundarbans

1.	Sundarbans Development Board was declared with the collaboration of the State Agriculture Department, Forest Department and Fishery Department, etc. for the developmental activities on the entire Sundarbanbs areas, within the 24 Parganas South and 24 Parganas North Districts	Since the Year 1973
2.	Sundarbans Tiger Reserve was declared within 2565.10 km² with the financial support of several International Organisations, *viz.* IUCN, UNESCO, UNDP & WWF	Since the Year 1973
3.	Sundarbans Crocodile Project was developed at Bhagbatpur of Namkhana Range with the financial support of several International Funds	Since the Year 1976
4.	Three Wildlife Sanctuaries were declared in Sundarbans at i) Sajnakhali Wildlife Sanctuary in an area of 362.40 km², within STR ii) Lothian Island Wildlife Sanctuary in an area of 38 km² outside the STR but within the mangrove forest and control of 24 Parganas Forest Division iii) Holiday's Island Wildlife Sanctuary in an area of 5.95 km² outside the STR but within the mangrove forest and control of 24 Parganas Forest Division	Since, June 1976
5.	Core Areas of the Sundarbans Tiger Reserve, covering an area of 1330.10 km² was declared as Sundarbans National Park	Since, May 1984
6.	Sundarbans Biosphere Reserve was declared in an area of 9630.00 km² with the financial support of several world organisations	Since the Year 1989
7.	Suindarbans Mangrove Forest Area was declared as the most important World Heritage Site by the UNDP, UNESCO, IUCN and WWF for having the unique Bio-diversity	Since the Year 1989

Photo Plate 19. Mangrove succession on newly silted up land- *Porteresia coarctata* followed by *Avicennia* sp.

Table 11 : Showing the Different Regions and the Project Areas in the Indian Sundarbans

Sundarbans Regions within India, showing different projects areas	**Area**
Sundarbans Biosphere Reserve, Indian Territory- covering an area	9630 km²
Mangrove forest zones, including estuarine rivers, canals, creeks, etc.	4266.66 km²
Only mangrove forest land areas (55%)	2347 km²
Only estuarine rivers, canals, creeks and lagoons (45%)	1920 km²
Sundarbans Tiger Reserve (STR) - Year of initiation since 1973	2585.10 km²
Core areas within STR declared as National Park - Year of initiation since 1984	1330.10 km²
Buffer areas within STR - outside the National Park	1255 km²
Sundarbans Wildlife Sanctuaries (3 numbers) declared during 1976	
1) Sajnakhali Wildlife Sanctuary (within the STR)	362.40 km²
2) Lothian Island Widlife Sanctuary (outside STR)	38.00 km²
3) Holiday's Island Wildlife Sanctuary (outside STR)	5.95 km²
Crocodile Breeding & Rearing Project at Bhagbatpur of Namkhana Range (outside STR), since 1975	32 ha
Mangrove Forest areas, rivers, creeks outside STR, *i.e.* in the western part of Sundarbans, W. B.	1682.00 km²
Mangrove Reclaimed human habitat zones in Indian Sundarbans (within 24 Parganas South & North)	5363.0 km²

Table 12 : Natural Calamities, *viz.,* Cyclones, Floods, Earth Quakes and Famines Recorded from Sundarbans

Year	Means of Natural Calamities	Year	Means of Natural Calamities	Year	Means of Natural Calamities
1582	One Severe Cyclone	1877	Three Severe Cyclones	1927	One Severe Cyclone
1688	One Severe Cyclone	1878	One Severe Cyclone	1928	One Severe Cyclone
1707	One Severe Cyclone	1880	One Severe Cyclone	1929	One Severe Cyclone
1737	Cyclone and Earth Quake	1881	Two Severe Cyclones	1932	One Severe Cyclone
1742	One Severe Cyclone	1882	Two Severe Cyclones	1934	One Severe Cyclone
1762	Earth Quake	1883	One Severe Cyclone	1935	One Severe Cyclone
1770	Famine	1884	One Severe Cyclone	1936	One Severe Cyclone
1791	Famine	1885	Flood	1937	One Severe Cyclone
1823	Flood	1887	Two Severe Cyclones	1940	One Severe Cyclone
1830	One Severe Cyclone	1888	Three Severe Cyclones	1941	One Severe Cyclone
1832	Three Severe Cyclones	1889	Two Severe Cyclones	1942	Famine
1833	One Severe Cyclone	1890	Several Floods	1943	One Severe Cyclone
1834	One Flood	1893	Two Severe Cyclones	1948	One Severe Cyclone
1830	One Severe Cyclone	1894	One Severe Cyclone	1956	One Severe Cyclone
1840	One Severe Cyclone	1895	Earth Quake	1960	One Severe Cyclone
1842	Once Earth Quake	1896	Five Severe Cyclones	1961	One Severe Cyclone
1844	One Severe Cyclone	1897	Earth Quake	1962	One Severe Cyclone
1848	One Severe Cyclone	1898	Four Severe Cyclones	1965	One Severe Cyclone
1850	One Severe Cyclone	1899	One Severe Cyclone	1966	One Severe Flood
1852	One Severe Cyclone	1900	One Severe Flood	1968	One Severe Cyclone
1856	Severe Flood	1901	Four Severe Cyclones	1970	One Severe Cyclone
1858	One Severe Cyclone	1904	Two Severe Cyclones	1973	One Severe Cyclone & Flood
1859	Two Severe Cyclones	1907	Two Severe Floods	1976	Severe Flood
1862	One Severe Cyclone	1909	One Severe Cyclone	1978	Severe Flood
1864	Several Cycones & Floods	1913	Two Severe Cyclones	1981	One Severe Cyclone
1865	Famine	1916	Two Severe Cyclones	1982	One Severe Cyclone
1867	One Severe Cyclone	1917	One Severe Cyclone	1985	One Severe Cyclone
1868	Severe Flood	1919	One Severe Cyclone	1988	One Severe Cyclone
1869	Severe Flood	1921	Famine	1991	One Severe Cyclone
1871	Severa Flood	1922	One Severe Cyclone	1994	One Severe Cyclone

Photo Plate 20. River bank *Rhizophora* sp. plant, showing tide water carrying silt mark on the leaves

Roles Played by these Mangroves in Indian Sundarbans

Mangrove litters release nutrients through mineralisation processes; in these mineralisation processes microbial organisms like, bacteria, fungus and other important mangrove dwelling fauna take active part. These organic nutrients help for the spontaneous growth of autotrophic algal species. Mangrove inhabiting fauna depends directly or indirectly on mangrove plants and other mangrove inhabiting consumer group of organisms. These Indian mangals are exploited very rapidly for fuel wood and timber harvest, grazing by domestic cattle, excavation and renovation of mangrove area to agriculture field, fish farms, salt pans, ports, harbour development, human habitations and also for discharge of sewage effluent, garbage from urban areas and industries. But, the important, role played by these mangrove ecosystems are unique; these are shown in Diagram - 1.

Diagram 1. Showing the Roles Played by the Mangrove Ecosystem of Sundarbans, India

4

MANGROVE ZONES AND SUCCESSION

Different mangrove species have different abilities of facing the specific tidal inundation along with specific salt tolerance. They can also grow and survive in specific soil physico-chemical properties, along with distinct soil nutrient status. Some of these mangrove species are distributed throughout the mangrove forest, while others have restricted distribution. This depends on the availability of good number of reproducing mother plants and specific soil physico-chemical properties. Based on the specific land strata, tidal nature, soil physico-chemical properties and distance from the coast or estuarine rivers different mangrove species grow in different areas. Sundarbans is the silt accreted delta, developed on different major estuarine river mouths or river-beds; which are raised 5.0 m - 8.0 m above MSL. During spring tides, most of these lands get inundated with saline sea-water, as spring tide water (HWSTL) rises up to 8.0 m above MSL Table 2, Graph 1. During average high tides (HWNTL), the tide water rises within 5.0 m- 6.0 m above MSL; during that time about 60% of the mangrove forest land get inundated with the saline water. During the average high tides, water reaches within 3.0 m-4.0 m from the MSL and at that time more than 40% of these mangrove forest lands in Sundarbans are inundated. During all the low tides most of these mangrove forest floor are exposed to air, as saline sea water recedes during low tide. These delta land and the mangrove forests are in dynamic stage; silt accretion, *vis-à-vis*, soil erosion and collapsing of lands in this area are the continuous processes. This tidal inundation and water flow caused change the soil layer frequently. Stages of land accresion, tidal inundation nature, distance or position from the coast line and availability of up-steam freshwater supply are the important factors for mangrove succession. As such, the soil types in different parts are very distinct; successions of these different mangrove species in different sub-tidal, inter- tidal and above tidal zones are also different (Diagrams 2).

Based on these above factors and causes, the following five categories of these Sundarbans mangrove zones and distinct mangrove successions are identified and studied critically by Naskar & Mandal (1999). These mangrove zones are - (1) Naked and vegetated river flats – 1st zone, ranges between MLWN and MTL Limit, *i.e.*, 2.14 m above MSL and 4.42 m above MSL; (2) Dense and ridge mangrove forests – 2nd zone, within 5.0 m above MSL, normal high tide water level; (3) Ridged with dominant tree mangrove forest species – 3rd zone, within 7.0 m above MSL; (4) Back mangrove area – 4th zone, beyond 7 m above MSL which very occasionally inundated with high tide water during SWHT and (5) Supra tidal zone- 5th zone, beyond 8.0 m above MSL which inundated with tide water during abnormal tide. Non-mangrove halophytes, mesophytes and xerophytic species replace the mangrove species in this zone - 5 and the mangroves decline in its climax stage. This climax phase of the mangrove succession really terminated at attaining its maturity with gregarious growth of the mangrove palm, *Phoenix paludosa* L.and fringes with the main mangrove associates, *viz., Excoecaria agallocha* L. and occasionally with *Heritiera fomes* Buch. Ham. In this mangrove succession stages, the most important mangrove grass species, *Porteresia coarctata* (Roxb.) is the pioneer and first of all Takeoka grows on the newly silted up soil of the riverbeds. While, the pioneer and important mangrove tree species, *Avicennia* spp. and *Sonneratia* spp. grow on the river flat lands after turning it more consolidated by the silt and with the help dense growth of *P. coarctata* (Roxb.) Takeoka.

A

B

Photo Plate - 21. (A) Succession of *P. coarctata* & *S. apetala.* (B) Secondary succession, *Bruguiera parviflora*

Diagram - 2. Inundation classes of different mangroves in the Ganga delta regions, Sundarbans in India

Species dominance in different regions

Zone-1. Dominated by sea-grass, *Ruppia maritima*

Zone-2. *Avicennia* spp., *Sonneratia* spp., *Rhizophora* spp., *Kandelia* sp., *Nypa* sp., *Sarcolobus* spp., *Suaeda* spp.

Zone-3. *Excoecaria* sp, *Avicennia* spp., *Lumnitzera* sp., *Ceriops* spp., *Aegialitis* sp., *Brownlowia* sp.,*Acanthus* sp. *Aegiceras* sp., *Scyphiphora* sp., *Excoecaria* sp., *Xylocarpus* sp.,*Bruguiera* spp.,*Sonneratia* spp., *Phoenix* sp.,*Heritiera* sp.

Zone-4. *Ipomoea pes-caprae*, *Salicornia* sp., *Tamarix* spp., *Acrostichum* sp., *Sesuvium* sp., *Aeluropus* sp., *Opuntia* sp., *Lumnitzera* sp., *Excoecaria* sp., *Heliotrophium* sp., *Acanthus* sp., *Derris* spp., *Pentatropis* sp., *Solanum* sp., *Sarcolobus* spp., *Premna* spp., *Phoenix* sp.,*Thespesia* spp., *Hibiscus* spp., *Clerodendron* spp.etc.

Zone-5. Non-mangrove, but few halophytic herbs, shrubs and introduced plants.

Zone-6. Mangroves fully reclaimed and turned human habitat zones, agricultural fields, mesophytes dominant.

The 1st succession zone gets inundated during all the high tides, at 30 days in a month and twice in a day; 2nd succession zone gets inundated with the tide water during 20 days in a month and twice in a day; 3rd zone gets inundated for 15 days and 4th zone gets inundated for 10 days in a months, in both the cases high tide inundates twice in a day. The 5th succession zone normally does not get inundated with frequent tide water, except during exceptional or abnormal natural calamities and bow spring tides. Due to different tidal inundation nature and position from the sea, along with available sand, silt and clay percentage and nutrient status of these different areas, different mangrove zones along with different mangrove species are formed (Naskar & Guha Bakshi, 1987). In these succession processes, the level of the silted up or accreted land rises up beyond the normal high tidal reach and the back mangrove species gradually appear. The 6th zone in Indian Sundarbans is the man made dense habitat non-saline area with dominant terrestrial non-halophytic floral assemblages. In these upper elevated zone, felling of trees and other deforestation activities usually accelerated 'saline blank formation'. This is caused due to capillary movement of salt water and successive evaporation of the surface water from the naked exposed soil surface. The water-dissolved salt remains on the surface soil and ultimately form the salt-crust. This formation of saline blank is also identified as major problem in the Sundarbans mangals. Felling of trees and deforestation in these zones is not advisable, as growth of these mangrove trees is very feeble and slow. This mangrove forest zones have several other dominant role rather than its timber and wood value. From the critical field investigations, it is noted that in the Indian Sundarbans mangal the supply of fresh water on the eastern part is comparatively more than that of the western part. This is resulted to form two distinct groups of mangrove habitats and the distributional difference between the mangrove species in the eastern and western Sundarbans. The central zone or islands form the transitory zone. These distinct mangrove distribution patterns of Indian Sundarbans are shown in Table - 13. The different and distinct soil physical and chemical properties of these distinct zones are also variable (Ghosh & Naskar, 2000).

Since 1975, along with more fresh water discharge from the Farakka Barrage, the mouth region of Hugli river became less saline, but that fresh water from the Hugli river mouth had little impact on desalinization on the river water of Sundarbans.

The Terms and Definitions of the Mangroves and the Mangal

The term 'mangrove' is derived from Portuguese and English words, *viz.*, *'mangue'* and *'grove'*, respectively. A mangrove tree or bush is called *'Mangue'* in Portugal, where as the community of trees and the bushes are known as *'grove'* in English (Naskar & Guha Bakshi, 1987). The mangroves may be defined as the *'coastal tropical forest formation, encircled or spread over by the tidal rivers and / or the sea water, flooded or bathed frequently."*

Naskar & Guha Bakshi (1987) have also collected informations, views of different scientists and based on those they have mentioned that *'mangroves do not necessarily mean by forest formation alone; mangroves may be either covered by herbs or shrubs or trees, i.e., herbaceous, bushy or arborescent, these occasionally lacking any such vegetation, viz., fringing waterways.*

Several world reputed mangrove workers have recognised and defined the term *'mangroves'* differently; few of these are as follows:

1. Davies (1910) defined the mangroves – *"plants which live in muddy loose wet soils in tropical tide waters"*.
2. Mac Nae (1968) defined the mangroves – *" trees or bushes growing between the level of high water of spring tide and level close to but above mean sea level"*. Mac Nae (1968) has also referred the term 'mangal' for referring the mangrove forest community, while the term 'mangrove' as individual kind of tree.
3. Aubreville (1970) defined the mangroves – *" as coastal tropical formations found along the border of sea and lagoons, reaching up to the edges of the river to the point where the water is saline, growing in swampy soils and covered by sea water during high tides"*.
4. Geriech (1973) reported that mangroves occur at edges of the tropical and sub-tropical sea, bays, lagoons and estuarine regions. Thus, *"Mangroves are trees of various species under several families, which grow only where they come into permanent contact with sea water or brackish water"*.
5. Blasco (1975 & 1977) defined mangroves–*"the mangrove is a type of coastal woody vegetation that fringes muddy saline shores as estuaries in tropical and sub-tropical regions"*.
6. Arroyo (1977) defined mangroves – *"A small group of true mangrove plants and associated species belonging to systematically unrelated families possessing similar physiological characteristics and structural adaptations with common preference to the inter-tidal habitat"*. The term 'Mangroves' used for the (a) forest ecosystem, (b) the component vegetation, (c) both forest ecosystem and the component vegetation.
7. Clough (1982) defined mangrove as – *"these are the only trees amongst relatively small group of higher plants that have been remarkably successful in colonizing the intertidal zone at the interface between land and sea"*.
8. Tomlinson (1986) defined mangroves – *"Either to the constituent of plants of tropical forest communities or forest community itself"*. Mangrove in the more limited sense may thus be defined-*"tropical trees restricted to intertidal and adjacent communities"* or *"mangrove is a community that contain mangrove plants"*.
9. Naskar & Guha Bakshi (1987) defined mangroves as–*"Coastal tropical forest formation encircled or spread by the tidal rivers and/or sea water, flooded frequently by the tidal water"*.
10. Mepham & Mepham (1984) have pointed out that though the term *'mangrove forest'* has been widely used and commonly accepted it is nothing but the synonym of tidal forest; they have also reported from the Indo-West Pacific tidal forests about 400 arborescent and 500 non-arborescent plant species. They highlighted to re-evaluate the old concept of the term *' Mangroves'*. According to them, it is now very much essential to clarify the term 'Mangroves' in relation to their characteristic plants and thus habitat status or morpho-anatomical features.

Photo Plate 22. Salt-crust on the surface soil of mangrove cleared inter-tidal habitat in Indian Sundarbans

Uniqueness of these Sundarbans Mangals

The Sundarbans mangal within the boundary of Bangladesh and India are unique, as these have the following under mentioned characteristic features and also for having there abundant natural wealth:

1. Sundarbans of both Bangladesh and India is the single largest mangrove chunk of the Globe, spreads over within the latitude 21° N – 22°31′ N and longitude between 88°10′ E and 92°15′ E;
2. This mangrove zone is the habitats for high bio-diversity, unique with interesting flora and fauna (Table-14 & 19);
3. Sundarbans mangal is the only Marshy mangrove Tiger- land′ of the World (Table - 17);
4. Tidal amplitude and fluctuations of the Sundarbans Mangals is very high (Table - 21);
5. Phyto- successions in the Sundarbans mangals are of unique types with varied and highest number of species diversity (Table - 14);
6. Sundarbans mangal supports coastal fisheries throughout the East Coast of India and in the Bay of Bengal (Table - 19E);
7. Sundarbans mangal itself is the important fish cradle;
8. Sundarbans mangal acts as the sink for mega - metropolitan pollutants.

For all these reasons, Sundarbans mangal was included in the ***World Heritage List***, since 1989.

Sundarbans mangal is very much potential for their renewable natural resources and this mangrove forest can also protect the coastal West Bengal from frequent tidal

Photo Plates 23. Fishing at Sundarbans by cast nets

flooding, surges arising in the bay and cyclone thrust, which very often originate and is concentrated in the Bay of Bengal (Table-12). Fosberg (1974) had mentioned this zone as very much prone to cyclones.

Finlayson and Moser (1991) have reported that the total mangrove areas of the World were about 14 million hectares. Out of these, the Old World tropical mangroves, *i.e.*, the Indo-West Pacific Tropical zones and Tropical Australia have the most dominant mangrove areas and the mangrove species abundance or diversity shows its unique succession pattern. Finlayson and Moser (1991) have also reported that about 1.48–1.73 million acres or 6,00,000 – 7,00,000 ha mangrove cover area was present in the southern part of the four coastal districts of Bangladesh and two coastal districts of India, *viz.*, Khulna, Barisal, Noakhali and Chittaganj of Bangladesh and 24 Parganas South & 24 Parganas North of India. These coastal zones popularly known as Sundarbans, where good number of mangrove species gregariously grow. In this undivided Sundarbans of India and Bangladesh mangrove species are estimated about 80% of total world mangrove species (Table–14, Mangroves of Indian Sundarbans).

In recent estimate it was shown that the total highest mangrove cover area is in Indonesia, which is about 42,510 km^2, situated within the latitude 6° N and 11°S and longitude between 95° E and 141°E. The 2nd highest mangrove cover area present in Brazil, which is about 25,000 km^2 and within the latitude 0° and 30°S and longitude between 30°W and 52°W. The 3rd dominant mangrove areas present in the tropical and sub-tropical Australia, within the latitude between 8° S and 38°45′ S and longitude between 115°E and 155° E, covering about 11,500 km^2 (Naskar & Mandal, 1999). On the other hand, the total coverage under the mangroves in Indonesia is spread over in several countries and thousands of islands, *viz.*, Java, Sumatra, Sulawesi, Irian Jaya, Kalimantan and several other small to large islands; these Indonesian mangroves also spread over both in the Indian and Pacific Oceans.

Mangrove areas in the New World zones at Brazil and the Old World zones of Australia and Indonesia are more than Sundarbans, but those zones are scattered and mangrove species diversity is less than Sundarbans mangal. This undivided Sundarbans mangroves in Bangladesh and India is the single largest mangrove chunk of the world, where the mangrove species diversity is also much more and important for harbouring several endemic and threatened plants and animal species. The mangrove area in the undivided Sundarbans of the Ganga–Brahamaputra deltas, *i.e.*, the Hugli-Padma-Meghna mouths, grows very compactly in hundreds of deltaic islands; which covers about 10,000 km^2 at the present time (Naskar & Guha Bakshi, 1987). But , all these mangrove species and mangrove lands in Sundarbans are in very much degraded conditions and threatened due to ever increasing human pressure and over exploitation.

The Unique Sundarbans Mangals for Its Highest Bio-diversity

During field survey and literature studies it was revealed that Sundarbans is the important mangrove zones for its abundant distribution of about 105 major and minor elements of mangroves, back mangrove species or mangrove associates (Table -14). Among these, true mangrove species in the Indian Sundarbans alone is about 40 (Naskar & Mandal 1998). The mangrove habitat faunal diversity in the Sundarbans is also unique and very rich (Table - 19). Several threatened, endangered and endemic species of plants are also reported from this Sundarbans (Table - 15); all these flora and fauna are also in very much threatened conditions for heavy human pressure and over exploitation (Das & Nandi, 1999). These need urgent measure for effective protection or conservation for the posterity of the future generations.

Table 13 : Different Groups and Numbers of Flora and Fauna Reported from Indian Sundarbans Mangal

Group/Class of Organisms	Order/Class Family	Reported Species	Reported by or Publications
1. Higher Groups of Plants	Major Mangrove Species	40 species	**Source :** Naskar, K.R. & R.N. Mandal (1999). Ecology & Bio-diversity of Indian Mangroves, Daya Publishing House, Delhi pp. 1-755 + 165 photo plates
	Minor Mangrove Species	32 species	
	Back Mangroves & Associates	30 species	
	Mangrove Habitat Ferns	3 species	
2. Lower Groups of Plants	Algal Species	150 species	**Source :** Sen Neera & K.R. Naskar (2003)
	Fungal species	Not assessed	————————
	Lichen Species	32 species	**Source :** Santra, et. al. (1999)
3. Higher Groups of Fauna	Phylum - 1. Chrodata	445 species	**Source :** 1. Das, A.K. & N.C. Nandi (1999). In : **Sundarbans Mangal** (Edts. Guha Bakshi, D.N., P. Sanyal & K. Naskar) pp. 417-427. 2. Nandi *et. al.* (1993). Rec. Zoological Survey of India 93 : 83-103.
(i) Kingdom : Animalla	Class - Chondrichthyes	22 species	
	Class - Osteichthyes	154 species	
	Class - Amphibia	8 species	
	Class - Reptilia	58 species	
	Class - Aves	163 species	
	Class - Mammalia	40 species	
4. Lower Group of Fauna	Phylum - 1. Sarcomastigophora	45 species	**Source :** 1. Das, A.K. & N.C. Nandi (1999). In : **Sundarbans Mangal** (Edts. Guha Bakashi, D.N., P. Sanyal & K. Naskar) pp. 417-427. 2. Nandi, *et al.* (1993). Rec. Zoological Survey of India 93 : 83-103. 3. Mukherjee, A.K. (1975). Fauna of Sundarbans. J. Bombay nat. Hist Soc. 73 : 2-20. 4. Mandal, A.K. & N.C. Nandi (1969), Fauna of Conservation Area Zoological Survey of India. 3 : 1-116. 5. Chaudhuri, A.B. & A. Choudhury (1994). Sundarbans. Part -I, IUCN. 1-247. 6. Anon (1995) West Bengal Estuarine Ecosystem Series. Zoological Survey of India. Part - 2, 1-542.
(ii) Kingdom : Protista	Phylum - 2. Aplicomplexa	24 species	
Sub-kingdom - Profozoa	Phylum - 3. Myxozoa	4 species	
	Phylum - 4. Clipphora	31 species	
	Phylum - 1. Porifera	1 species	
	Phylum - 2. Cnideria	33 species	
	Class - Hydrozoa	20 species	
(iii) Kingdom : Animalia	Class - Anthozoa	13 species	
	Phylum - 3. Ctenophora	2 species	
	Phylum - 4. Platyhelminthes	41 species	
	Class - Turbellaria	1 species	
	Class - Monogenera	21 species	
	Class - Trematoda	13 species	
	Class - Cestoda	6 species	
	Phylum - 5. Nemathelminthes	68 species	
	Phylum - 6. Acanthocephala	3 species	
	Phylum - 7. Nemertinea	2 species	
	Phylum - 8. Rotifera	4 species	
	Phylum - 9. Mollusca	143 species	
	Phylum - 10. Sipuncula	2 species	
	Phylum - 11. Echiura	3 species	
	Phylum - 12. Annelida	78 species	
	Class - Polycheta	69 species	
	Class - Oligochaeta	6 species	
	Class - Hirudina	3 species	
	Phylum - 13. Arthropoda	476 species	
	Class - Crustacea	240 species	
	Class - Insecta	201 species	
	Class - Arachnida	33 species	
	Class - Merostomata	2 species	
	Phylum - 14. Entoprocta	1 species	
	Phylum - 15. Bryozoa	3 species	
	Phylum - 16. Brachiopoda	1 species	
	Phylum - 17. Chaetognatha	4 species	
	Phylum - 18. Echinodermatra	20 species	
	Phylum - 19. Hemichordata	1 species	
Total Fauna/Animal Species Reported from Indian Sundarbans		**1434 species**	

5

FLORAL SUCCESSION IN THE MANGAL

Different and diverse habitats of the Indian Sundarbans mangals are - (i) the sea-face or beach forests, (ii) flora of the formative islands, (iii) flora of the reclaimed low-lying cultivated zones, (iv) flora of the river banks and (v) the swamp forests (Plate – 33). Naskar & Guha Bakshi (1987) had also categorized different mangrove associations in the Indian Sundarbans; these are as follows:

1. Deforested mangrove swamp, mostly in the western and central parts of the Indian Sundarbans;
2. Dense mangrove forests, presently only noted in the south-eastern part of the Indian Sundarbans;
3. *Heritiera fomes*, is now rare in the Indian Sundarbans and mostly restricted in the eastern part;
4. Brackish water mixed *Heritiera* sp. forests is very scattered and rare in the Indian Sundarbans and especially restricted in the north eastern part of the Indian Sundarbans;
5. Palm swamps dominated by *Phoenix paludosa*, which is distributed and present almost throughout the Indian Sundarbans.

Photo Plate - 24. A. Dense & tall mangrove Forest, B. Brackish water mixed *Heritiera* forest, C. Deforested mangrove swamp, D. Palm swamp, dominated by *Phoenix* sp. E. Brackish water *Nypa* sp. & *Heritiera* sp. forest, F. *Heritiera* fruits and G. Tall *Heritiera fomes* tree with buttress roots.

Table 14 : Common Mangroves and Mangrove Associated Plant Species of the Indian Sundarbans

Family Names (Mangroves & Mangrove Associates)	Genus under the Mangrove Family	Species & Genus, of Sundarbans Mangals, India	Habitat	Aerial/Air Root Pattern of the Mangroves	Germination/ Seeding Development	Remark
1. Rhizophoraceae	*1. *Rhizophora*	1. *R. mucronata*	3, 4	Prop roots & stilt roots	Viviparous Germination	+++
		2. *R. apiculata*	3, 4	Prop roots & stilt roots	Viviparous Germination	+
(The true and most dominant Mangrove Family in Indian Sundarbans; under this family only 4 genera and 9 major mangrove species were collected and identified from Sundarbans)	*2. *Bruguiea*	3. *B. gymnorhiza*	3, 4	Stilt root Broom Like	Viviparous Germination	+++
		4. *B. sexangula*	3, 4	Knee roots & Broom Like	Viviparous Germination	++
		5. *B. cylindrica*	3, 4	Knee roots & Brrom like	Viviparous Germination	++
		6. *B. parviflora*	3	Stilt root	Vivparous Germination	+
	*3. *Ceriops*	7. *C. tagal*	4, 5	Fused Stilt Foots	Viviparous Germination	++
		8. *C. decandra*	4, 5	Fused Stilt Roots	Viviparous Germination	+++
	*4. *Kandelia*	9. *K. candel*	3	Stilit Roots	Viviparous Germination	+++
2. Avicenniaceae	*5. *Avicennia*	10. *A. officinalis*	3, 4	Pneumatophore	Cyprto-viviparous	++++
(The 2nd dominant		11. *A. alba*	2, 3	Pneumatophore	Crypto-viviparous	++++
Mangrove family)		12. *A. marina*	2, 3	Pneumatophore	Crypto-viviparous	++++
3. Sonneratinceae	*6. *Sonneratia*	13. S. apetala	3, 4	Pneumatophore	Incipient Vivip.	+++
(The 3rd dominant		14. *S. caseolaris*	3	Pneumatophore	Non-viviparous	+
Mangrove Family)		15. *S. griffithii*	2, 3	Pneumatophore	Non-viviparous	++
4. Meliaceae	*7. *Xylocarpus*	16. *X. granatum*	3, 4	Root Buttress	Non-viviparous	+++
(The 4th dominant		17. *X. mekongensis*	3, 4	Root Buttress	Non-viviparous	++
Mangrove Family)	*8. *Aglaia*	18. *A. cucullata*	4	Pneumatophore	Non-viviparous	+
5. Myrisinaceae	*9. *Aegiceras*	19. *A. corniculatum*	3, 4	No-aerial roots	Incipient Viviparous.	+++
6. Aegialitidaceae	*10. *Aegialitis*	20. *A. rotundifolia*	4, 5	No Air Root Viviparous.	Incipient Viviparous	+++
7. Sterculiaceae	*11. *Heritiera*	21. *H. fomes*	4, 5	Blunt end peg-like	Incipient viviparous	++
8. Combretaceae	*12. *Lumnitzera*	22. *L. racemosa*	4	Looping aerial	Non-viviparous	++
9. Rubiaceae	*13. *Scyphiphora*	23. *S. hydrophyllacea*	3, 4	Wavy aerial root	Non-viviparous	+
	*14. *Hydrophyfax*	24. *H. maritima*	2	No aerial roots	Non-viviparous	+
10. Tiliaceae	=15. *Brownlowia*	25. *B. lanceolata*	5	No aerial roots	Non-viviparous	+
11. Arecaceae (Mangrove Palm)	*16. *Nypa*	26. *N. fruticans*	3	No aerial roots	Incipient viviparous	++
	*17. *Phoenix*	27. *P. paludosa*	5, 6	Pneumatothods	Incipient viviparous	++++
12. Euphorbiaceae	*18. *Excoecaria*	28. *E. agallocha*	4, 5	Gall on trunk	Non-viviparous	++++
13. Asclepiadaceae	+19. *Sarcolobus*	29. *S. globosus*	2, 3	No Aerial roots	Non-viviparous	++
(Mostly halophytic herbs		30. *S. carinatus*	2, 3	No Aerial roots	Non-viviparous	++
and twiners; mangrove	=20. *Pentatropis*	31. *P. capensis*	5, 6	No Areial roots	Non-viviparous	++
associates)	=21. *Tylophora*	32. *T. tenius*	6>	No Aerial roots	Non-viviparous	+
	=22. *Finlaysonia*	33. *F. obovata*	6>	No Aerial roots	Non-viviparous	+
	=23. *Hoya*	34. *H. parasifica*	Eph	No Aerial roots	Non-viviparous	+
14. Acanthaceae (Mangrove Associated)	+24. *Acanthus*	35. *A. ilicifolius*	4, 5	Stilit roots	Incipients viviparous	++++
		36. *A. volubilis*	4, 5	Stilt roots	Incipients viviparous	+
15. Papilionaceae	+25. *Cynometra*	37. *C. ramiflora*	4	Stilt roots	Non-viviparous	+
(Mangrove legumes;	+26. *Derris*	38. *D. indica*	6>	Stilt roots	Non-viviparous	++
associated back mangroves)		39. *D. scandens*	5	Stilt roots	Non-viviparous	++
		40. *D. trifoliata* ,	5	Stilt roots	Non-viviparous	++
	+27. *Dalbergia*	41. *D. spinosa*	4	Stilt roots	Non-viviparous	+

Notes : *Dodonaea viscosa* (L) Jack (Sapindaceae) & *Clerodendron nerifolium* var *macrocarpa* were collected from Sundarbans Tiger Reserve.

Family Names (Mangroves & Mangrove Associates)	Genus under the Mangroves Family	Species & Genus, of Sundarbans Mangala, India	Habitat	Aerial/Air Root Pattern of the Mangroves	Germination/ Seeding Development	Remark
16. Caesalpiniaceae	=28. *Caesaplinia*	42. *C. crista*	6>	Stilt roots	Non-viviparous	++
(Back Mangroves)		43. *C. bonduc*	6>	Stilt roots	Non-viviparous	++
17. Tamaricaeae	+29. *Tamarix*	44. *T. dioica*	6	Stilt roots	Non-viviparous	++
(Back Mangroves)		45. *T. gallica*	6	Stilt roots	Non-viviparous	++
18. Verbinaceae (Back	+30. *Clerodendrum*	46. *C. Inerme*	6	Stilt roots	Non-viviparous	+++
Mangroves	=31. *Premna*	47. *P. corymbosa*	6	Stilt roots	Non-viviparous	++
19. Malvaceae (Mostly	=32. *Thespesia*	48. *T. populnea*	6>	No Aerial roots	Non-viviparous	++
the Back Mangrove		49. *T. populneoides*	6>	No aerial roots	Non-viviparous	+
Plants or Mangrove		50. *T. lampus*	6>	No Aerial roots	Non-viviparous	+
associates)	=33. *Hibiscus*	51. *H. tortuosus*	5	No Aerial roots	Non-viviparous	++
		52. *H. tillaceous*	5, 6	No Aerial roots	Non-viviparous	++
20. Sapotaceae	=34. *Manikara*	53. *M. hexandra*	6>	No Aerial roots	Non-viviparous	+
21. Chenopodiaceae	+35. *Suaeda*	54. *S. nudiflora*	2, 3	No Aerial roots	Non-viviparous	+++
(Halophytic Herbs of the		55. *S. maritma*	2, 3	No Aerial roots	Non-viviparous	++
river flats)	+36. *Salicornnia*	56. *S. brachiata*	2	No Aerial roots	Non-viviparous	++
22. Aizoaceae	+37. *Sesuvium*	57. *S. portulacastrum*	2	No Aerial roots	Non-viviparous	++
23. Boraginaceae	+38. *Heliotrophium*	58. *H. curassavicum*	4, 5	No Aerial roots	Non-viviparous	++
24. Ruppiaceae	+39. *Ruppia*	59. *R. maritima*	Sub	No Aerial roots	Non-viviparous	++
25. Convolvulaceae	=40. *Ipomoea*	60. *I. pes-caprae*	6>	No Aerial roots	Non-viviparous	+++
26. Araceae	+41. *Crytocoryne*	61. *C. ciliata*	2	No Aerial roots	Non-viviparous	+
27. Amarylidaceae	+42. *Crinum*	62. *C. defixum*	2	No Aerial roots	Non-viviparous	++
28. Loranthaceae (Epiphytic	+43. *Viscum*	63. *V. orientale*	Epn	No Aerial roots	Non-viviparous	++
on Mangroves)	+44. *Dendropthoe*	64. *D. falcata*	Epn	No Aerial roots	Non-viviparous	++
29. Lauraceae	+45. *Cassytha*	65. *C. filiformis*	Epn	No Aerial roots	Non-viviparous	+
30. Cuscutaceae	+46. *Cuscuta*	66. *C. reflexa*	Epn	No Aerial roots	Non-viviparous	++
31. Poaceae	+47. *Porteresia*	67. *P. coarctata*	1, 2	No Aerial roots	Non-viviparous	++++
(Most Important	+48. *Myriostachya*	68. *M. wightiana*	1, 2	No Aerial roots	Non-viviparous	++
Pioneer Grass species	=49. *Phragmites*	69. *P. kakra*	2, 3	No Aerial roots	Non-viviparous	+
in the silted up saline	=50. *Aeluropus*	70. *A. lagopoides*	5, 6	No Aerial roots	Non-viviparous	+
soil of Sundarbans	+51. *Hemarthria*	71. *H. compressa*	2, 3	No Aerial roots	Non-viviparous	+
32. Bignoniaceae	=52. *Dolichandron*	72. *D. spathaceae*	6>	No Aerial roots	Non-viviparous	+
33. Birringtoniaceae	=53. *Barringtonia*	73. *B. acutangula*	6>	No Aerial roots	Non-viviparous	+
(Back Mangroves)		74. *B. racemosa*	6	No Aerial roots	Non-viviparous	+
34. Capparidaceae	=54. *Capparis*	75. *C. zeylanicas*	6>	No Aerial roots	Non-viviparous	++
(Sand Dune Flora)	=55. *Crataeva*	76. *C. roxburghii*	6>	No Areial roots	Non-viviparous	++
35. Opuntiaceae	=56. *Opuntia*	77. *O. dillenii*	6>	No Aerial roots	Non-viviparous	++
36. Cyperaceae	=57. *Scirpus*	78. *S. littoralis*	3-6	No Aerial roots	Non-viviparous	+++
37. Rutaceae	+58. *Atalantia*	79. *P. tectorius*	5, 6	Stit Roots	Non-viviparous	+
38. Pandanaceae	=59. *Pandanus*	79. *P. tectorius*	5, 6	Stit Roots	Non-viviparous	+
		80. *P. odoratissima*	5, 6	Stit Roots	Non-viviparous	+
		81. *P. foetidus*	5, 6	Stit Roots	Non-viviparous	+
39. Pteridaceae	+60. Acrostichum	83. *A. aureum*	5, 6	Fern-rhirzomes	Non-viviparous	++
Family = 39	**Genera = 60**	**Species : 83**				

Symbols Used : "++++" = denote - most dominate; "+++" = denote common; "++" = denote occasional; "+" = denote - rare. Asterisk (*) = Showing True Mangrove Genera; Plus (+) = Mangrove Associate; Parallel (=) = Showing Back Mangals. 1= Less Consolidated River Flat Grass Lands; 2 = Well consolidated Silted Up River Flat Forest Lands; 3 = Formative River Slope Land; 4 = Less Mature Flat Forest Land; 5 = Mature Forest Land = Beyond the Normal High Tide Level; 6 = Most Degraded Forest Land-Beyond The Spring High Tidal Reaches, Back Mangals; 6> = Back Mangals; Sub = Submerged; Eph = Epiphytic.

Table 15 : Rare, Threatened and Endangered Mangroves and Mangrove Associates in the Indian Sundarbans Mangal

Family Names	Scientific Names of the Rare/Thretened Species	Status in Sundarbans	Cause for Threating in Sundabans Mangrove Forest of Indian Part
Rhizophoraceae	1. *Rhizophora apiculata*	Occasional	Over exploitation and habitat lost
	2. *Bruguiera parviflora*	Occasional	Loss of natural habitat, rise of salinity
	3. *Ceriops decandra*	Occasional	Excessive exploitation and, habitat lost
	4. *Kandelia candel*	Occasional	Loss of natural habitat, rise of salinity
Meliaceae	5. *Aglaia cuculata*	Rare	Loss of natural habitat, rise of salinity
	6. *Xylocarpus mekongensis*	Threatened	Excessive exploitation and habitat lost
	7. *Xylocarpus granatum*	Threatened	Excessive exploitation and habitat lost
Sterculiaceae	8. *Heritiera formes*	Threatened	Excessive exploitation and habitat lost
Rubiaceae	9. *Scyphiphora hydrphyllacea*	Very Rare	Loss of natural habitat, rise of salinity
	10. *Hydrophyllax maritima*	Very Rare	Loss of natural habitat, rise of salinity
Tiliaceae	11. *Brownlowia lanceolata*	Occasional	Loss of natural habitat, rise of salinity
Arecaceae	12. *Nypa fruticans*	Occasional	Excessive expoliation and habitat lost
Acanthaceae	13. *Acanthus volubilis*	Very Rare	Loss of natural habitat, rise of salinity
Papilionaceae	14. *Cynometra ramiflora*	Rare	Loss of natural habitat, rise of salinity
	15. *Dalbergia spinosa*	Rare	Loss of natural habitat, rise of salinity
Sapotaceae	16. *Manlkara hexandra*	Rare	Loss of natural habitat, rise of salinity
Rutaceae	17. *Atalantia correa*	Very Rare	Loss of natural habitat, rise of salinity

Table 16 : Different Animal/Faunal Habitats in the Indian Sundarbans

Different Faunal Habitat	Dominant Animal Species	Existing Habitats
Estuarine and River Water	Crocodile, Dolphin, Water-Monitor, Shell-fish, Fin-fish, tortoise, turtle, terrapins, numbers of Aquatic Fauna	Dolphin and some of the shell-fish and fin-fish species are becoming rare and threateneed
Naked River Flats/ River Slopes	Different species of Mollusca, Crab, Annelid & Insect	Status of different species are not explored in Sundarbans.
Vegetated River Flats and River Slopes	Spotted Dear, Monkey, Wild Boar, Snakes, Frogs, Insects, Annelids, Mollusca, Crustaceans,	Status of micro and smaller animals are not assessed in Indian Sundarbans; but population, decreasing.
Dense Mangrove Forest Floor	Tiger, Fishing cat, Leopart cat, Jackal, Fox, Toddy cat, Civet, Wild boar, Spotted deer and few others...	Tiger population is said to be increasing after conservation measure by Sundarbans Tiger Reserve.
Naked Mangrove Forest Floor	Snake, Frog, Rodent, Lesser Cat, Mollusca, Crustacean, Insect, Annelid, etc...	Staus of different species are not explored in Sundarbans.
Tree Canopy	Monkey, Bat, Birds, snake, Insect, Mollusca and others.	Migratory and other bird species gradually decreasing
Amphibian Habitat	Most macro fauna, i.e., Tiger, Crocodile, Snakes, Water Monitor, good number of other fauna.	Heavy pressures threatened most of these fauna and their habitats in Sundarbans.

6

SUNDARBANS - THE MARSHY TIGER LAND

The Royal Bengal Tiger, *Panthera tigris tigris* has the world wide fame and glamour for their intelligence, body strength, quick movement ability, dangerous attacking techniques and their ferocious nature is unparalleled. Generally, the tiger is the dry forest land dweller, but in Sundarbans this Royal Bengal Tiger has to live in dense forest, entangled with dense aerial stilt roots, pneumatophores, tree trunks and the base soil is normally muddy. The pray animals of tiger in Sundarbans are wild boars, spotted deer, rhesus monkey, water monitors and even the large sized estuarine cat fish and crabs. Numerous medium to small rivers, creeks and canals spread and form close net work like mesh in these dense Sundarbans mangrove forests; there are about 70 small to large forest compartments within 15 forest blocks in Sundarbans Tiger Reserve (Table – 8). For these, the Bengal tiger has to swim and cross the wide rivers, canals, creeks very frequently for searching their prey animals, food, shelter and safe cub rearing place. Occasionally, they cross the river and enter the human habitat area. The tiger straying in and around the fringe villages of the Sundarbans mangal is one of the major problems. It is also seen that, this tiger can climb up the date palm tree *(Phoenix sylvestis)* in the village sides, which is not the normal behavior of the tiger. In Sundarbans, tiger has to drink the saline river water and it is assumed that for drinking of saline water, their kidney is affected and in turn they become most ferocious. The present day's tiger population in Indian Sundarbans is presented in Table - 17.

The concept of *'Keystone species'* was first introduced by Paine (1966 & 1969) to highlight the crucial role played by certain species, which directly or indirectly help many species to survived. As such, the loss of Keystone species must be resulted in the destruction of all other species or may bring the ecological disaster of an ecosystem. From this alarming point of view, scientists and naturalists denote the precise role of each and every species of an ecosystem that identify their significance as Keystone species and that need to be well accounting. The roles of Keystone species may be different or same that depends on the situation of the ecosystem. As such, it requires to be noted where and in which way the performances of Keystone species are prevailing? After this evaluation, *i.e.*, identifying the accurate role of Keystone species on the persistent of all other species of that ecosystem, the conservation programmes need to be undertaken on an urgent basis.

This mangrove swamp forest of Sundarbans are unique, that are characterised by diverse range of salt tolerant, unrelated trees, shrubs, herbs, grass, sedges, halophytic algae, fungi, lichens, bryophytes and few pteridophytes. These tidal swamps are also the habitats of large number of estuarine shell-fish, fin-fish, mollusks, amphibian, reptiles, birds and mammals. So that, these glorious habitats are maintaining the unique bio-diversity of both flora and fauna; among them, several threatened fauna, *viz.*, olive ridley, green ridley, few species of water monitors, fishing cats, local and seasonal migratory birds are the inhabitants here, though they are facing much pressure. The mangrove surrounded water-ways are the important spawning ground and nursery beds for several most economic and important marine and off-shore prawn, shrimp and fish species. Beside these, the forests check the devastating cyclones, sea-surges, tidal waves, and hurricanes and protect soil erosion considerably by building up of tree walls in these newly silted up delta land.

The Royal Bengal Tiger (*Panthera tigris tigris*) is the top consumer in the Sundarbans mangrove forest ecosystem. But unfortunately, these are threatened and are killed very often. Now several National projects and International funding organisations, agencies, scientific bodies, *viz.*, UNESCO, UNDP, WWF and IUCN have shown their interest to protect these Sundarbans mangrove forests tiger land; as a result, the Tiger Reserve Project was initiated since 1973 (Table-10). On the other hand the Sundarbans Biosphere Reserve Project had been established since 1989. Though it is late, presently several National and International Organisations have shown much interest and have also undertaken several conservation programmes for strict management of this Sundarbans mangals. But, identification of these problem is necessary and based on these problems conservation measures need to be undertaken on an urgent basis.

Important Roles Played by Tiger

The tiger plays very significant roles that may be considered under two categories, *viz.*, (a) Direct involvement and (b) Indirect involvement:

(a) Direct involvement means the interaction occurred in between tiger and interacted things, may be visible or invisible.

(b) Indirect involvement means no direct activities played by tiger, but the interactions come out through the activities, which are interfered by tiger indirectly.

The whole activities have been considered under 4 groups, *viz.,*

(i) *Tiger*-honey collector interaction;

(ii) Tiger- wood cutter interaction;

(iii) Tiger- fishermen interaction and

(iv) Tiger-deer-wild boar interaction.

(i) Tiger-Honey collector interaction

Honey for being the useful natural resources that are available in the Sundarbans mangrove forest and has attracted the honey collectors. For this, local honey collectors generally engage themselves for collecting honey, *etc.* During honey collection, major damages were made to tree branches and honey-bee hives. But the presence of tiger has to some extent restrict the large-scale entry of honey collectors in the forest. Based on this interaction, the following activities usually happen that are considered as indirect interaction or involvement. The incidents under indirect involvement are being taken place step by step as follows. Restriction of honey collectors means (a) less disturbance of honey-bees, which can (b) enhance the bee population. The increased population of honey-bees performs (c) the desirable pollination of mangrove flora, because of 80% of pollination of mangrove flora are performed by honey-bees. The effective pollination can only provide (d) large number of seed formation, seed maturation and seed germination under suitable environment that ultimately (e) maintain the regeneration of forest vegetation and by this way it save the virginity of f this threatened mangrove forest.

(ii) Tiger-Wood cutter Interaction

Several forest products like durable timbers, fuel wood, thatch leaves are the main articles of commerce to the local people. These valuable and highly priced wood, timber are being slain by the wood-cutters, time to time. But in many times, unlawfully entrance of woodcutters had become checked by the presence of tigers or had become prohibited by the tigers. This interactions in between the tigers and wood cutters considered as direct involvement. The following indirect interactions are the result of above direct actions that comes from the tiger. Tiger thus protected the virgin mangrove forests which are the gene pool of several biota of both flora and fauna.

(iii) Tiger – fisherman interactions

The mangrove surrounded waterways are the dwelling places of several aquatic fauna. The criss- cross canals, creeks, rivers and estuarine bodies are spawning ground and nursery beds of estuarine and offshore fish, prawn and crab species. For having these valuable resources of Sundarbans forests and waterways, large number of local people are engaged in fishing for every day's livelihood. But their movement to the shore or water courses of core forest areas have been checked as because of concealing movement of tiger in everywhere, even from island to island as tigers can swimming the vast rivers. This is the indirect interaction involved inbetween tiger and fishermen.

(iv) Tiger-Deer-Wild boar Interaction

Herbivorous deer and wild-boars are the first consumer in the trophic level of Sundarbans mangrove ecosystem and tiger has been maintaining the balance of the populations; the presence of tiger acts as the top consumer of the same tropic level. But the balanced population of deer is so far vital as because of protection of grass vegetation. So that, interaction between tiger and deer and wild boar are considered as indirect involvement. Depending on these interaction, the following incidents are involved. Due to the balanced populations of deer, grass (*Porteresia coarctata* Takeoka) vegetation is remained properly. This grass vegetation is not only treated as producer but also holds soil particles, that are coming with tide water and accelerates land formation towards offshore. The process of land formation occur where the grass vegetation covers the newly silted up soil by formation of immediate stolons and then protect it from soil erosion. Following the above succession process and direct and / or indirect interaction the next and real involvement of the tiger is protection role played in this mangrove forest and ecosystem. As a result the breeding times and spawning activities of different fauna is less distributed and the population ratio of aquatic fauna of different trophic levels are maintained. The discussion on Keystone species has opened so many factors, which are needed for further study. As per example, honey-bees play the important role that performs 80% of pollination of mangrove flora; as such, the loss of honey bees (*Apis dorsata*) will bring another disaster of mangrove vegetation as a whole by collapsing of total ecosystem. Similarly in aquatic ecosystem, crocodiles (*Crocodilus porosus*) protect partly aquatic bio-diversity as their presence check the rush of local people those who are engaged for fishing. Despite the wide performance of these known Keystone species in protecting the ecosystem, there are several factors, which have specific roles for the maintaining the biodiversity; but most of these are unknown to us and that should require to be investigated clearly. Now, time has come, while we have to take critical exploratory programmes along with awareness generation activities that will protect the sustainability of these forest.

Table 17 : Tiger Population in the Indian Sundarbans (Information taken from STR Reports)

Within the 2585 Km² Sundarbans Tiger Reserve Areas						Outside STR, 24 Pgs
Year of Census	Male	Female	Cub	Total	Remark	
1973	Sex was not determined			50	Census was incomplete	
1976	66	72	43	181	–	
1979	Sex was not determined			205	–	
1983	137	115	12	264	–	
1989	126	109	34	269	–	
1992	92	132	27	251	Computer programming was applied to count the number of tiger and determined sex and cub	26
1996	95	126	21	242	Plaster of paris casts of pugmarks were taken from field and also Computer programming was applied to count the number of tiger and determine sex and cub	35
1999	96	131	27	254	Pug-marks were taken from field	36
2001	93	129	23	245	–	

Table 18 : Tiger Straying Incidents (Source : STR Reports)

	SEX	
Year	Male	Female
1986	12	02
1987	16	04
1988	04	00
1989	03	01
1990	10	01
1991	01	00
1992	03	00
1993	00	01
1994	06	02

Photo Plate 25. Tiger released from cage at Indian Sundarbans (Source - STR)

Photo Plate 26. Swimming Tiger in Sundarbans river (Source : Sundarbans Tiger Reserve)

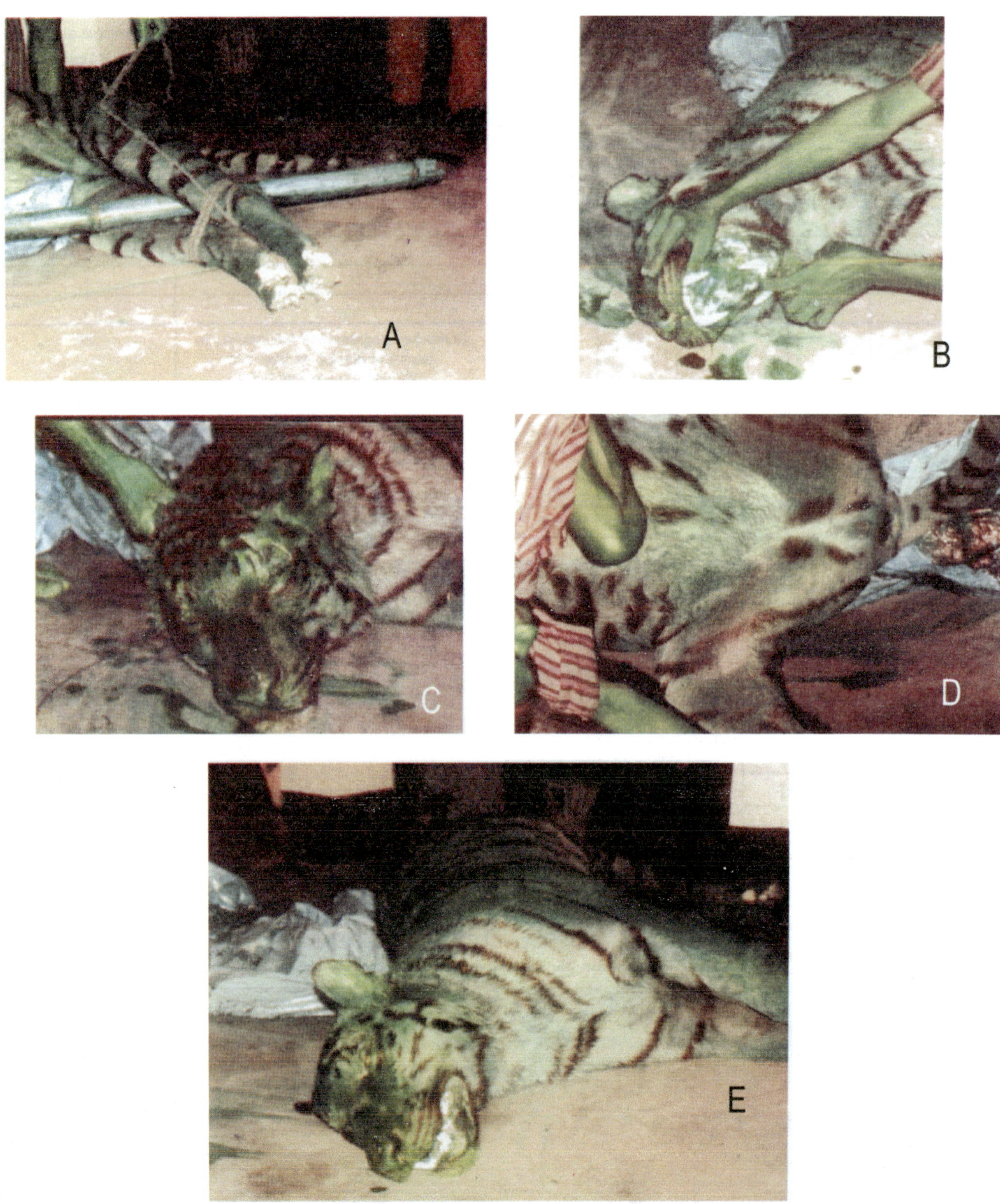

Photo Plate 27. Tiger- human conflict is the age old incidents in the Sundarbans area : Occasionally tiger enter in the mangrove reclaimed human settled/ habitat areas after crossing the river and sometime kill the domestic cattle; out of fear or grievance, local people occasionally kill such tiger brutally. The local people of the Sundarbans killed one such tiger - though they think little that the tiger actually protect the Sundarbans mangrove forest from human greediness.

Courtesy - Sundarbans Tiger Reserve, Forest Department, Government of West Bengal

Common and Higher Groups of Fauna of the Sundarbans

Sundarbans mangal is having diversified habitats, *i.e.*, terrestrial, inter-tidal, mud-flats, aquatic or estuarine, *etc.* (Table -16). As such, there are fauna like land dwellers/ terrestrials, amphibians, avian-fauna on tree canopy, aquatic, estuarine/off-shore species in this sea-land interphase mangrove habitats; they belong to different groups and vast numbers. Das & Nandi (1999) have reported 1434 fauna from this Sundarbans mangal, including Protozoa to Mammals, from the supra-tidal, inter-tidal and estuarine habitats (Table – 19). Out of these the Phylum – Chordate alone bears 890 species including 368 supra-tidal species, 346 estuarine species and 176 inter-tidal species (Das & Nandi (1999). Among these 445 species are Vertebrates; Das & Nandi (1999) further divided the faunal distribution in relation to their habitats (Table–16). In this context, mention may be made that these mangrove habitants or mangrove dwelling fauna are both residential type and also migratory or visiting time to time in different season or diurnal phases.

For bigger size ranges and comparatively easy identification possibilities, here in this treaties only the Phylum - Chordate and some very dominant Phylum - Arthropod and few others are mentioned in Table-19 A, B, C, D & E. But, for further detail studies and ready references mention may be made for some important books, monographs, periodicals, scientific papers, *etc.* Das & Nandi (1999) have also reported that about 1200 publications have come out on the fauna of Indian Sundarbans. Among these, after independence of India or in the later half of the 20th century the important and first comprehensive attempts towards completing the faunal list of Indian Sundarbans was made by Mukherjee (1975). The publications of Mandal & Nandi (1989) included also the big list of fauna of Indian Sundarbans under the title 'Sundarbans–A Socio-Biological Study'. Prior to this book, Naskar & Guha Bakshi (1987) have also published the book entitled 'Mangrove Swamps of Sundarbans – An Ecological Approach', where list of fish fauna and some other fauna of the Sundarbans were highlighted. Choudhury and his co-workers of the Marine Science Department, Calcutta University have also published good number of papers, articles, short notes and monographs on the lower group of fauna of Indian Sundarbans, especially from western fringe areas. The important publications made by IUCN on Sundarbans: Volume-I, which was compiled by Choudhuri & Choudhury (1994), where some information on the fauna of Sundarbans were mentioned. Besides these, the publications made by Chakraborty (1979, 1983 & 1984), Nandi & Misra (1987), Nandi, *et al.* (1993), Mandal & Ghosh (1989), Nandi & Pramanik (1994), Sanyal (1983,1999a & 1999b), Haldar & Misra (1999), Deuti & Roy Choudhury (1999), Naskar (1999) and several other publications are worth mentioning. But, Zoological Survey of India (1995) compiled and published the voluminous monograph on the Fauna of Hugli-Matla Estuary, where number of scientists and experts contributed their research papers on fauna.

In this context mention may also be made that N. Annandale and many others had contributed several sporadic publications on the fauna of Sundarbans during the early 20th century. The common Mammalian, Reptilian, Amphibian fauna and the Avis or birds of the Sundarbans are presented in Table – 19. The threatened, endangered and extinct fauna of the Sundarbans are also mentioned in Table- 20.

Regarding fish fauna of Sundarbans, several valuable contributions and scientific papers were published by the Scientists of Central Inland Fisheries Research Institute and Zoological Survey of India. The common and edible fish, shrimp, prawn and crab species in the Indian Sundarbans is more than 200; except Tetraodontiform species, most of these estuarine species are edible by local people (Table–19D).

Besides all these macro-fauna, microscopic zooplankters comprise of different taxonomic groups, comprising of copepodes, amphipods, ostracods, chaetognaths, mysids and hydromedusae are mentioned by Das & Nandi (1999). All these zooplankters depend on the mangrove detritus or as parasite on other fauna and act as the important food for the economically important fish and crustacean species. Along with these, both infauna and epifauna belonging to the benthic communities are important as the food for estuarine fish and crustaceans. The epifauna, *viz.*, gastropods are dominant in terms of number and bio-mass; barnacles, bivalves and sea anemones colony also remain attached with these mangroves. The infauna, like mangrove tree trunk borers and burrows are also important in these estuarine mangrove ecosystem.

Tiger (*Panthera tigris tigris*) is the top consumer in this mangrove ecosystem of Sundarbans, while the estuarine crocodile (*Crocodilus porosus*) is the top consumer in this aquatic ecosystem. The food chain from autotrophic algae-mangroves to the other heterotrophic fauna are shown in Flow Chart - 1 & Flow Chart - 2.

Photo Plate 28. - Tiger in the Sundarbans Tiger Reserve. (A) Tiger is being released from the cage by the forest staff; (B) Released Tiger run away out of fear; (C) Tiger crossing the vast river; (D) Tiger after release from the cage.

Courtesy **– Sundarbans Tiger Reserve, Forest Department, Government of West Bengal**

Photo Plate 29. Wild Animals of Sundarbans. A. Spotted Deer, *Cervis axis.*B.Rhesus monkey, *Macaca mulatta.* C. Jungle Fowl, *Gallus gallus.* D. Wild Boar, *Sus scrofa.* E. Olive Ridley, *Lepidochelys olivacea.* F. Estuarine Crocodile- *Crocodilus porosus*

***Courtesy* – Sundarbans Tiger Reserve, Forest Department, Government of West Bengal**

Photo Plate 30. Few Other Fauna of the Indian Sundarbans. A. Sea gull, B. Eggs of Marine Turtle. C. Snake Bird. D. Bones/ jaw of *Rhinoceros.* E. Snake. F. Local Bird. G.Water monitor, *Varanus salvator*

Table 19 : Common and Higher Group of Fauna of the Indian Sundarbans

A. Common Mammals : Class - Mammalia		
Aquatic Habitat Mammals in the Estuaries, Rivers and Offshore of the Indian Sundarbans		
Order - Cetacea	**Family**	**Common Name**
1. *Platinista gangetica* (Roxburgh)	Platinistidae	Gangetic Dolphin
2. *Neophocaena phocaenoides* (Cuvier)	Delphinidae	Little Porpoise/Black Finless Porpoise
3. *Sotalia plumbea* (Cuvier)	Delphinidae	Plumbous Dolphin
4. *Stanella malayana* (Lesson)	Delphinidae	Malay Dolphin
5. *Orcella brevirostris* (Owen)	Delphinidae	Irrawady Dolphin
Terrestrial Habitat Mammals in the Mangrove Forests of Indian Sundarbans		
Order - Carnivora		
6. *Panthera tigris tigris* (Linn.)	Felidae	Royal Bengal Tiger
7. *Felis viverrina* Bennett	Felidae	Fishing Cat
8. *Felis bengalensis* Kerr.	Felidae	Leopard Cat
9. *Fellis chaus* (Guldenstaedt)	Felidae	Jungle Cat
10 *Canis aureus* (Linn.)	Canidae	Jackal
11. *Vulpes bengalensis* (Shaw)	Canidae	Indian Fox
12. *Paraodoxurus hermaphroditus* Gray	Viverridae	Toddy Cat
13. *Vivericula indica* (Besmarest)	Viverridae	Small Indian Civet
14. *Herpestes edwardsir* (Geoffrey)	Viverridae	Common Grey Mingoose
15. *Herpestes auropunctatus* (Hodgson)	Viverridae	Small Indian Mongoose
16. *Herpestes palustris* Ghosh	Viverridae	Marsh Mongoose
Order - Mustelidae		
17. *Lutra perpicillata* Linn.	Mustelidae	Smooth Indian Otter
18. *Lutra lutra* (Linn.)	Mustelidae	Common Otter
Order - Insectivora		
19. *Suncus murinus* (Linn.)	Soricidae	House Shrew or Grey Musk Screw
Order - Artiodactyla		
20. *Sus scrofa* Linn.	Suidae	Wild Boar
21. *Cervus axis* (Erxleben)	Suidae	Spotted Deer
Order - Rodentia		
22. *Funambulus pennanti* Wroughton	Sciuridae	Five Stripped Squirrel
23. *Mus booduga* (Gray)	Muridae	Little Indian Field Mouse
24. *Mus musculus* Linn.	Muridae	House Mouse
25. *Rattus rattus* (Linn.)	Muridae	House Rat
26. *Bandicota indica* (Bechstein)	Muridae	Large Bandicoot Rat
27. *Bandicota bengalensis* (Gray)	Muridae	Lesser Bandicoot Rat
28. *Hystrix indica* Keer	Erethizontidae	Indian Crested Porcupine
Order - Chiroptera		
29. *Pteropus giganteus* (Brunnich)	Pteropidae	Indian Flying Fox
30. *Cynopterus sphinx* (Vahl)	Pteropidae	Shortnosed Fruit Bat
31. *Pipistrellus mimus* (Wroughton)	Vespertilionidae	Pigmy Pipistrelle
32. *Scotophilus kuhli* Leach	Vespertilionidae	Lesser Yellow Bat
33. *Megaderma lyra* (Geoffroy)	Megadermatidae	Indian False Vampire
34. *Megaderma spasma* Linn.	Megadermatidae	Vampire
35. *Rhinopoma hardwickii* (Gray)	Rhinopomatidae	Lesser Rat Tailed Bat
36. *Hipposideros bicolor* Temminck	Rhinopomatidae	Bicoloured Leafnosed Bat
37. *Hipposideros lankadiva* Kelaart	Rhinopomatidae	Cave Bat
38. *Taphozous longimanus* Hardwicke	Emballonuridae	
39. *Rhinolophus lepidus* Blyth	Emballonuridae	
Order - Primate		
40. *Macaca mulatta* (Zimmermann)	Cercopithecidae	Rhesus Monkey
Order - Pholidota		
41. *Manis pentadactyla* Linn.		Indian Pangolin

Contd...

B. Common Reptiles : Class - Reptilia		
Aquatic Habitat Reptiles in the Estuaries, Rivers, Freshwater Pools and Offshore of the Indian Sundarbans		
Order - Chelonia (Turtles, Tortoises & Terrapins)	**Family**	**Common Name**
1. *Batagur baska* (Gray)	Emydidae	Common Batagur River Terrapin
2. *Geomyda tricarinata* Blyth	Emydidae	Three Keeled Terrapin
3. *Kachuga kachuga* (Gray)	Emydidae	
4. *Kachuga tecta* Tentbridge	Emydidae	
5. *Geoclemys hamiltoni* (Gray) Terrapin	Emydidae	Spotted pond turtle
6. *Morenia ocellata* (Dumerid & Bibron)	Emydidae	Bengal Eyed Terrapin
7. *Trionyx gangeticus* Cuvier	Trionychidae	Freshwater Turtles
8. *Trionyx hurum* Gray	Trionychidae	
9. *Lissemys punctata* (Bonnaterre)	Trionychidae	Pond Turtle
10. *Pelochelys bibroni* (Owen)	Trionychidae	Coast Soft Shell Turtle
11. *Chitra indica* (Gray)	Trionychidae	Chitra Turtle
12. *Lepidochelys olivacea* (Eschscholtz)	Chelonidae	Ridley Turtle
13. *Chelonia myrdus*	Chelonidae	Green Turtle
14. *Eretmochetys imbricata*		Hawk's Bill
Terrestrial Habitat Reptiles of the Indian Sundarbans		
Order - Squamata :		
Sub-Order - Lacertilia (Lizards, Geckos & Monitors)		
15. Gekko gecko (Linn.)	Gekkonidae	
16. *Hemidactylus frenatus* Schelgab	Gekkonidae	House Gecko
17. *Hemidactylus brooki Gray*	Gekkonidae	
18. H*emidactylus leschenaulti Dumeril & Bibron*	Gekkonidae	
19. *Hemidactylus flaviviridis* Rupell	Gekkonidae	House Gecko
20. *Calotes versicolor* (Daudin)	Agamidae	Garden Lizard
21. *Chamaeleon zeylanicus* Laurenti	Chamaeleontidae	Indian Chamaeleon
22. *Mabuya carinata* (Schneider)	Scincidae	
23. *Mabuya dissimilis* (Hallowell)	Scincidae	
24. *Riopa punctata* (Daudin)	Scincidae	
25. *Riopa albopunctata* Gray	Scincidae	
26. *Varanus salvator* (Laurenti)	Varanidae	Water Monitor
27. *Varanus flavescens* (Gray)	Varanidae	Monitor Lizard
28. *Varanus bengalensis* (Daudin)	Varanidae	Monitor Lizard
Sub Order - Ophidia (Snakes)		
29. *Typhlops porrectus* Stoliczka	Typhlopidae	Blind Snake
30. *Typhlops braminus* (Daudin)	Typhlopidaei	Common Blind Snake
31. *Python molurus* Linnaeus	Boidae	Indian Python
32. *Acrochordus granulatus* Schneider	Boidae	Thicket Snake
33. *Eryx conicus* (Schneider)	Boidae	Rusell's Sand Boa
34. *Elaphe helena* (Daudin)	Boidae	Trinket Snake
35. *Ptyas mucosus* Linnaeus	Boidae	Rat Snake
36. *Oligodon arnensis* Shaw	Boidae	Kukri Snake
37. *Ahaetulla nasutus* (Lacepede)	Boidae	Green Whip Snake
38. *Ahaetulla myctarijans* (Linnaeus)	Boidae	Bronze Back
39. *Lycodon aulicus* Linnaeus	Boidae	Common Wolf Snake
40. *Natrix stolata* Linnaeus	Boidae	Checkered Keelback
41. *Atretium schistosum* Daudin	Boidae	Olivaceous Keelback
42. *Dendrelaphis ahaetulla* (Linnaeus)	Boidae	Bronze Back Snake
43. *Dendrelaphis tristis* (Daudin)	Boidae	Common Indian Bronze Back
44. *Enhydris enhydris* Schneider	Boidae	Common Smooth Water Snake
45. *Cerberus rhynchops* Schneider	Boidae	Dog-faced Water Snake
46. Gerardia frevostiana Eydous& Gervais	Boidae	
47. *Bungarus caeruleus* Schneider	Elapidae	Common Indian Krait
48. *Bungarus fasciatus* Scheneider	Elapidae	Banded Krait
49. *Naja naja* Linnaeus	Elapidae	Indian Cobra

Contd...

Order - Squamata :		
Sub Order - Ophidia (Snakes)	**Family**	**Common Name**
50. *Ophiophagus hannah* (Cantor)	Elapidae	King Cobra
51. *Vipera russelli* Shaw	Viperidae	Rusell's Viper
52. *Trimeresurus erythrurus* Cantor	Viperidae	Pit Viper
53. *Praesculata viperina* Schmidt	Hydrophidae	
54. *Microcephalophis gracilis* Shaw	Hydrophidae	
55. *Microcephalophis cantoris* Gunther	Hydrophidae	
56. *Enhydrina schistosa* Daudin	Hydrophidae	
57. *Hydrophis nigrocinctus* Daudin	Hydrophidae	
58. *Hydrophis obscurus* Daudin	Hydrophidae	
59 *Hydrophis caerulescens* Shaw	Hydrophidae	
Order - Crocodilia (Crocodiles)		
60. *Crocodylus porosus* (Schneider)	Crocodidae	Estuarine Crocodile
C. Common Amphibians of the Indian Sundarbans		
(i) Aquatic Amphibians		
1. *Rana cyanophlyctis* Schneider	Ranidae	Skipping Frog
2. *Rana tigerina* Daudin	Ranidae	Indian Bull Frog
3. *Rana limnocharis* Wiegmann	Ranidae	Paddy Field Cricket Frog
4. *Rana hexadactyla* Lesson	Ranidae	Pond Green Frog
5 *Microhyla ornata* Dumeril & Bibron	Microhylidae	Ornate Microhylid
(ii) Terrestrial Amphibians		
6. *Bufo melanostictus* Schneider	Bufonidae	Common Indian Toad
7. *Rhacophorus maculatus* (Gray)	Rhacophoridae	Tree Frog
D. Class - Avis (Common Birds / Avi-Fauna of the Indian Sundarbans)		
Order - Podicipediformes (Permanent Resident in the Mangrove Forest Area & Localities of Sundarbans)		
1. *Podiceps ruficollis* (Pallas)	Podicepididae	Dabchick
Order - Ciconiformes.		
2. *Ardea cinerea* (Linn.)	Ardeidae	Grey Heron
3. *Ardea purpurea* (Linn.)	Ardeidae	Purple Heron
4. *Egretta alba* Linn.	Ardeidae	Large Egret
5. *Egretta intermedia* (Wagler)	Ardeidae	Smaller Egret
6. *Egretta grazetta* (Linn.)	Ardeidae	Little Egret
7. *Ardeola grayi* (Sykes)	Ardeidae	Pond Heron
8. *Ardeola striatus* (Linnaeus)	Ardeidae	Little Green Heron
9. *Bubulcus ibis* (Linn.)	Ardeidae	Cattle Egret
10. *Nycticorax nycticorax* (Linn.)	Ardeidae	Night Heron
11. *Ixobrychus minutus* (Linn.)	Ardeidae	Little Bittern
12. *Ixobrychus cinnamomeus* (Gmelin)	Ardeidae	Chestnut Bittern
13.Ixobrychus flavicollis (Latham)	Ardeidae	Black Bittern
14. *Anastomus oscitans* (Boddaert)		Openbilled Stork
15. *Threskiornis melanocephala* (Latham)	Ardeidae	White Ibis
16. *Leptoptilos dubius* (Gmelin)	Ardeidae	Adjutant Stork
17. *Xenorhynchus asiaticus* (Latham)	Ardeidae	Blacknecked Stork
Order - Pelicaniformes		
18. *Anhinga rufa* (Daudin)	Anhingidae	Darter
Order - Anseriformes		
19. *Nettapus coromandelianus* (Gmelin)	Anatidae	Cotton Teal
Order - Falconiformes		
20. *Milvus migrans* (Boddaert)	Accipitridae	Pariah Kite
21. *Haliastur indus* (Boddaert)	Accipitridae	Brahminy Kite
22. *Haliaetus leucogaster* (Gmelin)	Accipitridae	Whitebellied Sea Eagle-
23. *Haliaetus leucoryphus* (Pallas) Pallas's	Accipitridae	Fishing Eagle-

Contd...

Species Name	Family	Common Names
24. *Spilornis cheela* (Latham)	Accipitridae	Crested Serpent Eagle
25. *Icthyophaga ichthyaetus* (Horsefield)	Accipitridae	Greyheaded Fishing Eagle
26. *Accipiter badius* (Gmelin)	Accipitridae	
Order - Galliformes		
27. *Francolinus gularis* (Temminck)	Phasianidae	Swamp Patrifge
28. *Gallus gallus* Linn.	Phasianidae	Red Jungle Fowl-
Order - Pelicaniformes		
29. *Phalacrocorax niger* (Vieillot)		Little Cormorant
Order - Columbiformes		
30. *Treron phoenicoptera* (Latham)	Collumbidae	Green Pigeon
31. *Streptopelia decaocto* (Frivaldszky)	Collumbidae	Ring Dove
32. *Streptopelia chinensis* (Scopoli)	Collumbidae	Spotted Dove
Order - Psittacieormes		
33. *Psittacula krameri* (Scopoli)	Psittacidae	Roseringed Parakeet
Order - Cuculiformes		
34. *Eudynamys scolopacea* (Linn.)	Cuculidae	Koel
35. *Clamator jacobinus* (Boddaert)	Cuculidae	Pied crested Cuckoo
36. *Cuculus varius* (Vahl)	Cuculidae	Brainfever Bird
37. *Cuculus micropterus* Gould	Cuculidae	Indian Cuckoo
38. *Centropus sinensis* (Stephens)	Cuculidae	Crow Pheasant
Order - Stigiformes		
39. *Tyto alba* (Scopoli)	Strigidae	Barn Owl
40. *Otus scops* (Linn.)	Strigidae	Scops Owl
41. *Otus bakkamoena* Pennant	Strigidae	Collared Scops Owl
42. *Bubo zeylonensis* (Gmelin)	Strigidae	Brown Fish Owl
43. *Bubo bubo* (Linn.)	Strigidae	Eagle Owl
44. *Athene brama* (Temminck)	Strigidae	Spotted Owlet
Order - Caprimulgiformes		
45. *Caprimulgus indicus* Latham	Caprimulgidae	Jungle Nightjar
Order - Apodiformes		
46. *Cypsiurus parvus* (Lichtenstein)	Apodidae	Palm Swift
Order - Coraciformes		
47. *Coracias benghalensis* (Linn.)	Coraciidae	Indian Roller
48. *Alcedo atthis* (Linn.)	Alcedinidae	Common Kingfisher
49. *Pelargopsis amauroptera* (Pearson)	Alcedinidae	Brownwinged Kingfisher
50. *Halcyon pileata* (Boddaert)	Alcedinidae	Blackcapped Kingfisher
51. *Halcyon smyrnensis* (Linn.)	Alcedinidae	Whitebreasted Kingfisher
52. *Halcyon chloris* (Boddaert)	Alcedinidae	Whitecollared Kingfisher
53. *Ceryle rudis* (Linn.)	Alcedinidae	Lesser Pied Kingfisher
54. *Merops orientalis* Latham	Meropidae	Green Bee-eaters
55. *Upupa epops* Linn Hoope	Upupidae	
Order - Piciformes		
56. *Megalaima haemacephala* (Muller)	Capitonidae	Coppersmith
57. *Micropternus brachyurus* (Vieillot) Rufous	Picidae	Woodpecker
58. *Dinopium benghalense* (Linn.)	Picidae	Goldenbacked Woodpecker
59. *Chrysocolaptes lucidus* (Scopoli)	Picidae	Goldenbacked Woodpecker
Order - Charadriformes		
60. *Metopidius indicus* (Latham)	Jacanidae	Bronze-winged Jacans
61. *Hydrophasianus chirurgus* (Scopoli)	Jacanidae	Pheasant tailed Jacana
62. *Vanellus indicus* (Boddaert)	Charadriidae	Redwattled Lapwing
63. *Charadrius dubius* Scopoli	Charadriidae	Little Ringed Plover
64. *Rostratula benghalensis* (Linn.)	Ros tratulidae	Painted Snipe

Contd...

Species Name	Family	Common Names
Order - Gruiformes		
65. *Rynchops albicollis* (Swainson)	Laridae	
66. *Sterna albiforns* (Pallas)	Laridae	Little Tern
67. *Sterna acuticauda* J.E.Gray	Laridae	Blackbilled Tern
68. *Sterna bergii* (Lichtenstein)	Laridae	Largecrested Tern
69. *Sterna bengalensis* (Lesson)	Laridae	Indian Lessercrested Tern-
70. *Sterna aurantia* J.E. Gray	Laridae	Indian River Tern -
71. *Amaurornis phoenicurus* (Pennant)	Rallidae	Water-hen
72. *Porphyrio porphyrio* (Linn.)	Rallidae	Purple Moorhen-
73. *Gallinula chloropus* (Linn.)	Rallidae	Moorhen
74. *Fulica atra* (Linn.)	Rallidae	Coot
Order - Passeriformes		
75. *Alauda gulgula* Franklin	Alaudidae	Eastern Skylark
76. *Lanius schach* Linn.	Laniidae	Blackheaded Shrike
77. *Oriolus xanthornus* (Linn.)	Oriole	
78. *Dicrurus adsimilis* (Bechstein)	Dicruridae-	Black Drongo
79. *Artamus fuscus* (Vieillot)	Artamidae-	Ashy Swallow Shrike
80. *Sturnus malabaricus* (Gmelin)	Sturnidae-	Greyheaded Myna
81. *Sturnus contra* Linn.	Sturnidae	Pied Myna
82. *Acridotheres tristis* (Linn.)	Sturnidae	Common Myna
83. *Acridotheres ginginianus* (Latham)	Sturnidae	Bank Myna
84. *Acridotheres fuscus* (Wagler)	Sturnidae	Jungle Myna
85 *Dendrocitta vagabunda* (Latham)	Corvidae	Tree Pie
86. *Corvus splendens* Vieillot	Corvidae	House Crow
87. *Corvus macrorhynchos* Wagler	Corvidae	Jungle Crow
88. *Pycnonotus jocosus* (Linn.)	Pycnonotidae	Redwhiskered Bulbul-
89. *Pycnonotus cafer* (Linn.)	Pycnonotidae	Redvented Bulbul
90. *Turdoides striatus* (Dumont)	Timaliidae	Jungle Babbler-
91. *Muscicapa parva* Bechstein	Muscicapidae	Redbreasted Flycatcher
92. *Terpsiphone paradisi* Linn.	Muscicapidae	Paradise Flycatcher-
93. *Pachycephela grisola* (Blyth)	Muscicapidae	Mangrove Whistler-
94. *Orthotomus sutorius* (Pennant)	Sylviidae	Tailor Bird-
95. *Acrocephalus dumetorum* Blyth's	Sylviidae	Read Warbler-
96. *Phylloscopus fuscatus* (Blyth)	Sylviidae	Dusky Leaf Warbler
97. *Phylloscopus collybita* (Viellot)	Sylviidae	Chiffchaff.
98. *Acrocephalus agericola* (Jerdon)		Paddy-field Warbler
99. *Copsychus saularis* (Linn.)	Turdidae	Magpie Robin
100. *Saxicola caprata* (Linn.)	Turdidae	Pied Buchechat
101. *Parus major* Linn.	Paridae	Grey Tit
102. *Dicaeum erythrorhynchus* (Latham)	Dicaedae	Tickell's Flower pecker
103. *Nectarinia zeylonica* (Linn.)	Nectarinide	Purplerumped Sunbird
104. *Nectarinia asiatica* (Latham)	Nectarinidae	Purple Sunbird-
105. *Zosterops palpebrosa* (Temminck)	Zosteropidae	White-eye
106. *Passer domesticus* (Linn)	Ploceidae	House Sparrow
107. *Ploceus philippinus* (Linn.)	Ploceidae	Baya
108. *Lonchura striata* (Linn.)	Ploceidae	Whitebacked Munia
109. *Lonchura punctulata* (Linn.)	Ploceidae	Spotted Munia-
Order- Ciconiformes (Migratory Bird Species in the Indian Sundarbans Mangrove Forests & Human Settled Areas)		
110. *Ardea goliath* Cretzschman	Ardeidae	Giant Heron
Order - Pelecaniformes		
111. Pelecanus philippensis Gmelin	Pelecanidae	Spotted - billed Pelecan-
Order - Anseriformes		
112. *Aythya ferina* (Linn.)	Anatidae	Common Poachard
113. *Anas crecca* Linn.	Anatidae	Common Teal

Contd...

Species Name	Family	Common Names
114. *Anas acuta* (Linn.)	Anatidae	Pintail
115. *Anas strepera* Linn.	Anatidae	Gadwall
116. *Aythya nyroca* (Guldenstadt)	Anatidae	White-eyed Poachard
117. *Aythya marila* (Linn.)	Anatidae	Scaup Duck
118. *Tadorna ferruginea* (Pallas)	Anatidae	Brahminy Duck
119. *Dendrocygna javanica* (Horsfield)	Anatidae	Lesser Whistling Teal
120. *Netta rufina* (Pallas)	Anatidae	Redcrested Pochard
Order - Falconiformes		
121. *Falco tinnunculus* Linn.	Falconidae	Kestrel
122. *Falco peregrinus* Tunstall	Falcondae	Peregrine
123. *Pandion haliaetus* (Linn.)	Accipitridae	Osprey
124. *Circus aeruginosus* (Linn.)	Accipitridae	Marsh Harrier
Order - Gruiformes		
125. *Larus argentatus* (Pontoppidan)	Laridae	Herring Gull
126. *Larus ridibundus* (Linn.)	Laridae	Blackheaded Gull
127. *Larus brunnicephalus* (Jerdon)	Laridae	Brownheaded Gull
128. *Hydroprogne caspia* (Pallas)	Laridae	Caspian Tern
129. *Sterna hirundo* (Linn.)	Laridae	Common Tern
130. *Sterna fuscata* (Linn.)	Laridae	Sooty Tern
131. *Gelochelidon nilotica* (Gmelin)	Laridae	Gullbilled Tern
132. *Chlidonias hybrida* (Pallas)	Laridae	Whiskered Tern
Order - Charadriiformes		
133. *Vanelius cinereus* (Blyth)	Charadriidae	Greyheaded Lapwing
134. *Calidris tenuirostris* (Horsfield)	Charadriidae	Eastern Knot
135. *Calidris minuta* (Leisler)	Charadriidae	Little Stint
136. *Calidris testacea* (Pallas)	Charadriidae	Curlew Sandpiper
137. *Charadrius dubius* Scopoli	Charadriidae	Little Ringed Plover
138. *Lemnodronus semipalmatus* (Blyth)	Charadriidae	Snipebilled Godwit
139. *Calidris alpinus* (Linn.)	Charadriidae	Dunlin
140. *Arenaria interpres* (Linn.)		
141. *Numenius phaeopus* (Linn.)	Scolopacidae	Whimbrel
142. *Numenius arquata* (Linn.)	Scolopacidae	Curlew
143. *Limosa limosa* (Linn.)	Scolopacidae	Blacktailed Godwit
144. *Tringa totanus* (Linn.)	Scolopacidae	Common Red Shank
145. *Tringa glareola* (Linn.)	Scolopacidae	Wood Sandpiper
146. *Tringa terek* (Latham)	Scolopacidae	Terek Sandpiper
147. *Tringa ochropus* Linnaeus	Scolopacidae	Green Sandpiper
148. *Tringa hypoleucos* Linn.	Scolopacidae	Common Sand Piper-
149. *Gallinaga gallinago* (Linn.)	Scolopacidae	Fantail Snipe
150. *Gallinago stenura* (Bonaparte)	Scolopacidae	Pintail Snipe
151. *Tringa nebularia* (Gunner)	Scolopacidae	Green Shank
Order - Columbiformes		
152. *Streptopelia orientalis* (Latham)	Columbidae	Rufoustailed Dove
Order - Piciformes		
153. *Jynx torquilla* Linn.	Picidae	Wryneck
154. *Hirundo rustica* Linn.	Hirundinidae	Common Swallow
Order - Passeriformes		
155. *Phoenicurus ochruros* (Gmelin)	Turdidae	Black Redstart
156. *Saxocola torquata* (Linn.)	Turdidae	Collard Bush Chat
157. *Monticola solitarius* (Linn.)		Blue Rock Thrunk
158. *Erithacus svecicus* (Linn.)	Turdidae	Blue Throat
159. *Motacilla alba* Linn.	Motacillidae	White Wagtail
160. *Motacilla citreola* (Pallas)	Motacillidae	Yellowheaded Wagtail
161. *Anthus novaeseelandiae* (Gmelin)	Motacillidae	Paddy Field Pipit
162. *Anthus hodgsoni* Richmond	Motacillidae	Indian Tree Pipit-

E. Common Fish Fauna of the Rivers, Estuaries and Off-shore of the Indian Sundarbans

Class - Chondrichthyes - Cartilagenous Fishes

(Shark, Tiger Shark, Hammer Headed Shark, Saw Fish, Guitar Fish, Electric Fish, Ray Fish, *etc.*).

(a) Common Marine and Estuarine Fish, Shark Species

1. *Chiloscyllium griseum* Muller & Henle-Tiger Shark
2. *Stegostoma fasciatum* (Hermann)-Tiger Shark
3. *Carcharhinus limbatus* Valenciennes
4. *Glyphis gangeticus* (Muller & Henle)
5. *Sphyma blochii* (Cuvier)
6. *Pristis microdon* (Latham)-Saw Fish
7. *Rhinobatas annandalei* (Norman)-Guitar Fish
8. Dasyatis bleekeri (Blyth)
9. *Dasyatis Zugei* (Muller & Henle)
10. *Dasyatis marginatus* (Blyth)
11. *Dasyatis uamak* (Forsskal)
12. *Dasyatis sephen* (Forsskal)
13. *Aetobatus narinari* (Euphrasen)-Eagle Rays

Class-Osteichthyes (Bonny Fish)

(b) Common Marine/Estuarine Fish

14. *Pelona ditchela* Valenciennes
15. *Scatophagus argus* (Linnaeus)-Butter Fish
16. *Streinateus sinensis* Day-Pomfret
17. *Pampus chinensis* (Euphrasen)
18. *Pampus argentateus* (Euphrasen)
19. *Cynoglossus cynoglossus* (Hamilton)
20. *Cynoglossus lingua* (Hamilton-Buchanan)
21. *Cynoglossus brevis* (Gunthar)
22. *Pama pama* (Hamilton)
23. *Daysciaena albida* (Cuvier)
24. *Nibea soldada* (Lacepeda)
25. *Otolithoides biauritua* (Cantor)
26. *Johnius belangerii* (Cuvier)
27. *Johinus coitor* (Hamilton)
28. *Tenualosa* (*Hilsa*) *ilisha* (Hamilton)
29. *Tenualosa* (*Hilsa*) toli (Valenciennes)
30. *Tenualosa* (*Hisha*) *elongate* (Bennett)
31. *Nematolosa nasus* (Bloch)
32. *Raconda russeilianna* Gray
33. *Anodonstoma chaeunda* (Hamilton)
34. *Leiognathus blochii* (Valenciennes)
35. *Leiognathus equulus* (Forsskal)
36. *Leiognathus fasciatus* (Lacepeda)
37. *Garras oyena* (Forskal)
38. *Gerreomorpha setifer* (Hamilton)
39. *Ceilia dussumeiri* Valencienkes
40. *Coilia neglecta* Whitehead
41. *Coilia ramcarati* (Hamilton)
42. *Coilia raynaidi* Valenciennes
43. *Setipinnia phasa* (Hamilton)
44. *Stepinnia laty* (Valenciennes)
45. *Corica saboma* (Hamilton)
46. *Gudusia chapra* (Hamilton)
47. *Pelona elongata* (Bennett)
48. *Anchovialli indica* (va Hass)
49. *Trissocles hamiltonii* (Gray)
50. *Trissocles purava* (Hamilton)
51. *Chysocentrus dorab* (Forskal)
52. *Boarius boarius* (Hamilton)
53. *Tachisurus iella* (Day)
54. *Tachisurus sona* (Hamilton)
55. *Thryssa purava* (Hamilton)
56. *Thryssa hamiltonei* (Gray)
57. *Megalops cyprinoides* (Broussoner)
58. *Anguilla bengalensis* (Gray and Hardwicke)
59. *Chanos chanos* (Forskal)
60. *Acrichthys aor* (Hamilton)
61. *Mystus cavasius* (Hamilton)
62. *Mystus gulio* (Hamilton)
63. *Mystus villatus* (Bloch)
64. *Rita rita* (Hamilton)
65. *Pangasius pangasius* (Hamilton)
66. *Plotesus angultarius* (Bloch)
67. *Plotesus angultarius* (Bloch)
68. *Arius arius* Day
69. *Arius gagora* (Hamilton)
70. *Arius jella* Day
71. *Arius sagor* (Hamilton)
72. *Arius sona* (Hamilton)
73. *Batrachocephalus nino* (Hamilton)
74. *Oesteogeniosus militaris* (Linnaeus)
75. *Harpodon nehereus* (Hamilton)
76. *Harpodon tumbil* (Bloch)
77. *Batrichthys gruuiens* (Linnaeus)
78. *Rhynchoramphus georgii* (Valeneiennes)
79. *Strongylura leira* (Bleeker)
80. *Sillaginopsis panijus* (Hamilton)
81. *Sillage sihama* (Forsskal)
82. *Therapon jarbora* (Forsskal)
83. *Lates calcarifer* (Bloch)
84. *Valamugil speigleri* (Bleeker)
85. *Valamugil cunnesius* (Valenciennes)
86. *Awaonichthys menoni* Chatterjee
87. *Bathygobins orbicularis* (Visweswara Rao)
88. *Boleophthalmus boddarert* (Pallas)
89. *Brachygobius nunus* (Hamilton)
90. *Periophthalmodon schlosserri* (Valenciennes)
91. *Periophthalmus koelreuteri* (Pallas)
92. *Periophthalmus chrysospiles* Bleeker
93. *Periophthalmus malaccensis* Eggart
94. *Periophthalmus vulgaris* Eggert
95. *Periophthalmus weberi* Eggert
96. *Pseudapocryptes lanceolatus* (Bloch & Schneider)
97. *Scartelaos histophorus* (Valencienes)
98. *Stigmatogobius sadanundio* (Hamilton)
99. *Trypauchen vagina* (Bloch & Schneider)
100. *Kurtus indicus* Bloch
101. *Trichiurus haumala* (Forskal)
102. *Trichiurus gangeticus* Gupta
103. *Trichiurus muticus* (Gray)
104. *Trichiurus savala* (Cuvier)
105. *Lepturacanthus pantulai* (Gupta)
106. *Lepturacanthus savala* (Cuvier)
107. *Rhinomugil corsula* (Hamilton)
108. Polydactylus indicus (Shaw)
109. *Ambassis nama* (Hamilton)
110. *Ambassis baculis* (Hamilton)
111. *Ambassis ranga* Hamilton
112. *Caranus carangus* (Bloch)
113. *Megalesis cordyla* (L.)
114. *Lutianus argentimaculatus* (Forskaal)
115. *Gazza minuta* (Block)

116. *Drepane punctatus* (L.)
117. *Etroplus suratensis* (Bloch)
118. *Cybium commersonii* (Lacepede)
119. *Odontamblyopus rubicundus* (Hamilton)
120. *Thycenophrya indicus* (L.)
121. *Platycephalus insidiator* (Forskal)
122. *Colea bleekeri* (Bouleenger)
123. *Triacanthus brevirostris* (Temm & Sch.)
124. *Oryzias melastigma* (Mc Chelland)
125. *Amphipnous chuchia* (Hamilton)
126. *Nundus nundus* (Hamilton)
127. *Elops saurus* (L.)
128. *Corcia sabomia* (Hamilton)
129. *Psiodonopsis boro* (Hamilton)
130. *Chanos chanos* (Forsskal) - Milk Fish
131. *Zenarchopterus dispar* (Valenciennes)
132. *Strongylura strongylura* (van Hassalt)
133. *Xenontodon cancila* (Hamilton)
134. *Ambasis commersoni* (Valenciennes)
135. *Terapon jarbura* (Frosskal)
136. *Lutjanus johni* (Bloch)
137. *LIza parsia* (Hamilton)
138. *Liza tade* (Frosskal)
139. *Liza macrolepis* (Smith)
140. *Mugil cephalus* (Linnaeus)
141. *Mugil oligolepis* (Bleeker)
142. *Eleutheronema tetradactylum* (Shaw)
143. *Polynemus paradiseus* Linn.
144. *Apocryptes bato* (Hamilton)
145. *Glossogobius giuris* (Hamilton)
146. *Gobiopterus chuno* (Hamilton)

(Besides these common edible or economic fish several other fish species are also available in the estuaries of Sundarbans)

F. Class-Echinoderms (Marine or Estuarine Species)

Sub-Class-Asteroidea (Starfish)

1. *Astropecten curyacanthus* L.

Sub-Class-Ophiuroidea (Brittle Snakes)

2. *Amphioplus hastatus* (L. Jungman)
3. *Amphioptus ireois* (Lyman)
4. *Ophiactis delagoa* Balinky
5. *Ophiactics modesta* Broch
6. *Ophioenenuis marmorata* (Lamark)

Sub-Class-Echinoidea (Sea Urchins)

7. *Temnopleurus toreumaticus* (Leske)
8. *Clypeaster rarispinus* de Meijere
9. *Laganum decagonala* (de Blain ville)
10. *Echinodiscus auritus* Leske

Sub-Class-Holothuroidea (Sea Cucumber)

11. *Stolus rapak* (Kochler & Vaney)
12. *Thorsonia investigatoris* (Kochler & Vaney)
13. *Acaudina molpadioides* (Semper)
14. *Paracaudina australis* (Semper)

G. Class-Arthopods (Shrimps, Prawns and Crabs)

Sub-Class-Crustacea (Prawns and Shrimps)

1. *Caridina gracilipes* de Man
2. *Macrobrachium rosenburgi* (de Man)
3. *Macrobrachium malcomsonii* (Milne-Edw)
4. *Macrobrachium lamarrei* (Milene-Edw)
5. *Macrobrachium ruda* (Heller)
6. *Macrobrachium mirabili* Kemp.
7. *Macrobrachium scrobiculum* (Heller)
8. *Macrobrachium javanicum* (Heller)
9. *Metapenaeus affinis* (Milne-Edw)
10. *Metapenaeus brevicornis* (Milne-Edw)
11. *Metapenaeus dobsonii* (Miers)
12. *Metapenaeus monoceros* (Fabricius)
13. *Palaemon styliferus* (Milne-Edw)
14. *Palaemon tenuipes* (Henderson)
15. *Parapenaeopsis sculptilis* (Heller)
16. *Parapenaeopsis stylifera* (Milne-Edw)
17. *Penaeus indicus* (Milne-Edw)
18. *Penaeus monodon* (Fabricius)
19. *Penaeus semisulcatus* De Mann
20. *Acetes indicus* (Milne-Edw)
21. *Acetes erythraeus* Nobilis

Order-Decapoda (Crabs)

1. *Paratelphusa spinigera* (Wood-Mason)
2. *Paratelphusa jacquemontii* (*Rothbun*)
3. *Paratelphusa hydrodromus*
4. *Varuna litterata*
5. *Scylla serata* (Forsskal)
6. *Matuta plauipes* (Fabricius)
7. *Portunus pelagicus* (Linnaeus)
8. *Portunus sanguinolentus*
9. *Chrybdii rostata* (A.M. Edw.)
10. *Chrybdii orientales* (Herbst)
11. *Chrybdii ornata* (A.M. Edw)
12. *Chrybdii merguiensis* (de Man)
13. *Sesarma bidens* (de Man)
14. *Sesarma impressa* (M. Edw.)
15. *Sesarma taeniolatum* (White)
16. *Sesarum tetragona* (*Fabricius*)
17. *Sesarum edwardsi* de Man
18. *Sesarum longipes* Krauss
19. *Sesarum quadrata* (Fabricius)
20. *Sesarum smithii* Edwards
21. *Metaplax crenulata* (Gerstaecker)
22. *Metaplax dentipes* (Heller)
23. *Metaplax distincta* H.M. Edwards
24. *Metaplax intermedia* de Man
25. *Metapograpsus maculatus* (Edw.)
26. *Metapograpsus messor* (Forsskal)
27. *Matuta victor* (Fabricius)
28. *Matuta lunaris* (Herbst)
29. *Calappa lophos* (Herbst)
30. *Dorippa facchino* (Herbst)
31. *Leucosia craniolaris* (Herbst)
32. *Ethusa indica* (Alcock)
33. *Hymenicus wood-masoni* (Alcock)
34. *Hymenicus inachoides* (Alcock)
35. *Philyra globosa* (Fabricius)
36. *Philyra globulosa* (Edwards)
37. *Doclea japonica* (Ortman)
38. *Doclea canalifera* (Stimpson)
39. *Uca acutus* Stimpson
40. *Uca lactea annulipes* (M. Edw.)
41. *Uca dussumieri* (M. Edw.)
42. *Uca triangularis* (M. Edw.)
43. *Macropthalmus pectinipes* (Guerin)
44. *Illyoplas gangeticus* (Kemp)
45. *Dotilla blanfordi* Alcock
46. *Dotillopsis brevitarsis* (de Man)
47. *Ocypoda macrocienra* (M Eww.)
48. *Scopimera globosa* (de Hann)

Hermit Crabs

1. *Coenobita cavipes* Stimpson
2. *Diogenes avarus* Heller
3. *Clibanarius padavensis* de Man

H. Mollusca (Brackishwater Species)

Class - Grastropoda

Order - Archaeogastropoda

1. *Potamacmaea fluviatilis* (Blanford)
2. *Cyclostrema innocens* Preston
3. *Nertia articulata* Gould
4. *Neritina smithii* Wood
5. *Neritina violacea* Gmelin
6. *Psendonerita sulculosa* von Martens
7. *Littorina carinifera* (Meuke)
8. *Littorina melanostama* Gray
9. *Littorina scabra* (Linn.)
10. *Littorina undulata* Gray
11. *Gangetica miliacea* Nevill

I. Branchiopods

1. *Artemia* sp.
2. *Cladocerus* sp.
3. *Daphnia* sp.
4. *Ceriodaphnia* sp.
5. *Moina* sp.
6. *Bosmina* sp.
7. *Bosminopsis* sp.
8. *Maerothris* sp.
9. *Triopa* sp.

J. Coepods

1. *Diaptomus sp.*
2. *Pseudodiaptomus* sp.
3. *Cyclops* sp.
4. *Microstella* sp.
5. *Acartiella* sp.
6. *Calanus* sp.
7. *Parcalanus* sp.
8. *Oithona* sp.
9. *Pentella* sp.

K. Lepidoptera

1. *Spodoptrera mauritia* (Boisduval)

L. Coleoptera

1. *Haltiea cyanea* Weber
2. *Epicauta* sp.
3. *Monanus concinnulus* (Walker)

M. Hymenoptera

1. *Apis dorsata*
2. *Apis cerena*
3. *Sima rufonnigra*
4. *Menaforia indica* Gupta & Saxena
5. *Enicospilus nocturnus* (Kohl)

N. Diptera

1. *Psychoda bengalensis* Burncm
2. *Geranomyia circipunctata* Brunctti
3. *Geranomyia faviocosta* Brun
4. *Geranomyia tridens* Brun
5. *Phlebotomus argentipes* Ann. & Brun
6. *Phlebotomus bengalensis* Brun
7. *Phlebotomus minutus*
8. *Aedes nivens* (Ludlow)
9. *Anopheles subpictus* Grassi
10. *Aedeeomycia catas*b
11. *Culex vishnui* Theob.
12. *Lutzia fuscans* Wied.
13. *Tacniorhynchus annuliferus* Theob.
14. *Tacniorhynchus uniformis* Theob.
15. *Odontomyia rubrithorax* Macq.
16. *Odontomyia dorsoangulata*
17. *Chrysops dispa*r Fabr.
18. *Haematopota javana* Wied.
19. *Tabanus striata* Fabr.
20. *Sepsis indica* Wied.
21. *Sarcophaga dux* Thoms.
22. *Sarcophaga ruficornis* Fabr.
23. *Musea crassirostris* Stein
24. *Musea pattonia* Aust.
25. *Musea seniorwhite* Patton
26. *Musea vicia* Macq.
27. *Stomoxys calcitrans* Linn.
28. *Stomoxys indica* Pic.
29. *Curicea pumila* Wied.
30. *Cyrtoneura stobulans* Fil
31. *Lipsa pumila* Wied.

O. Class - Arachnida

a) Acarina

1. *Bakeriella ocimis* Chakrobarty & Mandal
2. *Eriophyes pongamiae* (Channabasavanna)
3. *Aculops abutiloni* Mandal & Chakroborty

1. *Leneage decorata* Blackwell
2. *Araneus mitifica* (Thorell)
3. *Ecuta javanica* Thorell
4. *Hippasa* sp.
5. *Oxopes* sp

c) Xiphosura

1. *Leneage decorata rotundicanda*
2. *Tachypleusgigas*

P. Class - Polychaeta (Bristle Worms)

Order Phyllodieida

1. *Phyllodoce madeivensis* Langerhans
2. *Etenoni borantollae* Fauvel
3. *Gattyana daludens* Fauvel
4. *Gattyana doludens* Fauvel
5. *Glycera alba* (Muller)
6. *Glycera lancadive* Schmarda
7. *Glycera rouxi* Audouin & Milne - Edw.
8. *Glycera emerita* Audouin & Milne - Edw.
9. *Nephlyx oligobranchia* Southern

10. *Areistrosyllis constricta* Southern
11. *Tahlesapia annaudeli* Fauvel
12. *Lepidonotus lenuisetsus* (Gravier)
13. *Dedrobereuder geterioida* Southern
14. *Namalyeastis indica* (Southern)
15. *Lycastonereis indica* Rao
16. *Nereis cricagnatha* Ehlers
17. *Nereis chingrighattensis* Fauvel
18 *Perineis nigropnnetuta* Horst

Order - Capitellida

19. *Heteromastus similis* (Southern)
20. *Mastrobranchus indicus* Southern

Order - Sternaspida

21. *Sternaspis scutate* (Ranjai)

Order - Spionida

22. *Polydora kempi* Southern
23. *Prionospio cirrifera* Wiren
24. *Spia bengalensis* Willey
25. *Sabellaria pectinata* var. *intermedia* Fauvel

Order - Eunicida

26. *Diopatra cuprea* (Bose)
27. *Diopatra neopolitana* Delle Chiaji
28. *Marphysa macintoshi* Crossland
28. *Marphysa sauguinea* (Montagu)
29. *Lumbrinereis heteropoda* Marenze llar
30. *Lumbrinereis notocirrata* (Fauvel)
31. *Lumbrinereis polydesma* (Southern)

Order Terebellida

32. *Laoimia medusa* (Saigony)

Order - Sabellida

33. *Laonome indica* Southern

Table 20 : The Threatened, Endangered and Extinct Fauna of the Sundarbans Mangal

Sl. No.	Scientific Names of the Fauna Extinct/Threatened	Common Name	Status in the Sundarbans Mangal
Threatened Mammalian species in Indian Sundarbans			
1.	*Felis bengalensis*	Leopard Cat	Threatened species in Sundarbans
2.	*Felis viverrina*	Fishing Cat	Threatened species in Sundarbans
3.	*Panthera tigris tigris*	Tiger	Threatened species in Sundarbans
4.	*Platinista gangetica*	Gangetic Dolphin	Threatened species in Sundarbans
5.	*Orcaella brevirostris*	Irrawady Dolphin	Threatened species in Sundarbans
6.	*Neophocaena phocaenoides*	Little Indian Porpoise	Threatened species in Sundarbans
7.	*Manis pentadactyla*	Chinese Pangoin	Threatened species in Sundarbans
Extinct Mammalian species in Indian Sundarbans			
1.	*Cervis duvauceli*	Swamp Deer	Extinct Species from Sundarbans
2.	*Bubalus bubalis*	Wild Buffalo	Extinct Species from Sundarbans
3.	*Rhinoceros sondaicus*	Indian Rhinoceros	Extinct Species from Sundarbans
4.	*Rhinoceros unicornis*	*Great 1 - Horned Rhinoceros*	Extinct Species from Sundarbans
Threatened Reptilian species in Indian Sundarbans			
1.	*Crocodylus porosus*	Estuarine Crocodile	Threatened species in Sundarbans
2.	*Lepidochelys olivacea*	Olive Ridley Turtle	Threatened species in Sundarbans
3.	*Batagur baska*	Batagur Turtle	Threatened species in Sundarbans
4.	*Lissemys punctata*	Indian Flapshelled Turtle	Threatened species in Sundarbans
5.	*Trionyx gangeticus*	Indian Softshelled Turtle	Threatened species in Sundarbans
6.	*Trionyx hurum*	Peacock-marked Turtle	Threatened species in Sundarbans
7.	*Kachuga tecta*	Indian Tent Turtle	Threatened species in Sundarbans
8.	*Varanus bengalensis*	Common Indian Monitor	Threatened species in Sundarbans
9.	*Varanus flavescens*	Yellow Monitor	Threatened species in Sundarbans
10.	*V aranus salvator*	Water Monitor	Threatened species in Sundarbans
11.	*Python morulus*	Indian Rock Python	Threatened species in Sundarbans
Threatened Birds species in Indian Sundarbans			
1.	*Ardea goliath*	Great Heron	Threatened Bird species in Sundarbans
2.	*Pelecanus philippensis*	Dalmatian Pelican	Threatened Bird species in Sundarbans
3.	*Leptotilos dubius*	Lesser Adjutant Stock	Threatened Bird Specie in Sundarbans

Source : Das & Nandi (1999).

Mangal Support Coastal and Estuarine Fisheries

The Sundarbans mangal is a dense mangrove dominant coastal tidal influenced vegetation in the Lower Ganga Delta. This coastal mangal release an average of 6000 t/ ha mangrove litters per year. After decomposition of these mangrove litters through mineralisation processes, enormous organic nutrients are released in the coastal ecosystem; which help to grow the plankton and other fish food organisms. As such, good number of estuarine fish species, prawns, shrimps, crabs are present in this estuarine water for having abundant natural food. Besides these, dense mangrove forests also provide shelter and nursery ground for the fry, fingerlings, juveniles of the estuarine and off-shore fish and shrimp species. The shell-fish and fin-fish species collected and identified from the Indian Sundarbans coast are shown in Table - 19 E, F & G.

Fish, fisheries and fishing in the Sundarbans tidal estuaries, rivers, creeks and vast estuarine wet-lands are very important in terms of rural economic set up and also supply of cheaper protein food. These shrimp and fish are transported to the local, as well as, to the International markets. Rural people are engaged in the fishing activities and fisheries, which are the substantial source of their earning. Irrigation is not much developed in this area; as such, the traditional rain-fed rice cultivation is the main agricultural practice here. In this coastal area, the climate is very much hostile and frequent natural calamities (Table - 12) cause considerable damage the agricultural crop. Very often the farmers have to face all such critical problems for frequent flood due to collapsing of the river dykes during cyclones and tidal thrusts. In this perspective, these rural people have to depend on fishing in these estuarine rivers, canals, creeks and also in the off-shore fishing during winter months and also other period of the year. Shrimp seed (*Penaeus monodon*) collection is also now the common occupation for thousand of rural people. The fish species collected from these estuarine water of Sundarbans are listed in Table – 19 E. Besides these fishing activities, about 40,000 ha brackish-water fisheries in coastal Sundarbans areas are also available for traditional fish culture. Some of the prawn farmers from outside are now engaged in the culture of export quality shrimp through semi-intensive practices. These in turn pollute the environment for discharging the untreated polluted fishery water in the side by canals and rivers. For the export market demand of shrimp (*P. monodon*), culture, *vis-à-vis*, collection of shrimp seed ultimately damage other less priced fish and prawn seeds. All these had created tremendous detrimental effects and problems in these natural ecosystem. Besides these, capture of gravid mud crab (*Scylla serrata*) and export of the same to the foreign markets has also adverse effect, which depleting the natural stock. This mud crab used to take shelter in the mud holes in the mangrove forest floor and play important role in this ecosystem. But for high demand for crab in the International markets exploitation of this wild mud-crab are now threatening the natural balance. Till date, artificial cultural practices or fattening of mud crab was inadequately done in Sundarbans. Netting with fine mesh nylon net of about 2 to 3 km length (*Char pata jal*) on river side forest lands have also very adverse effect, as it kills or damages the juveniles of fish, prawn and shrimp species, indiscriminately. All these harmful, detrimental ecological damage require to be stopped on an urgent basis. Act are there against exploitation and damage of this coastal mangrove ecosystem, but the question of implementation still remains less attended.

Photo Plate 31. Aquatic Food Chain - Detritus Food Chain of Estuarine Fish and Shrimp Species in the Sundarbans Mangal

Tropic level on the Sundarbans mangrove ecosystem

Mangrove ecosystem is no way exception than the other typical tropical ecosystem. Here the basic and the producer group of organisms are planktonic, benthic, periphytic algae and the inter-tidal mangrove plants. The primary consumers are zooplankters, zoobenthos, large group of microbes, mollusca, crabs and crustacean; all these micro-fauna graze on these alga and the decomposed mangrove detritus. Furthermore, these alga, mangrove detritus, smaller group of micro - fauna, mollusca and crustaceans act as the natural food for the secondary group of consumers, these are particularly the smaller fish, amphibians, reptiles and others. List of other fauna of the ecosystem is presented (Table - 19).

Predator fish, cat fish, different species of sharks, dolphins are the secondary and tertiary consumers and the crocodile, *Crocodilus porosus* in the estuarine water of Sundarbans is the top consumer. The herbivorous deer, wild boar and monkey graze on these mangrove seedlings and tender leaves or grass, *Porteresia coactata*. All these herbivorous primary consumer group are the important food for the secondary, tertiary and top consumers, *viz*., different types of wild cats including tiger. Through the Flow Chart -1, the Tropic Level in the Sundarbans mangals is shown.

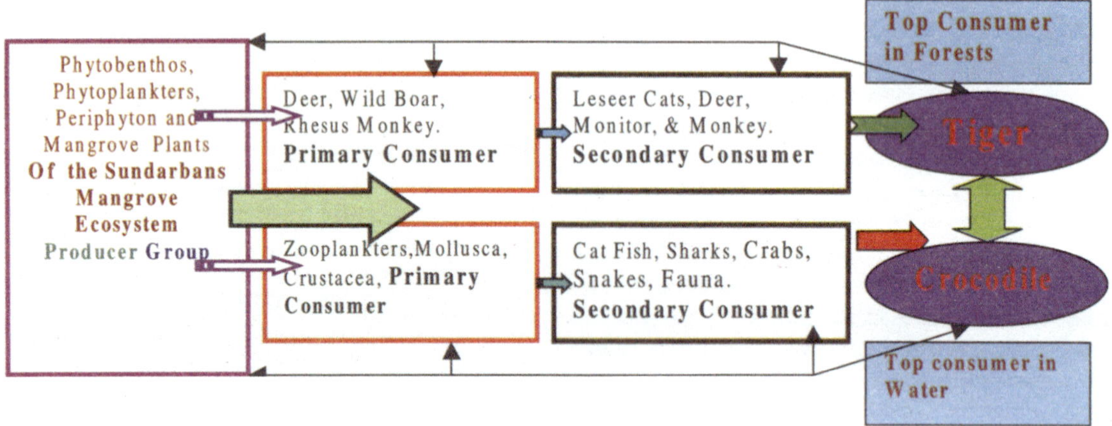

Flow Chart 1. Tropic Level in Indian Sundarbans

Photo Plate 32. Mangrove Litter Collection by litter traps for bio-mass estimation.

Tidal amplitude/fluctuations of the Sundarbans mangal

Tidal amplitude, especially during the high water spring the tide attains up to 7.0 m from the MSL; occasionally it is up to 9.0 m from the MSL. In the Sundarbans within the Indian territory, tidal fluctuation is very high and is next to the Gujarat coast (Table – 21). Due to this high tidal fluctuation and high tidal amplitude, most of the mangrove forest in Sundarbans is very frequently inundated with the saline river water. Therefore, diverse groups of mangrove species get their suitable or preferred and natural habitats for varied succession.

Table 21 : Tidal Amplitudes in the Indian Sundarbans

Tidal Amplitude Level	Height (Metre)	
Spring Low Level (S.L.L.)	0.0	Low Tide
Mean Low Water Spring (M.L.W. S.)	1.0	0.0 m - 2.5 m
Mean Low Water Neap (M.L.W.N.)	2.5	-
Mean Tidal Level (M.T.L.)	4.5	Mid Tidal
Mean High Water Neap (M.H.W.N.)	6.5	2.5 m - 6.7 m
Mean High Water Spring (M.H.W.S.)	8.0	-
Spring High Level (S.H.L.)	9.0	High Tidal
Supra Tidal Level (Supra T.L.)	9.1	6.5 m (9+) m

Graph - 1. Tidal Amplitude in the Mangals of Indian Sundarbans During Different Tidal Phases

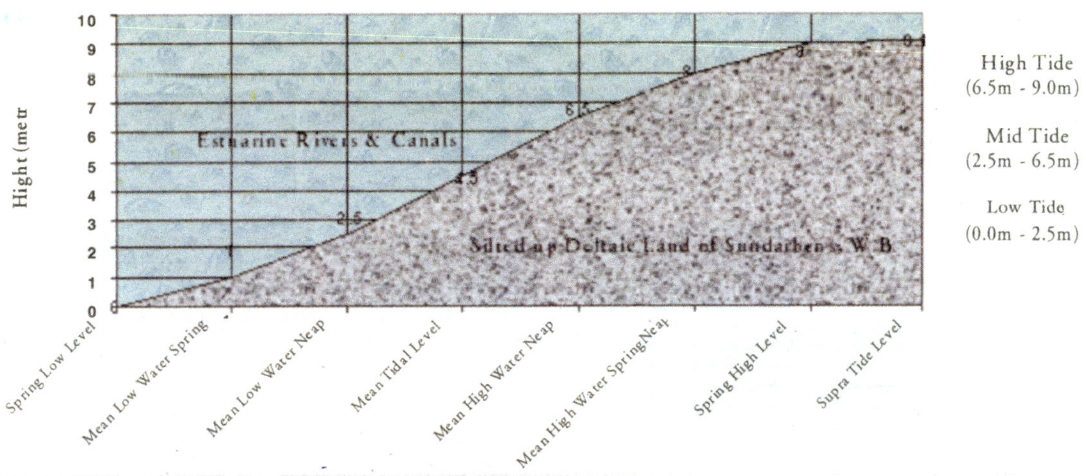

Photo Plate 33. (A) Dense River Bank Mangrove Forest. (B) Tidal inundated *Rhizophora* sp. forest, Sundarbans

Photo Plate 34. Fishermen along with their poor fish catch

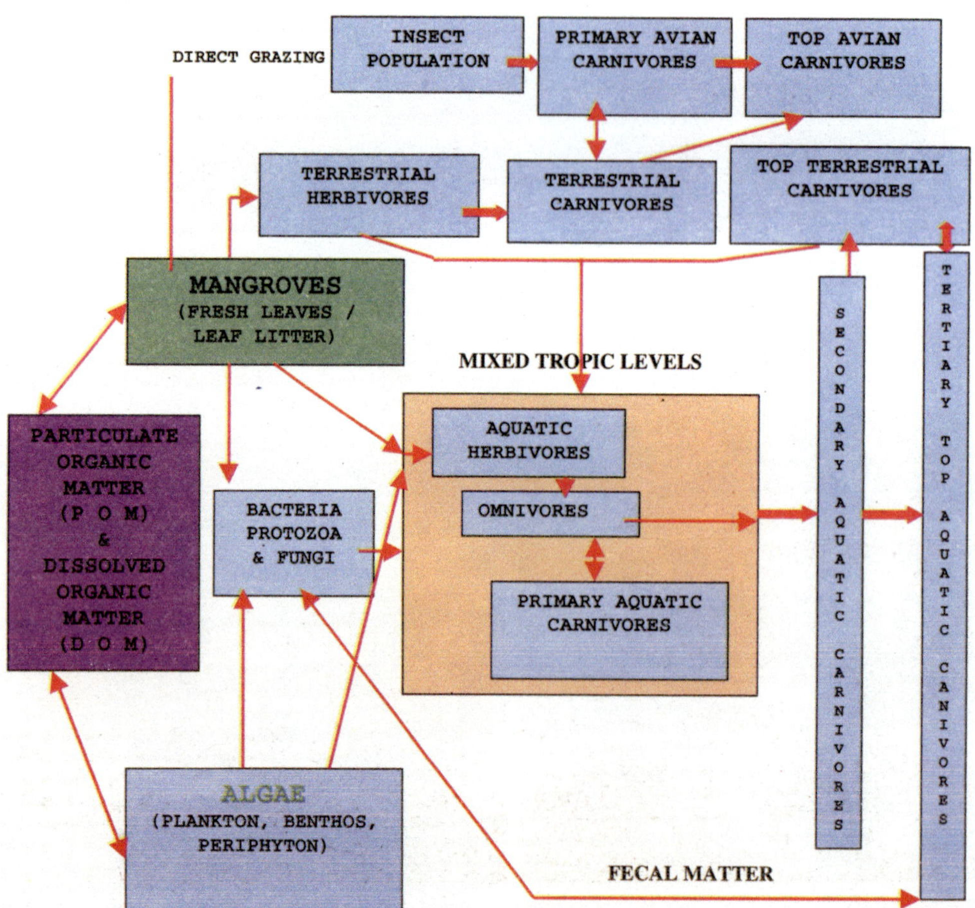

Flow Chart 2. Compartmental Model of the Food Chains Operating in the Sundarbans Mangal

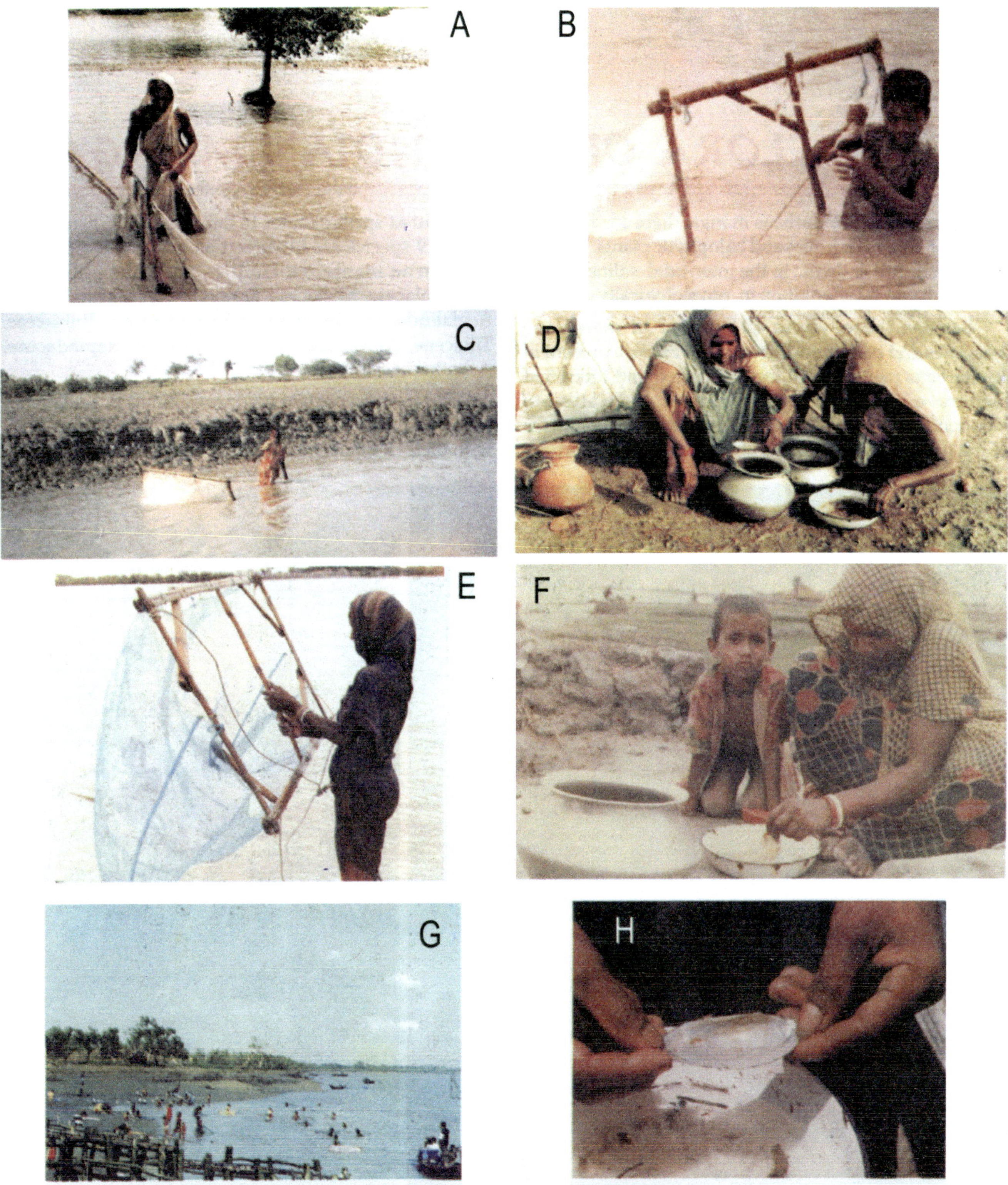

A. Netting out of shrimp and other fish juveniles by old women. B. Netting by young girl. C. Netting for shrimp juveniles by girl. D.Screening of shrimp seeds by rural women. E. Showing the net and the rural women. F. Screening of shrimp seeds . G. Netting for shrimp juveniles collection by rural men, women, boys or girls. H. Segregation of shrimp juveniles.

Photo plate 35. Large scale exploitation of shrimp seeds and damage of other fish, prawn and aquatic fauna (juveniles) is alarming at Sundarbans

7

FLORA OF THE INDIAN MANGALS

In the Indian Sundarbans mangal there are about 60 species of true mangroves and halophytic plants. Most of these plants are endemic in the inter-tidal saline deltaic areas with special adaptation in physiologically dry soil.

Besides these, about 40 number of mangrove associates and back mangrove species are also present in Sundarbans mangals. All these collected and identified mangrove plants and the mangrove associates are mentioned in Table–14. The major and minor mangroves or the true mangrove plants of the Indian Sundarbans are about 41 species under 16 families and 26 genera and about 19 species are identified as mangrove associates. These true mangrove families are Rhizophoraceae, Avicenniaceae, Meliaceae, Sonneratiaceae, Myrsinaceae, Aegialitidaceae, Sterculiaceae, Combretaceae, Rubiaceae, Tiliaceae, Arecaceae, Euphorbiaceae, Asclepiadaceae, Acanthaceae, Papilionaceae and Rutaceae.

The most important mangrove families, genera and species of the Indian Sundarbans are mentioned here along with their characteristic features.

Photo Plate - 36. A. Plantation of *Avicennia* spp. at river flat. B. Natural succesion of *Porteresia* sp. followed by *Avicennia* spp. C. *Rhizophora mucronata* Lamk. in its natural habitats at Sundarbans mangal.

Table 22 : Key Characters of the Dominant Indian Mangroves and Mangrove Associated Families

No.	Families	Characters of the Family
1.	**Rhizophoraceae** Genera - 4, Species - 9	Trees or shrubs, aerial stilt roots, prop roots, root - knees, root - buttress and short to long viviparous germinated hypocotyles well developed.
2.	**Avicenniaceae** Genus - 1, Species - 3	Small to medium trees, aerial roots, pencil like pneumatophores; crypto-viviparous germination; ventral surface of leaves pale green to whitish.
3.	**Sonneratiaceae** Genus - 1, Species - 4	Medium-tall trees, leaves partially pendulous-iso-bilateral; woody, conical, pneumatophores, 1 m high; fruits with persistent pointed calyx lobe.
4.	**Combretaceae** Genus - 1, Species - 2	Small to medium tree, pneumatophores looping, from lateral underground roots, trunk inconspicuous, dense tomb shaped canopy, no vivipary.
5.	**Arecaceae** Genera - 2, Species - 2	Palms, pinnately compound leaves, either arising from rhizomatous stem/tufted on stem apex, with/without sharp spines on petiole base/leaf tip.
6.	**Meliaceae** Genera - 2, Species - 4	Medium tree, stout trunk bases, stem buttress/buffalo horn like pneumatophores; fruit globose, round, generally 5 cm-15 cm diameter.
7.	**Euphorbiaceae** Genera - 2, Species - 2	Medium tree, aerial roots absent, gall on trunk bases, spreading roots horizontally exposed on erosion; milky latex, leaves yellow on maturity.
8.	**Myrsinaceae** Genus - 1, Species - 1	Small tree, no aerial roots, trunk base thick with basal roots, flowers in simple umbels, white, fragrant; fruits long, curved, conical pointed apex.
9.	**Aegialitidaceae** Genus - 1, Species - 1	Small tree, trunk base swollen, spongy, leaves with long clasping petiole, flowers white, fruits elongated, cryptoviviparous germination, long ribbed.
10.	**Sterculiaceae** Genus - 1, Species - 2	Tall tree, prominant buttress, flat blind root suckers and conical woody pneumatophores, ventral surface of leaves pale green, fruits winged.
11.	**Rubiaceae** Genus - 1, Species - 1	Shrub-small tree, tomb shaped canopy, inconspicuous wavy aerial roots, flowers in condensed cymes, reddish white, fruits with persistent calyx.
12.	**Acanthaceae** Genus - 1, Species - 3	Viny shrub/twining, small stilt roots, leaf margins with spines/succulent, flowers clustered on raceme spike, violet-white, fruits capsule, 4-seeded.
13.	**Apocynaceae** Genus - 1, Species - 1	Small-medium tree, on sandy soil, leaves whorls on branch apex, veins prominent, flowers fragrant, fruits 3-5 cm diam. smooth surface, long stalk.
14.	**Asclepiadaceae** Genera - 4, Species - 5	Twining herb to shrub, occasional climbers, succulent leaves, milky latex, fruits in pairs, follicles, seeds numerous, flattened.
15.	**Bignoniaceae** Genus - 1, Species - 1	Medium tree, leaves pinnately compound, flowers large, capsule long, two from a single stalk, seeds rectangular with corky wings, in back mangals.
16.	**Malvaceae** Genera - 2, Species - 5	Smell-medium tree, mucilaginous juice, fused stamens from single staminal column, mangrove associate or back mangal.
17.	**Tiliaceae** Genus - 1, Species - 2	Small tree, leaves similar to *H. fomes* with twining elongated branches, cyme mixed inflorescence, flowers ebracteate, fruits capsule, small.
18.	**Rutaceae** Genus - 1, Species - 1	Shrub, swollen trunk bases, leaves with paired lemon like spines, leaves with lemon like smell, fruit small lemon like, ribbed.
19.	**Poaceae** Genera - 4, Species - 4	Intertidal habitat grass, grow on the river-flat; the dominant species *Porteresia coaractata* (Roxb.) Takeoka looks similat to rice plant.
20.	**Verbinaceae** Genera - 2, Species - 3	Stragling much branched shrub, dark green, twiggy on river bank; inflorescence cymes, axillary, fruits pyriform drupe, woody; seeds four.
21.	**Tamaricaceae** Genus - 1, Species - 3	Shrubs or small trees with deliquiscent branches; leaves minute but fleshy and ornamented, triangular; catkin inflorescence, pendulous.
22.	**Solanaceae** Genera - 4, Species - 9	Shrubs with erect or twing habit, stem and leaves with yellow arms; flowers violet; fruits globose, round typical *Solanum* type.
23.	**Sapotaceae** Genus - 1, Species - 1	Tall evergreen tree with dense leafy crown above the tidal reach; flowers in axillary clusters; fruits round about 1.0 cm long; grow above tidal ridges.

Contd...

No.	Families	Characters of the Family
24.	**Sapindaceae** Genera - 2, Species - 2	Shrubs generally grow beyond the inter-tidal zones; leaves alternate or opposite; fruits capsule with or without arillus; albumen rarely present; embryo usually thick.
25.	**Amaryllidaceae** Genus - 1, Species - 1	Rhizomatous bulbous herb, attain the height upto 1.5 m; simple lamina up to 75.0 cm, sheathing at base, umbel scape axillary, erect, cylindrical
26.	**Araceae** Genus - 1, Species - 1	Erect herb on the inter-tidal river flats or canal side in the inner estuary and also in the marshy waste land; with white flowers on long stalk
27.	**Aizoaceae** Genus - 1, Species - 1	Decumbent or ascending much branched herb on the river flats; leaves opposite sessile, linear-spathulate, entire, glabrous; flowers-pinkish
28.	**Boraginaceae** Genus - 1, Species - 1	Prostrate, much spreading branched herb on the salt crusted soil, above the frequent tidal reaches; inflorescence unipodial helicoid cyme
29.	**Caesalpiniaceae** Genus - 1, Species - 2	Erect or climbing shrubs, bipinnate, on thickets, beyond the frequent tidal reach; fruit coat smooth or with dense spines; seeds flattened or globose
30.	**Fabaceae** Genera - 3, Species - 5	Leaves unipinnate; flowers papilionaceous, hermaphrodite; vexillary aestivation; pods flattened orbicular, not terminated into beaks.
31.	**Chenoipodiaceae** Genus - 2, Species - 3	Prostrate or erect herbs on the river flats within frequently tidal inundated zones; stem and leaves dark green to reddish with much succulent leaves.
32.	**Clusiaceae** Genus - 1, Species - 1	Medium tree beyond frequent tidal reaches; leaves dark green leathery and round fruit on elongated long stalk
33.	**Convolvulaceae** Genus - 1, Species - 1	Frequent on sandy beaches beyond tidal reach; leaves thick, with bilobed apex and large typical bell shaped pink white flowers; fruits globose.
34.	**Cyperaceae** Genera - 2, Species - 2	Root-stalk stout, hard with fibrous roots; leaves membranous, acuminate, sheath soft fleshy with long typical inflorescence
35.	**Flacourtiaceae** Genera - 4, Species - 9	Shrub; leaves spirally arranged; blade simple, ovate; grow beyond the inter-tidal zones and in the river side thickets.
36.	**Lauraceae** Genera - 1, Species - 1	Leafless twining parasitic herb, stem solid, cord like, much branched; flowers small, sessile, bracteoles short; fruit white, globose, smooth.
37.	**Loranthaceae** Genus - 3, Species - 3	Parasitic shrubs on the mangrove trees, viz., *Excoecaria* sp., *Avicennia* spp. and *Xylocarpus* spp., leaves fleshy green; flowers green small or red.
38.	**Barringtoniaceae** Genus - 1, Species - 2	Medium tree with deliquiscent much branched; leaves dense in terminal part of the branches; lamina oblong-elliptic; receme inflorescence
39.	**Opuntiaceae** Genus - 1, Species - 1	Purely xerophytic on sandy and dry soil; shrub with jointed phylloclades, flowers yellow, berries purple, pear shaped, tubercled
40.	**Pandanaceae** Genus - 1, Species -1	Small tree, dichotomously branched, up to 5.0 m tall, stemless, palm like, sub-erect, supported by stilt or prop-roots; leaves upto 2.0 m tall serrated
41.	**Pteridaceae** Genus - 1, Species - 2	Erect fem with frequent stilt root like rhizoids; rhizome globose, horizontal, caudex, circinate vernation, young rachis covered by gregarious hairs like.
42.	**Rupiaceae** Genus - 1, Species - 1	Submerged, delicate, soft, green, much branched herb in the brackish water fisheries and stagnant pools or canals; flowers minute.
43.	**Asteraceae** Genus - 1, Species - 2	Herbs on the coastal sandy soils but beyond the frequent tidal reaches; inflorescence characteristic capitulum type.
44.	**Cuscutaceae** Genus - 1, Species -1	Stem slener, brown or golden-yellow, leafless; parasitic on *Excoecaria* tree and beyond the frequent tidal reaches

Family - 1. Mangrove Rhizophoraceae

Plants under this family - Rhizophoraceae Blume (Mangrove- Rhizophoraceae) are usually trees or shrubs; these plant species grow gregariously in the inter-tidal silted up river slopes, river flats, ridges and dense inter-tidal mangrove forests. In Sundarbans, 4 genera including 9 species under the family - Mangrove Rhizophoraceae were collected and reported by Naskar & Guha Bakshi (1987) and Naskar & Mandal (1999). Their above ground aerial prop roots, stilt roots, broom like stilt roots, root buttresses, root knees and other different types of aerial breathing roots are typical for adapting best in these inter-tidal, compact and highly saline, physiologically dry soil. Viviparous germination is also interesting and unique characteristic feature of these members of the family - Mangrove Rhizophoraceae. When their fruits hang on their mother plants, their seeds start germination. Their seeds mature completely in the mother plants and without taking much dormant phase seed germination starts on the mother plant itself; which is known as viviparous germination. These are unique adaptations of the mangrove plants for their better survivability and regeneration of their offspring in these hostile tide affected, high saline delta or estuarine environment. Their newly developed slender hypocotyles and seedling plants are quite different in shape and size from species to species.

The family name – Rhizophoraceae Blume had been derived from the Tribe Genus – *Rhizophora* L. Sp. Pl. 443.1753. This pantropical family has about 16 genera and 120 species of shrubs and trees, distributed throughout the tropical and sub-tropical regions of the Globe. But, all these species under the family - Rhizophoraceae are not true mangroves; only the members which belong to the Tribe - Rhizophoreae (Mangrove - Rhizophoraceae) are true or major mangroves and the species under these four genera, *viz.*, – *Rhizophora* L., *Bruguiera* Lamk., *Ceriops* Arnold and *Kandelia* Wight & Arnold are the major mangroves. The Indian mangrove zones in general and the Sundarbans mangal in particular have 9 true mangrove species under these above 4 mangrove genera; these are *Rhizophora mucronata* Lamk., *R. apiculata* Blume, *Bruguiera gymnorhiza* (L.) Lamark, *B. sexangula* (Lour.) Poir., *B. cylindrica* (L.) Blume, *B. parviflora* W. & A. *ex* Griffith, *Ceriops tagal* (Perr.) Robinson, *C. decandra* (Griff.) Ding Hou and *Kandelia candel* (L.) Druce. Besides these, in Andaman & Nicobar Islands and Mahanadi delta, other mangrove species, *viz.*, *R. stylosa* Griff., *R.* X *lamarckii* Montr., *etc.* were also reported by Banerjee, *et al.*(1989) and Dagar, *et al.*(1991). One more species under the family - Rhizophoraceae is also distributed in the Andaman & Nicobar Islands, *viz.*, *Carallis brachiata* (Lour.) Merr., but this species is not a mangrove. However, in this treaties the key characters of all these common Indian mangrove species under the family - Mangrove Rhizophoraceae are presented for easy field identification. These members under the family - Mangrove Rhizophoraceae can be easily identified by their characteristic viviparous germinated hypocotyles and conspicuous aerial stilt roots, proof roots, root knees, broom like stilt roots, root buttress, *etc.*

Table 23 : Differences among these four Genera of the Family - Mangrove Rhizophoraceae

Characters	*Rhizophora* spp.	*Bruguiera* spp.	*Ceriops* spp.	*Kandelia* sp.
Aerial Roots	Profuse, distinct tall, looping stilt roots.	Broom like stilt roots fused on trunk base.	Fused stilt roots form conical trunk base.	Fused stilt roots form trunk base.
Habits	Globose, tall trees.	Pyramidal, tall tree	Trees or shrubs.	Diffused- small tree.
Leaves	Acute, mucronate, dark green, glossy wax coated leaves.	Acute, reddish - dark green, grossly leathery leaves.	Leaves, ovate/ round shaped, reddish green less glossy.	Leaf lanceolate, not much acute end, not much glossy.
Flowers	Tetra-merous, not lobed, ovary 2 - celled.	Tetra-merous, lobed, ovary 2-4 celled.	Penta-merous, ovary 3-celled.	Penta-merous, petals, bilobed, ovary1-celled.
Hypocotyles	Not ridged, up to 75 cm long, thick, not much pointed end	Ridged, up to 20 cm long, persistent calyx remain longer period.	Ridged, 10 to 25 cm long, not much pointed.	Smooth, up to 40 cm long, green with sharp pointed.
Germination	Viviparous germination	Viviparous germination	Viviparous germination	Viviparous germination

The Genus - *Rhizophora* L. has two species in the Indian Sundarbans, *viz., Rhizophora mucronata* Lamk. and *R. apiculata* Blume; other species under the genus - *Rhizophora* L. are *R. stylosa* Griff. and *R.* X *lamarchii* Montr.; these later two species were reported from mangrove forests of the Andaman & Nicobar Islands and Mahanadi delta in Orissa Banerjee, *et al.* (1989) & Dagar, *et al.* (1991).

Table 24 : Difference between the two species of *Rhizophora* L. in Sundarbans

Characters	*Rhizophora apiculata* Blume	*Rhizophora mucronata* Lamk.
Leaf	Petiole - 4.0 cm long; lamina with acute apex; lamina 13.0 cm long x 6.0 cm broad; black spots on the dorsal surface of lamina.	Petiole - 4.0 cm long; lamina with mucronate apex; lamina 18.0 cm long x 7.5 cm broad; black spots on on the dorsal surface of lamina
Inflorescence	2-flowers in each peduncle; peduncle straight; flowers below leaf cluster.	2-flowers in each peduncle; peduncle bent slightly; flowers within the leaf cluster.
Flowers	Sessile, small to medium, 0.8 cm long; petals - 7cm long; glabrous margin; stamens 11-12, 0.6 cm long; filament sessile; ovary 0.25 cm long.	Short pedicellate, medium, 1.7 cm long; sepals - 1.2cm long; petals - 1.0 cm long; hairy margin; stamens - 8, 0.7cm long; filament - short, ovary - 0.30cm long.
Fruits	2.0cm - 4.0cm long x 5.0cm 6.0cm periphery	5.0cm-7.0cm long x 8.5cm 9.5cm periphery
Occurrence	Frequent in the Indian Sundarbans	Common in the Indian Sundarbans
Germination	Truly viviparous germination	Truly viviparous germination

The Genus - *Bruguiera* Lamk. has 4 species in the Sundabans and other mangrove habitats in India. These are *Bruguiera gymnorhiza* (L.) Lamark, *B. sexangula* (Lour.) Poir., *B. cylindrica* (L.) Blume, *B. parviflora* W. & A. *ex* Griffith. The species, *B. exaristata* Ding Hou and *B. hainesii* Rogers also present and reported from the Old World tropics.

Table 25 : Difference between the four species of the Genus – *Bruguiera* Lamk. of the Indian Sundarbans

Characters	*B. gymnorhiza*	*B. sexangula*	*B. cylindrica*	*B. parviflora*
Tree Colour	Reddish green	Yellowish green	Dark green	Yellowish green
Lamina	Upto 20.5 cm x 6.5 cm	Upto 13.0 cm x 6.5 cm	Upto 12.0 cm x 5.0 cm	Upto 10.0 cm x 5 cm
Inflorescence	Solitary, big sized, reddish flower	Solitary, big sized, yellowish flower	3-flowers group, small greenish flower	4-7 flowers in a group, small reddish flower.
Flower	3.5 cm long	2.0 cm long	1.3 cm long	0.85 cm long
Fruit	3.8 cm long	3.2 cm long	1.8 cm long	1.0 cm long
Hypocotyle	16.0cm long, 2.0-2.5cm diam., ribbed with reddish persistent calyx-tube; green hypocotyles like ladies finger.	10.0cm long, 2.0-2.5cm diam., ribbed with yellow persistent calyx tube greenish yellowish	16.0 cm long, 1.5cm diam., smooth with green inversely folded calyx tube and with curved hypocotyle apex.	25.0 cm long, 1.0 cm diam., smooth, hypocotyle; with yellow straight persistent long calyx tube
Occurrences	Common in Sundarbans	Rare in Sundarbans	Occasional Sundarbans	Occasional Sundarbans
Germination	Viviparous germination	Truly viviparous	Viviparous germination	Truly viviparous

The Genus - *Ceriops* **Arnold** has only two species in the Old World Tropics, *viz., Ceriops decandra* (Griff.) Ding Hou and *Ceriops tagal* (Perr.) Robinson; both these species are present in the Indian Sundarbans. These are the true mangrove species and common almost throughout in Sundarbans.

Table 26 : Difference between the two species of *Ceriops* Arnold in Indian Sundarbans

Characters	*Ceriops decandra* (Griff.) Ding Hou	*Ceriops tagal* (Perr.) Robinson
Habitat	Shrub, diffuse branched, 3m-5m tall	Small-medium tree, upto 8 m tall
Inflorescence	Within and below the leafy cluster, straight, 6-10 peduncles on each shoot.	Within leafy cluster, pendulours, 2.4 peduncles on each shoot.
Fruit	Capsule 3.0 cm long	Capsule 2/5 cm long
Hypocotyle	Upto 12.0 cm long, distinctly ridged with blunt end, greenish yellow.	Upto 25.0 cm long, distinctly ridged with sharp end, green
Occurrence	Common in the Sundarbans mangals	Comparatively less in the Sundarbans
Germination	Truly viviparous germination	Truly viviparous germination

The mangrove species, ***Kandelia candel* (L.) Druce** is a mono-specific member under the Genus - *Kandelia* Wight & Arnold of Mangrove - Rhizophoraceae and very occasionally found in the Sundarbans mangal.

Distinct characters of the hypocotyles of the Members of Mangrove - Rhizophoraceae

The viviparous germination is one of the most important and conspicuous character among the members of the family Mangrove- Rhizophoraceae. The specific characters of the seedling plants and the hypocotyles of the Rhizophoraceae are presented below for easy identification in the fields during seedling stages of these mangroves.

Table 27 : Different Characters of Viviparous Germinated Hypocotyles of Mangrove - Rhizophoraceae

Name of the Mangrove Species of Rhizophoraceae	Length of the Hypocotyles	Cross section of Hypocotyles	Cotyledon Leaves/Foliages	Seed Germination
Rhizophora mucronata Lamk	Upto 75.0 cm	Upto 2.0 cm	Acute leaf apex	Epigeal, viviparous
Rhizophora apiculata Blume	Upto 55.0 cm	Upto 2.0 cm	Acute leaf apex	Epigeal, viviparous
Bruguiera gymnorhiza Lamk	Upto 16.0 cm	Upto 2.0 cm	Acute leaf apex	Epigeal, viviparous
Bruguiera sexangula Poir	Upto 10.0 cm	Upto 1.5 cm	Acute leaf apex	Epigeal, viviparous
Bruguiera parviflora W. & A	Upto 25.0 cm	Upto 1.0 cm	Acute leaf apex	Epigeal, viviparous
Bruguiera cylindrica Blume	Upto 16.0 cm	Upto 1.0 cm	Acute leaf apex	Epigeal, viviparous
Ceriops decandra Ding Hou	Upto 12.0 cm	Upto 1.0 cm	Round leaf apex	Epigeal, viviparous
Ceriops tagal Robinson	Upto 25.0 cm	Upto 1.0 cm	Round leaf apex	Epigeal, viviparous
Kandelia candel (L.) Druce	Upto 45.0 cm	Upto 1.5 cm	Slightly acute apex	Epigeal, viviparous

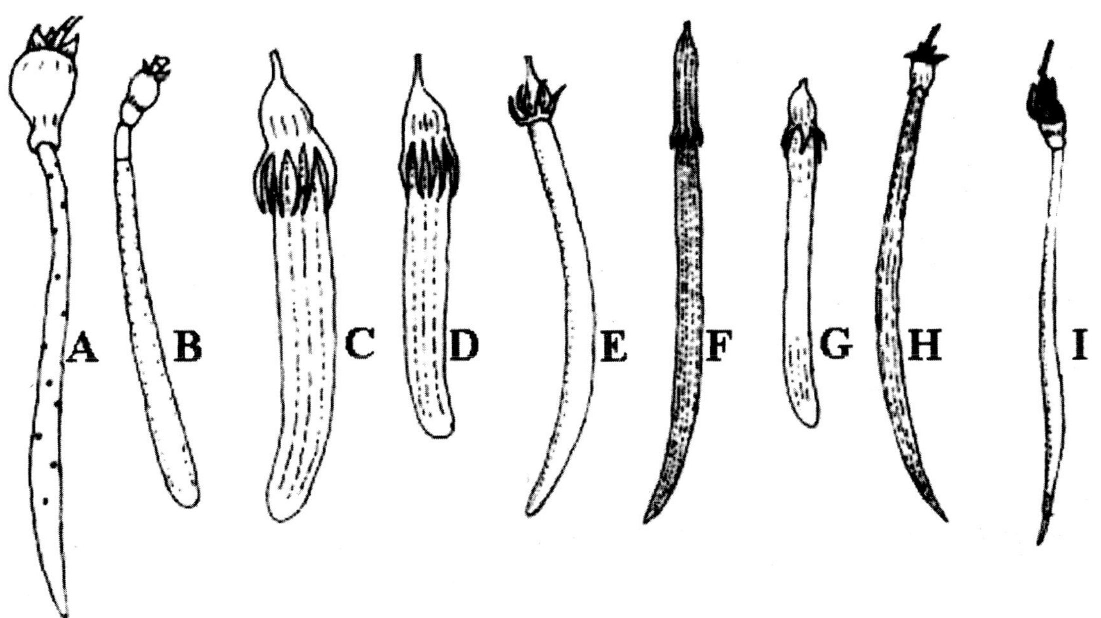

Sketches- 1. Different Hypocotyles of the Species under Mangrove - Rhizophoraceae

A. *Rhizophora mucronata*, B. *R. apiculata*, C. *Bruguiera gymnorhiza*, D. *B. sexangula*, E. *B.cylindrica*, F. *B. parviflora*, G *Ceriops decandra*, H *C. tagal* & I. *Kandelia candel*

Notes : Plant species under the family Mangrove - Rhizophoraceae are considered the major and most important in the inter-tidal mangrove habitats of the tropical and sub-tropical zones of both the Old World and the New World. In India there are eleven species under this family mangrove - Rhizophoraceae, distributed in Sundarbans mangals and the mangrove habitats of Andaman & Nicobar Islands, Bhitarkanika and other mangrove areas of both the east and west coasts of India. The other New World species under the family- Mangrove Rhizophoraceae are *Rhizophora mangle* L., *R. racemosa* Meyer, *R. samoensis* (Hochr.) Salvoza, *R.* X *selala* (Salvoza) Tomlinson, *R.* X *harrisonii* Leechm (Tomlinson, 1986). Besides these, the Genus - *Bruguiera* Lamk. has two other species, *viz., B. exaristata* Ding Hou and *B. hainesii* Rogers, but these two species are not present or reported from Indian mangals.

Table 28 : Key to the Genera under the Family - Mangrove Rhizophoraceae (Tribe - Rhizophoreae)

A	Medium to tall trees with conspicuous broom like stilt roots, up to 4.0 m or prop roots and viviparous germinated hypocotyles, hypocotyles ranging between 50.0 cm to 75.0 cm long, hanging downwards, plants grow densely on the inter-tidal river slopes or ridges of the forest	See B & BB
AA	Medium to small trees with broom like still roots and viviparous germinated hypocotyles; hypocotyles ranging between 10.0 cm to 25.0 cm long, hanging downwards or erect, commonly grow on the inter-tidal river slopes or ridge forests	See C & CC
B	Medium to large tree, with wide spreading branches; aerial stilt roots or prop roots conspicous, up to 4.0 m high or more; leaves ovate-elliptic, mucronate and acute apex; 16.0 - 20.0 cm X 7.5 cm leaf size; viviparous germinated hypocotyles generally up to 60.0 cm long with small dot like warts on surface and tapering end but not very pointed; very common in Indian Sundarbans	1. *Rhizophora* Linn.
BB	Small to medium tree, with spreading branches; aerial stilt roots fused on the trunk bases, less conspicuous, not more than 50.00 cm high; leaves oblong lanceolate with obtuse end, leaves 10.0 - 14.0 cm X 4.0-5.0 cm; viviparous germinated hypocotyles generally up to 45.0 cm long with smooth glabrous surface and sharp pointed end; this species is less common in the Indian Sundarbans	4. *Kandelia* Wight & Arnold
C	Medium to small tree, with less spreading branches; aerial stilt roots fused on the trunk bases, form conical broom like appearance, conspicuous, knee roots or fused stilt roots to buttress, more than50.00 cm high; leaves oblong lanceolate with obtuse end, 16.0 - 20.0 cm X 7.5 - cm leaf size; viviparous germinated hypocotyles generally up to 25.0 cm long with ribbed or smoothglabrous surface, sharp or blunt end, common in Indian Sundarbans	2. *Bruguiera* Lamk.
CC	Small tree or shrub, with less spreading branches; aerial stilt roots fused on the trunk bases, form conical broom like appearance, conspicuous, knee roots or fused stilt roots to buttress, flanges densely encircle the trunk bases; more than 30.00 cm high; leaves ovate oblong, obtuse or with round oval shape and round apex, 9.0 cm X 4.0 cm leaf size; viviparous germinated hypocotyles generally up to 20.o cm long with ribbed, glabrous surface, sharp or blunt end; common in the Indian Sundarbans	3. *Ceriops* Arnold

Table 29 : Key to the species under Genus - *Rhizophora* L. Sp. Pl. 443. 1753.
(Species distributed in Indian Sundarbans)

A	Leaves with mucronate apex; flowers 4 on each peduncle; mature flowers blossom within the leafy cluster, peduncle long, up to 0.5 cm slightly bent; flowers medium up to 1.7 cm long, pedicellate; petals hairyin margin; stamens 8; filaments short	1. *R. mucronata* Lamk.
AA	Leaves with acute apex; flowers 2 on each peduncle; flowers blossom below the leafy cluster; peduncle 2.0 cm long, straight; flowers smaller up to 1.2 cm long, sessile, petal glabrous margin; stamens 11 - 12, sessile	2. *R. apiculata* Blume

Besides these two above species of *Rhizophora* L., *R. stylosa* Griff. and *Rhizophora X lamarckii* Montr. were also reported from the Gahirimata estuary of Orissa coast and Andaman & Nicobar Islands by Banerjee, *et al.* (1989) and Dagar, *et al.* (1991). Tomlinson (1986) has stated distribution of *R.stylosa* Griff. from S. E. Asia and Australia only, while Tomlinson & Womersley (1976) have also mentioned that *R.* X *lamarckii* Montr. is a sterile F-1 hybrid between *R. apiculata* and *R. mucronata* and also considered this hybrid as endemic to N. W. coast of New-Caledonia.

Table 30 : Key to the Species under the Genus - *Bruguiera* Lamk. Encycl. Meth. Bot. 4 : 696. 1797-8

A	Inflorescence with single flower; flower up to 4.0 cm long, reddish or yellowish, persistent calyx tube in hypocotyles; calyx tube - cup shapped with 12 - 16 pointed calyx lobes	See B & BB
AA	Inflorescence with 3-7 flowers in group in each peduncle; flowers smaller up to 2.0 cm long, yellowish or dark green with persistent calyx tubes, not cup shaped and calyx lobes - 8	See C & CC
B	Sepal lobes 13 - 16, reddish; petals terminated into 4 cilia; stamens 26 32; hypocotyles up to 18.0 cm long X 2.5 cm diam. with reddish persistent cup shaped calyx tube	1. *B. gymnorhiza* (L.) Lamk.
BB	Sepal lobes 10 - 12, yellowish; petals terminated, not ciliated; stamens 20 - 34; hypocotyles up to 12.0 cm long X 2.0 cm diam. with yellowish persistent cup- shaped calyx tube	2. *B. sexangula* (Lour.) Blume
C	Three flowers in each group; flowers up to 1.3 cm long; petals white; hypocotyles up to 15.0 cm long X 1.5 cm diam. with dark green persistent calyx tube, calyx lobes / tips folding inversely	3. *B. cylindrica* (L.) Blume
CC	Four to seven flowers in each group; flowers up to 1.0 cm long; petals yellowish; hypocotyles up to 25.0 cm long X 1.0 cm diam. with yellowish persistent calyx tube, calyx lobes small, straight	4. *B. parviflora* W. & A

Table 31 : Key to the Species under the Genus - *Ceriops* Arnold. Ann. Nat. Hist. 1 : 363. 1838

A	Small tree, branches pyramidal appearance, leaves elliptic, peduncle with 2-4 flowers, within leafy cluster; petals terminally 3 - clavate appendages, anther shorter than filament length; hypocotyle up to 25.0 cm long with sharp end and ridged surface	1. *C. tagal* (Perr.) Robinson
AA	Shrub, deliquescent branching; leaves ovals, elliptic, peduncle 6-10 flowers, within and below leafy cluster; petals terminally ciliated; anther equal to the filament length; hypocotyle up to 12.0 cm long, apex blunt, club shaped and ridged surface	2. *C. decandra* (Griff.) Ding Hou

Notes : Dagar, Mongia & Bandopadhyay (1991) had reported that -

1. *Rhizophora* X *lamarckii* Mont, Mem. (Acad. Sci. Lyon 10 : 201) is moderate sized tree, height attain up to 10.0 m; bark - grey to brown; leaves elliptic, rarely ovate; inflorescence in the leaf axils like that of *R. mucronata* Lamk., peduncle usually with 2-flowers, rarely 2-3 flowers; pedicels 6.0-9.0 mm long; calyx deeply lobed, ovate, 12.0 to 15.0 mm long, pale yellow; petals lanceolate, white, 10.0 to 15.0 mm long with hairy margin; stamens 12 to 15, arranged in two whorls. It is considered as the hybrid between *R. apiculata* X *R. stylosa*.
2. The other Indian Species, *viz.,Rhizophora stylosa* Griff. (Nat. Pl. As. 4 : 665. 1854.) is moderately tall tree, height attain up to 20.0 m; bark grey to brown black; stem with forking stilt roots at base; branches produce aerial slender roots; leaves opposite, clustered towards ends of the branches; petiole slender; flowers in sub-terminal axillary cymes on slender recurved peduncles of about 3.0 cm long; calyx glabrous, tube shortly turbinate, four lobed, valvate, persistent and recurved in fruit; petals white, broadly ovate, long silky pilose on margins; stamens 8; ovary inferior, 2-celled; style slender. 2-lobed; fruit pyriform, 2.0 3.0 cm long, hypocotyle slender, up to 40.0 cm long. Not very common in Andaman Islands; reported from Chiriatapu of Andamans.

Table 32 : Comparison between the four Indian species of *Rhizophora* L. Sp. Pl. 443.1753

1. *Rhizophora mucronata* Lamk.	Flowers generally 4 on slender peduncles, borne within leafy crown, leaf blade broad, 20.0 c X 10.0 cm; petals hairy on the margin; stigma sessile; seedling warted, 50.0-70.0 cm long.
2. *Rhizophora apiculata* Blume	Flowers generally 2 or rarely 4 on stout peduncles, about 10.0 mm long, borne below leafy crown; each flower sessile; leaf blade broad, 20.0 cm X 10.0 cm; petals glabrous; stigma almost sessile; seedling warted, 40.0-50.0 cm long
3. *Rhizophora stylosa* Griff.	Flowers generally 4, 8 or more on slender peduncles, borne within leafy crown; leaf blade narrow and short, 12.0 cm X 7.0 cm; stigma on a slender style, 4.0 - 5.0 mm long; seedling smooth, not exceeding 30.0 cm
4. *Rhizophora lamarckii* Mont	Flowers generally 4, sometime 2 on short peduncles, about 1.5 cm long and rather stout, borne within leafy crown; leaf blade narrow, short 12.0 cmX7.0 cm, stigma on a slender style; 4.0 - 5.0 mm long; seedling not formed; Trees usually sterile and do not produce seedling.

Photo Plate 37. *Rhizophora mucronata* Lamk. A. Full grown *R. mucronata* Lamk. tree in the inter-tidal river flat land at Sundarbans, India. B. Stilt roots and viviparous germinated hypocotyles hang from the mother plant. C. Full grown hypocotyles in hand, Mayabandar, Andaman. D. Twigs with two full grown hypocotyles of *R. mucronata* Lamk. at Sundarbans. E. Flowering Twig of *R. mucronata* Lamk. Sundarbans.

1. *Rhizophora mucronata* Lamk. *Encycl. 6 : 189. 1804.*

[*Rhizophora mucronata* Lamk *var. typica* Schimp. 1838 (non Linn.) and *R. candelaria* DC.]

(Chromosome No. 2n = 36, Subramaniam,1980)

Common Name : Long fruited – stilted mangrove. Local Name (Bengali): Garjan or Khamu

Frequently on the bank of tidal creeks and narrow rivers; medium tree, up to 15.0 m tall in Indian Sundarbans; with dense wide spreading deliquescent horizontal branches; dense canopy, globose; trunk normally not conspicuous, supported by aerial stilt and prop roots, hang from trunk bases and mature branches, up to the height 3.0 – 4.0 m, 2.5 cm – 7.5 cm diam, corky, woody, slender, cylindrical. Leaves clustered in shoot apex, cyme, opposite decussate, ebracteate, leaf scar prominent elliptic, simple, exstipulate, petiole 4.0 cm long, acute - mucronate apex; petiole 16.0 cm long X 8.0 cm broad, dark green, dorsiventral, coriaceous dorsal surface shining, ventral surface pale green. *Inflorescence on the axis of clustered of existing leaves, up to 5.2 cm long, 4 - flowers on each peduncle, cyme opposite, ebracteate, 6-8 inflorescence in each branch, dichotomously branched, flower ebracteate, up to 2.0 cm long, with long stalk,* glabrous, regular hermaphrodite, erect, calyx persistent in fruit, sepals 4, polysepalous, ovate, entire, acute; corolla white, petals 4, polypetalous, hairy in margin; stamens 8, free, 4 - stamens opposite to petals and 4 stamens opposite to sepals, superior, filament short, 1.0 cm long, round, white, basifixed, anther bilobed, introrse; carpels 2, syncarpous, ovary globose, inferior, 0.3 cm long, 2-chambered, 2 ovules in each chamber. Capsule 5.0 cm - 7 cm long, 7.5 cm periphery, with 4 persistent sepal teeth. Seed one, covered by thick indehiscent pericarp; after maturation the seed bifurcated transversely within fruit; seed 1.8 cm long X 2.2 cm diam., germinate when the fruit hang in mother plant, mature viviparous germinated hypocotyle attained the length up to 75.0 cm and 3.0 cm diam. Hypocotyles fall on the soft soil and immediately forms seedling plant. Flowering and fruiting normally during late winter to monsoon months. Stem bark contain about 20% tannin, wood used for construction purposes and as fuel; bark has astringent properties and can be medicinally used in the treatment of hemorrhage, hematuria and angina.

Photo Plate 38. ***Rhizophora apiculata*** **Blume. A. Full grown tree of *R. apiculata* Blume at Sundarbans. B. View of branches with flowers and viviparaous germinated hypocotyles. C. Close view of branch with flowers. D. Branch with the initiation of viviparous germinated hypocotyles. E. Branch with full grown hypocotyles in hand.**

2. *Rhizophora apiculata* Blume *En. Pl. Jav. 1 : 91. 1827.*

(*Rhizophora conjugata* Arn. Ann. Mag. Nat. Hist. 1 : 363. 1838 (non Linn.) and *R. candelaria* DC.)

(**Chromosome No.** 2n = 36, Subramaniam,1980)

Common Name : Long fruited – stilted mangrove **Local Name** (Bengali): Garjan or Bhara

Occasionally on the bank of tidal creeks and narrow rivers; medium tree - up to 12.0 m tall in Indian Sundarbans, with dense wide spreading deliquescent horizontal branches; dense canopy, globose; main trunk normally not conspicuous, supported by aerial stilt and prop roots,prop roots hang from trunk bases and mature branches, up to the height 3.0 m × 2.5 cm – 7.5 cm diam, corky, woody, slender, cylindrical. Leaves clustered in shoot apex, cyme, opposite decussate, ebracteate, leaf scar prominent elliptic, simple, exstipulate, petiole 4.0 cm long, acute apex; leaves 15.0 cm long × 7.0 cm broad, dark green, dorsiventral, coriaceous dorsal surface shining, ventral surface pale green with distinct spots. *Inflorescence on the axis of fallen leaves, 2 flowers on each peduncle*, cyme opposite decussate, ebracteate, 6-8 inflorescence in each branch, dichotomously branched, flower ebracteate, up to 2.0 cm long, sessile, glabrous, regular hermaphrodite, erect, calyx persistent in fruit, sepal 4, polysepalous, ovate, entire, acute; corolla white, petal 4, polypetalous, hairy in margin; stamens 11-12, sessile, free, 4- stamens opposite to petal and 4 stamens opposite to sepal, superior, filament short, 1.0 cm long, round, white, basifixed, anther bilobed, introse; carpels 2, syncarpous, ovary globose, inferior, 0.25 cm long, 2- chambered, 2 ovules in each chamber. Capsule 2.0 cm- 4.5 cm long, 6.5 cm periphery, with 4 persistent sepals teeth. Seed one, covered by thick indehiscent pericarp; after maturation the seed bifurcated transversely within fruit; seed 1.5 cm long × 2.0 cm diam., germinate when the fruit hang in mother plant, mature viviparous germinated hypocotyle attain length up to 55.0 cm and 3.0 cm diam. Hypocotyles fall on the soft soil and immediately form gregarious seedling plants. Flowering and fruiting normally during late winter to monsoon months. Stem bark contain about 20% tannin, wood used for construction purposes and as fuel wood; bark astringent and can be medicinally used in hemorrhage, hematuria and angina.

Photo Plate 39 : *Bruguiera gymnorhiza* (L.) Lamk. A. Full grown tree of *Bruguiera gymnorhiza* (L.) Lamk at Sundarbans. B. *B. gymnorhiza* (L.) Lamk. within tidal water. C. View of broom like fused stilt roots. D & E. Close view of branch with flowers. F. Branch with viviparous germinated hypocotyles. G. Newly germinated seedling plant.

3. *Bruguiera gymnorhiza* (L.) Lamk. Encyl. Meth. Bot. 4:696.1797-1798.

(*R. gymnorhiza* L.Sp. Pl. 443.1753; *B. rheedii* Blume En.Pl.Jav. 92.1827; *B. conjugata* Merr.Philipp. J.Sci.Bot. 9: 118.1914.)
(**Chromosome No**. 2n = 26 Patil, 1958 & Sidhu, 1968)

Common Name : Large – Leafed Orange Mangrove **Local Name** (Bengali): Kankra

Common on the inter-tidal forest, bank of tidal creeks and narrow rivers; very common throughout in Sundarbans; medium tree - up to 12.0 m tall, with dense canopy, compact branches, pyramid shaped; main trunk conspicuous, supported by basal fused stilt roots on the trunk base and form conical base with fused aerial roots, occasionally form buttresses and horizontal knee roots. Leaves clustered in shoot apex, opposite decussate, simple, cauline, exstipulate, elliptic, petiole 5.0 cm long, acute apex; leaves up to 20.0 cm long X 8.5 cm broad, dark green, dorsiventral, dorsal surface shining, ventral surface pale green. Inflorescence solitary, single flower in each peduncle; flowers ebracteate, pedicellate, pedicel 2.0 cm long, round, glabrous, complete, bisexual, regular, spiral, pendulous, 3.5 cm long, calyx persistent in fruit, sepals 13 - 16, polysepalous, pointed sepal lobes above but cup or tube-shaped base, reddish, glabrous, ridged, velvet, 1.5 cm long, persistent in fruit and viviparous germinated hypocotyle; petal 13-16, polypetalous, each petal folded into two parts, white; stamens 26 - 32, free, two stamens pair within one petal, filament up to 0.25 cm long; basifixed, superior, anther bilobed; carpels 3, syncarpous, inferior, cup-shaped , 2-4 locules, 2 ovules in each chamber. Fruit capsule, pendulous with persistent calyx; seed one, hypocotyle 18.0 cm long, 2.5 – 3.0 cm diam., dark green, longitudinally narrow ridges, showy with reddish cup shaped persistent calyx. Seed germinate when the fruit hang in mother plant, mature viviparous germinated hypocotyle attain length up to 18.0 cm long and 3.0 cm diam. on maturity; viviparous germinating hypocotyles fall on the underneath soft soil and immediately anchor on the soft soil and form seedling plants. Flowering and fruiting normally during late winter to monsoon months. Stem bark contain higher percentage of tannin, wood hard - used for construction purposes, fuel wood; bark is said to be useful in diarhea; leaves and hypocotyles occasionally cooked as vegetable, a famine food.

Photo Plate 40 : *Bruguiera sexangula* (Lour.) Poir. A. Full grown tree of *Bruguiera sexangula* (Lour.) Poir. at Sundarbans. B. Close view of branch with flowers. C. Close view of branches with viviparous germinated fruits. D. Close view of branch with flowers. F. Branch with viviparaous germinated hypocotyles. G. Close view of branch with persistent twisted calyx tube.

4. *Bruguiera sexangula* (Lour.) Poir. in Lamk. Encyl. 4 : 262.1816.

(*Bruguiera eriopetala* Wight & Arnold ex Arn., Ann. Mag. Nat.Hist.1:368.1838; *B. malabarica* F. Vill. (non Arn., New Appl. 79. 1880 non Arnold & *Rhizophora sexangula* Lour.)

(**Chromosome No.** 2n = 26, Sidhu, 1968; G Gill. 1978)

Common Name : Large – Leafed Orange Mangrove **Local Name** (Bengali): Kankra

Rare in the inter-tidal forest; occasionally grow in the north-eastern part of Indian Sundarbans; small to medium tree, up to 10.0 m tall, much branched, spreading or diffused; yellowish to dark green foliages; branches deliquiscent, dense canopy; trunk base conspicuous, supported by aerial fused stilt roots, form buttresses and horizontal knee roots. Leaves clustered in shoot apex, opposite decussate, simple, cauline, exstipulate, elliptic, petiole 1.8 cm long, acute apex; leaves up to 13.0 cm long X 6.5 cm broad, dark green, dorsiventral, dorsal surface shining, ventral surface pale green. Inflorescence solitary, single flower in each peduncle; flowers ebracteate, pedicellate, pedicel 1.0 cm long, bisexual, regular, spiral, pendulous, 2.0 cm long, calyx persistent in fruit, sepals 10-12, polysepalous, pointed sepal lobes above but cup shaped base, yellowish, glabrous, ridged, 1.5 cm long, persistent in fruit, viviparous germinated hypocotyle; petals 10-12, polypetalous, yellowish; stamens 20-24, free, 2 stamen pairs within one petal, filament up to 0.25 cm long; basifixed, anther bilobed; carpels 3, syncarpous, ovary 0.5 cm long, superior, cup shaped, 2-4 celled, 2 ovules in each chamber. Fruit capsule, pendulous with persistent calyx; seed one; *seed germinates when the fruit hang in mother plant, mature viviparous germinated hypocotyle attain the length up to 12.0 cm long and 2.5 cm diam.*; hypocotyles fall on the underneath soil and immediately anchor on soft soil and form gregarious seedling plants. Flowering and fruiting during summer to monsoon months. Stem bark contain higher percentage of tannin, wood hard - used for construction purposes, fuel wood; bark is said to be useful in diarrhea.

Photo Plate 41. *Bruguiera cylindrica* (L.) Blume. A. Close view of branches with flowers of *Bruguiera cylindrica* (L.) Blume at Sundarbans. B. Close view of branch with viviparous germinated hypocotyles. C. Close view of branches with viviparous germinated hypocotyles. D. Close view of branches with viviparous germinated hypocotyles. E. Seedlings of *Bruguiera cylindrica* (L.) Blume

5. ***Bruguiera cylindrica*** (L.) Blume Ency. Pl. Jav. 1 : 93. 1827.

(*Rhizophora cylindrica* L.Sp.Pl.443.1753; *R. caryophylloides* Burm.f. Fl. Ind.109.1768; *B. malabarica* Arn.Ann.Mag. Hist. 1 : 369.1838.)

Common Name : Small – Leafed Orange Mangrove **Local Name** (Bengali): Bakul Kankra

Occasionally in the inter-tidal forest; throughout in Indian Sundarbans; small to medium tree, up to 12.0 m tall, much branched, spreading or diffused; dark green foliages; branches irregularly arranged, trunk base conspicuous, supported by aerial fused stilt roots. Leaves clustered in shoot apex, opposite decussate, simple, cauline, exstipulate, elliptic, petiole 2.0-4.0 cm long, acute apex; leaves up to 12.0 cm long X 5.0 cm broad, dark green, ovate lanceolate, dorsiventral, dorsal surface shining, ventral surface pale green. Inflorescence -cyme, three to many flowers in each peduncle, peduncle long, terete, green, glabrous; flowers ebracteate, pedicel short, 2.0 cm long, greenish, bisexual, regular, complete, pendulous or erect during anthesis, medium sized -1.3 cm long, 0.5 cm across; calyx tube 0.4 -0.5 cm long x 0.2 cm broad; sepals- 8, polysepalous, 8 sepal lobes, funnel like tube base, surface smooth, lobes short, 0.5 cm long, pointed apex, green; persistent in fruit; calyx teeth bend backward; petal 8, polypetalous, 0.35 cm long × 0.15 cm broad, slightly lobed, each petal folding, *i.e.*, two parts, white, thin herbaceous; androecium -16, free but 8 groups, two stamens in each petal, filament unequal length, soft, white, basifixed, 0.15 cm - 0.25 cm; carpels 2, 3, syncarpy, ovary 0.25 cm long, cup-shaped, 2-4 celled, one ovule in each chamber, axile placentation, style -1, terminal, 0.75 cm long, round, white, glabrous, stigma bi-fid, hairy. Fruits berry, pendulous, persistent calyx, lobes bent towards pedicels; hypocotyle up to 16.0 cm long, 1.0 cm diam. end blunt, bending, green, slightly ridged. Seed germinates when the fruit hang in mother plant, mature viviparous hypocotyles fall on the underneath soil and immediately anchor on soft soil and form gregarious seedling plants. Flowering and fruiting during late winter to monsoon months. Stem bark contain higher percentage of tannin, wood hard - used for construction purposes and fuel.

Photo Plate 42. *Bruguiera parviflora* W. & A. ex Griff. A. Long view of *Bruguiera parviflora* W. & A. ex Griff. at Sundarbans. B. Close view of branch with flowers. C. Close view of branches with viviparous germinated fruits. D. Close view of seedling plant

6. *Bruguiera parviflora* W. & A. ex Griff.Trans. Med. Phys. Soc. Cal. 8:10. 1936.

(**Chromosome No**. 2n = 36, Sidhu, 1961 & 1968)

Common Name : Small – Leafed Orange Mangrove **Local Name** (Bengali): Bakul Kankra

Occasional in the inter-tidal forest; south-eastern part of Indian Sundarbans; medium tree up to 15.0 m tall, much branched, pyramidal, looks most beautiful plant among the mangroves; yellowish green foliages looks it golden in sunny weather, trunk base conspicuous supported by aerial fused stilt roots. Leaves simple, opposite, decussate, cauline, exstipulate, petiolate, petiole 4.0 cm long, yellowish green, lamina elliptic oblong, entire acute up to 10.0 cm long X 5.0 cm broad, glabrous, dorsiventral, dorsal side shiny, coriaceous, unicostate reticulate venation. *Inflorescence cyme 4-7 flowers in each peduncle*, 2.0 cm – 2.5 cm long, glabrous, terete, yellowish; flowers ebracteate, pedicel short, 0.8 cm long, terete, glabrous, smooth, yellowish, solid, bi-sexual regular, complete, pendulous, small, 0.85cm long; lobes above, below cylindrical tube, acicular, pointed apex, persistent greenish, regular, complete, pendulous or erect during anthesis; sepals- 8, 8 lobes above, below cylindrical tube, acicular, pointed apex, persistent, 0.25cm long, equal length; corolla – 8, polypetalous, 0.1 cm – 0.2 cm long × 0.5 cm broad, yellowish, petals folded in two parts, at the terminal part 3 long cilia, petal margin slightly hairy, thin, herbaceous, glabrous, each petal contain two stamens of unequal filament length; stamens – 16 free, but 8 groups, 2 stamens in each folding petal, glabrous, soft, terete, white, basifixed, anther bilobed, longitudinal dehiscent, introrse, inserted, yellowish white; carpels 3, syncarpy, ovary up to 0.15 cm long but on maturity become 1.5 cm long, attached fruit, cylindrical, 2-celled, axile placentation, style one 0.15 cm –0.25 cm long, white, soft, stigma 2, bi-fid, short; fruit berry, pendulous, up to 2.0 cm long, 0.4-0.6 cm diam. hypocotyle yellowish green, sepal lobes short, persistent and long calyx tube, end slight pointed, up to 20.0 cm. Flowering and fruiting during summer to monsoon months. Stem bark contain higher percentage of tannin, wood hard - used for construction purposes, fuel wood.

Photo Plate 43. *Ceriops decandra* (Griff.) Ding Hou. A. Full grown plant at Sundarbans. B. Close view of branch with flowers. C. Close view of branches with viviparous germinated fruits. D. Viviparous germinated hypocotyles. E. Viviparous germinated hypocotyles hang in the branches. F. Viviparous germinated hypocotyles.

7. *Ceriops decandra* (Griff.) Ding Hou, Fl. Males. Ser. I, 5(4): 471.1958.

(*Bruguiera decandra* Griff. Trans. Med. Phys. Soc. Calc. 8 : 10. 1836; *C. roxburghiana* Arn., Ann. Mus. Nat. Hist. 1 : 364. 1838 & *Rhizophora decandra* Roxb., Hort. Beng. 36.1814, *nomen ex* Griff. Natul 4: 663. 1854).

(**Chromosome No.** 2n = 36, Sidhu, 1968)

Common Name : Yellow Mangrove. **Local Name** (Bengali): Jale Goran or Jhamti Goran

Very common in Indian Sundarbans, evergreen shrub or small tree, pyramidal or with diffused branches; up to 5.0 m tall, in the inter-tidal forest floor, slender or erect and bushy habitat in degraded forest; broom like and fused stilt roots, conical on the trunk base, up to 30.0 cm upward aerial fused stilt roots on trunk base. Leaves simple, opposite decussate, cauline, exstipulate, petiolate, petiole 2.5 cm long, round, yellowish green, glabrous, pulvinous, lamina ovate or ovate-elliptic, oblong, entire, obtuse or round apex, 9.5 cm long x 5.5 cm broad, green, mid-vein yellowish, dorsiventral, glabrous, coriaceous, unicostate reticulate venation. Inflorescence 1.0 cm long, having single stalk on both sides of the nodes, umbel, 1 peduncle having 16 flower buds, but 4-6 of them are only matured; 6-10 peduncles on each shoot apex, within or below the leafy cluster; 16 flower buds aggregate on each head, bracteate, 2 bracts adjoining with 4 buds, each of these 2 buds have another bracteoles; flowers short, sessile, bisexual, regular, pentamerous, complete, glabrous, green, 0.75 cm long × 0.6 cm across, rosaceous, bracts 0.2 cm long × 0.2 cm broad; sepals 5, polysepalous, 0.4 cm long, 0.15 cm broad, persistent, whitish yellow, fleshy thick, elliptic, entire, acute, mature sepals turn brown; petals 5, polypetalous, oblong, entire, 0.28 cm long × 0.08 cm broad, white, terminally ciliated, numerous; stamens 10, free, filament length equal to anther, stamens 0.25 cm long, white filament, 0.14 cm long, whitish, soft; anther dorsifixed, 0.10 cm long, bilobed, longitudinally dehiscent, introrse, inserted, each stamen in equal length, shorter than petal; carpels - 3, syncarpy, ovary 0.2 cm long × 0.2 cm diam, 2-celled axile placentation, style 1, 0.15 cm-0.25 cm long, white, soft, stigma 2, bifid, short; fruit capsule, 3 cm long with 5 persistent calyx, brown with rough surface, blunt apex; seed germination epigeal, viviparous; hypocotyle erect with blunt end, *club shaped and attain the length up to 12.0 cm.* Flowering and fruiting during summer to monsoon months, but viviparous hypocotyles present on the mother plant almost throughout the year. Stem used as fuel wood and cottages.

Photo Plate 44. *Ceriops tagal* (Perr.) Robinson. A. Full grown small tree at Sundarbans. B. Close view of branch with flowers. C.Long view of tree on the river bank. D. Viviparous germinated hypocotyles.

8. *Ceriops tagal* (Perr.) Robinson Philipp. J. Sci. Bot. 3 : 306. 1908.

(*Rhizophora tagal* Perr. Mem. Soc. Linn. Paris 3 : 138.1824; *C. candalleana* Arn. Ann. Mag. Nat. Hist. 1 : 364.1838; *C. timoriensis* Domin, Bibl. Bot. 89. 444. 1928 & *C. boiviniana* Tulasne, Ann. Sci.Nat. 6 : 112.1856).

Common Name : Yellow Mangrove. **Local Name** (Bengali): Mat Goran or Jat Goran

Frequent in Indian Sundarbans, evergreen, small tree, much branched, pyramidal up to 8.0 m tall, in the inter-tidal forest floor, slender or erect and bushy habitat in degraded forest; broom like and fused stilt roots form buttress on the trunk base, up to 50.0 cm upward; leaves simple, opposite decussate, cauline, exstipulate, petiolate, petiole 2.7 cm long, round, green, glabrous, pulvinous, lamina elliptic, obtuse, entire, tapering, 9.4 cm long × 4.4 cm broad, coriaceous, dorsiventral, dorsal side green, shiny in comparison to ventral side, glabrous, unicostate, reticulate venation, vein inconspicuous. Inflorescence 2.0 cm long, within leafy cluster, cyme, opposite decussate, ebracteate, peduncle develop from leaf base, peduncle erect but bent downwards, 1.0 cm long, glabrous, green, flattened, 2-4 peduncles, peduncle terminate in 4-6 short branches, each branch have 2-3 flowers; flowers short pedicellate, 0.1 cm long, terete, glabrous, green, bisexual, complete, regular, 0.7 cm long X, 0.6 cm diam., rosaceous, pentamerous, minute, 10-12 flowers on single peduncle; sepals - 5, polysepalous, superior, elliptic- lanceolate, entire, thick, green, acute, 0.45 cm long X 0.25 cm broad at base; persistent, tapering end, woody, equal length, upper surface same level, petals 5, polypetalous, valvate, petals white, thin, herbaceous, inconspicuous venlets.; petals 0.3 cm long; stamens 10, free, superior, filament 0.25 cm long, terete, soft, white, anther bilobed, 0.1 cm long, sagittate, introrse, inserted, filaments unequal length; carpels 3, syncarpous, ovary inferior, elliptic globose, 2-chambered, axile placentation, 2 ovules in each chamber, but 1 survives and 3 degenerate, 0.3 cm diam., style - terminal, 0.3 cm long, terete, short, greenish white, glabrous, stigma 3, shortly tri-fid; fruit capsule, 2.5 cm long with 5 persistent calyx, brown with rough surface, sharp pointed apex; seed germination epigeal, viviparous; hypocotyle up to 25.0 cm long pendulous and sharp end and ridged. Flowering and fruiting during summer to monsoon months, but viviparous germinated hypocotyles present on the mother plant almost throughout the year. Stem barks contain higher percentage of tannin, wood hard - used for construction purposes, fishing stalks, poles for cottages, high calorie value for fuel wood.

Photo Plate 45. *Kandelia candel* (l.) Druce. A. Full grown small tree at Sundarbans. B. Close view of branch with flowers. C. Long view of flowers. D. Viviparous germinated hypocotyles in hand. E. Close view of viviparous germinated hypocotyles in hand. F. Viviparous germinated hypocotyles in hand

9. *Kandelia candel* (L.) Druce, *Rep. Bot. Exch. Club. Brit. Isles 1913 (3) : 420.1914.*

(*K. rheedii* Wight & Arnold, Prodr. 311. 1834.)

Common Name : Candle Mangrove. **Local Name** (Bengali): Goria

Occasional in Indian Sundarbans, evergreen small or medium tree, much branched, spreading deliquescent branches up to 7.5 m tall, in the inter tidal forest, fused broom like stilt roots form buttresses on the trunk base. Leaves simple, opposite decussate, cauline, stipulate, petiolate, petiole 2.0-cm long, terete, green, glabrous, pulvinous, lamina oblong lanceolate, acute, entire, tapering, glabrous, coriaceous, slightly fleshy, dorsiventral, dorsal side dark green and slightly shining, unicostate. Inflorescence cyme in axils, opposite decussate, bracteolate, bracteoles cup shaped, main peduncle 4.4 cm long, dichotomously branched, each peduncle with 4 - white flowers, peduncle green, glabrous, 0.2 cm broad, slightly flattened, pulvinous, terminal portion of the peduncle bracteolate, cup-shaped and grooved, 0.4 cm in diam., each shoot having 4-8 inflorescence stalks. Flowers complete, regular, bisexual, pedicellate, pedicel 10.2 cm long, 2-flowered generally remain in pair; 1.8 cm long; sepals – 5, polysepalous, slender, sepal lobes above upper surface, thick fleshy, green, glabrous, imbricate, persistent;; petals 5, white, polypetalous, equal length alternate to sepals, oblong lanceolate, 1.2 cm long, thin coriaceous, apical papilate; carpels 3, syncarpous ovary, minute, 0.35 cm long; inferior, globose, unilocular, 6 ovules, placentation axile; style-1, terminal, 1.0 cm long; stigma 3, short, tri-fid. Fruit capsule, up to 2.0 cm long, with twisted or folded persistent calyx lobes; seed 1; viviparous hypocotyle up to 43.0 cm long, green smooth surface and pointed end; germination epigeal. Stem bark contain higher percentage of tannin, wood hard - used for construction purposes, fishing stalks, poles for cottages, high calorific value for fuel wood. Said to be used medicinally for the treatment in diabetes.

Table 33 : Other New World Mangrove species under the Family - Mangrove Rhizophoraceae

1. *Rhizophora mangle* L. *Sp. Pl.*(ed.) **1** : 443. 1753.	Leaves smaller - 10.0cm x 5.0 cm; petiole up to 2.0 cm, pale green to almost white below, apex blunt; older shoot light brown; bracteoles conspicuous, 1.0 mm or longer; flower buds pointed at apex; fruits elongated, 2.0 - 3.0 cm long, asymmetric, tapered at apex, olive green, rough surface; seedlings 15.0 cm - 20.0 cm long, blunt without conspicuous lenticels. Distributed in tropical America on both the Atlantic and Pacific coasts.
2. *R. racemosa* Meyer Prim. *Fl. Esseq.* 185. 1818.	Inflorescence much branched, with up to 6 orders of bifurcation, flower number up to 128 on one axillary flowering branch; bracteoles thick, short, rounded; flower buds rounded; flowers often 32-64 per inflorescence; pointed and with small flowers; sepals 8.0 - 10.0mm long. Distributed in West Africa and Atlantic coast of tropical America.
3. *R* . X *harrisonii* Leechm. *Kew Bull.* **1918** : 8. 1918.	A probable hybrid between *R. mangle* and *R. racemosa.* Many flowered inflorescence with blunt or round flower buds; sepal 10.0 mm long. West African, Atlantic and Pacific coasts of tropical America.
4. *R. samoensis* (Hochr.) Savoza *Bull. Nat. Appl. Sci. Univ. Philipp.* **5** : 220.pl. 6. 1936.	Very much similar to *R. mangle* L.; only differ with round or blunt flower buds and inconspicuous bracteoles, less than 1.0 mm long.. New Caledonia, New Hebrides, Fiji and probably in W. Pacific coast
5. *R.* X *selala* (Salvoza) Tomlinson *Arnold Arbor.* **59** : 159. 1978.	Peduncle 2.5 - 3.0 cm long, but often longer, 2.6 - 3.2 mm wide, often with more than 2 orders of branching; flowers 2 - 9 per inflorescence; mature flower buds white. A probable sterile hybrid between *R. samoensis X R. stylosa.* Distributed in New Caledonia. Fiji and the New Hebrides

Photo Plate-46. A.& B. *Rhizophora mucronata* Lamk. with dominant stilt roots in Port Blair, Andaman (Bay Islands)

Family – 2. Avicenniaceae

(Grey Mangrove Family)

Next to the family Rhizophoraceae, the 2nd important mangrove family in the Indian Sundarbans is the Avicenniaceae. Under this family there is only one Genus - *Avicennia* L., along with its 3 species and 1 variety; these are very common in the Indian Sundarbans, *viz., Avicennia alba* Blume, *Avicennia officinalis* L. *Avicennia marina* (Forsk.) Vierh. and *A. marina* (Forsk) Vierh. var. *marina* (Forsk.) Vierh.

These are mostly tall to medium trees, dominant plants in the Sundarbans mangal; these species are distributed throughout the tropical and the sub-tropical zones. Plants under this family Avicenniaceae have pencil like, up to 30.0 cm high slender negatively geotropic air breathing roots, the pneumatophores; occasionally form slender stilt roots from the trunk bases. Pneumatophores of these members of *Avicennia* L. have no secondary growth; small pores are scatteredly present on the surface of these aerial roots for performing effective root zone respiration in this compact inter-tidal saline soil. Sometimes the fibrous, 0.5 cm – 1.0 cm diam. stilt roots develop and hang from the trunk bases in case of *Avicennia* spp. Viviparous germination like the members of Rhizophoraceae is absent for the members of Avicenniaceae. Seeds of the *Avicennia* spp. germinate soon after their fruits fall from the mother plants. In this case, their fruits mature in the mother plant itself and develop the seeds to their optimum condition for effective germination. This phenomenon is known as pseudo- viviparous or incipient - viviparous seed germination.

Table 34. The distinct characteristic features between three species of *Avicennia* L. of Sundarbans

Characters	*Avicennia alba* Blume	*Avicennia marina* Vierh	*Avicennia officinalis* L.
Bark	Gray black or brown	Light yellow, guava like	Gray to light, whitish
Leaves	Elongated long lamina, apex acute or acuminate	Elliptic-oblong, lanceolate, elliptic-ovate, apex acute	Ovate oblong, bigger in size, apex round or obtuse
Inflorescence	Raceme compact spikes, peduncle 5.0 cm long	Raceme compound spike, terminal capitate, peduncle upto 10.0cm	Compound and much branched spike, peduncle up to 25.0 c
Flowers	Small, 0.5 cm - 0.8 cm long, ovary dumbbell shaped	Medium, 0.6 cm - 1.0 cm long, ovary dumbbell shaped	Comparatively larger flower; 0.8 cm - 1.2 cm long, ovary conicae
Fruits	Conical, extended with terminal beak, up to 2.7 cm	Spherical to ovoid, with short apical beak, flattened, 2.5 cm	Flattened with apical beak, up to 3.0 cm long and 2.8 cm width
Seeds Germination	Crypto- viviparous seedling germination	Crypto- viviparous seedling germination	Crypto- viviparous seedling germination

Photo Plate 47.(A & B). *Avicennia officinalis* L. - with pneumatophores and stilt roots, in Sundarbans mangal.

Avicennia L. Sp. Pl. 1: 110.1753 is the only and Type – Genus of the Family – Avicenniaceae; this family was segregated from the family – Verbenaceae by Tieghen (1898) based on the characteristic wood and leaf anatomy, *i.e.*, aerial roots, axile placentation, pseudo-viviparous or incipient-viviparous seedling germination. This genus – *Avicennia* L. was originally described (*l.c.*) based on its type species, *Avicennia officinalis* L. Sp. Pl. 1: 110. 1753 and the Generic epithet was coined after the name of the famous Arabian Philosopher, *'Iban Sina'* (Tan & Keng, 1977).

Tomlinson (1986) described 10 species under the Genus- *Avicennia* L., *viz.*, *A. officinalis* L., *A. marina* (Forsk.) Vierh., *A. alba* Blume, *A. lanata* Ridley, *A. eucalyptifolia* (Zipp. Ex Miq.), Moldenke, *A. balanophora* Stapf & Mold., *A. germinanas* (L.) Stearn, *A. africana* Palisot de Beauv., *A. bicolor* Standley and *A. schaueriana* Stapf. & Leechman ex Moldenke along with 7 taxonomical putative varieties or ecotypes. Out of these 10 distinct species, 6 species and few varieties belong to the Indo-West Pacific Regions (Old World Tropics and Sub-Tropics), while the rest 4 species and few other putative varieties or ecotypes are present in the New World Tropics.

In the Indian coastal and Islands mangals only 3 species, *viz.*, *A. officinalis* L., *A. marina* (Forsk.) Vierh., *A. alba* Blume and few other putative varieties are quite significant as major or true mangrove flora; these are found almost throughout the mangrove ecosystem in the inter-tidal areas. These mangrove species are commonly known as the 'Grey mangroves' or 'White mangroves' for their whitish grey barks.

Table 35 : Key to the Species under the Genus – *Avicennia* L. Sp. Pl. 1: 110. 1753

A	Medium to tall tree; pneumatophores longer than the other members of Avicennia L; profuse fine stilt roots often developed from the bent trunks; stem-bark whitish, smooth; leaves obovate-oblong, round apical, not much silvery beneath; flowers 1.0 cm-1.5 cm across; fruit 2.5cm-3.0cm long with short narrow beak, almond shaped	1. *A. officinalis* L.
AA	Shrub, small to medium tree; pneumatophores shorter than the above; fine stilt roots often developed from the bent trunks of *A. alba* Blume; stem-bark blackish or guava stem like peeling, smooth; leaves lanceolate or linear with pointed apex, silvery or whitish beneath, flowers less than 0.5 cm across; fruit with or without norrow beak, oval or kidney shaped	See – B & BB
B	Leaves lanceolate or linear, upto 12.5 cm long; ventral surface bright silvery; stem bark turning blackish with aging; flowers dull yellow, less than 0.4 cm across; fruit upto 4.0 cm long, conical, tapering towards the apical portion	2. *A. alba* Blume
BB	Leaves ovate - lanceolate, upto 10.0 cm long; ventral surface not much silvery nor much white like *A. alba*; stem bark guava stem like peeling; flowers orange yellow, less than 0.4 cm-0.5 cm across; fruit upto 4.0 cm long, conical, tapering towards the apical portion most common mangrove plants in the inter tidal zones throughout tropical and sub-tropical zone in the old world.	3. *A. marina* (Forsk.) Vier

Avicennia spp, are the common mangrove flora in diverse or varied ecological conditions and different saline zones; this genus is common both in the Old World and in the New World mangals and occasionally in the temperate zone at about 38° S latitude in S. E. Australian coast; as such, the members of *Avicennia* L. are the cosmopolitan in distribution.

Banerjee, *et al.* (1989) have reported that two varieties, *viz.*, (i) *Avicennia marina* (Forsk) Vierh. var. *marina* (Forsk.) Vierh and (ii) *Avicennia marina* (Forsk) Vierh. var. *acutissima* Stapf & Mold. from the Indian mangrove habitats. The variety, *A. marina* (Forsk) Vierh. var. *marina* (Forsk.) Vierh has elliptic-oblong or elliptic-ovate leaves and apiculate fruit; while, *A. marina* (Forsk) Vierh. var. *acutissima* Stapf & Mold. is with sessile and sharply acuminate leaves. The 1st variety is said to be distributed in the mangals of both East coast and West coast of India, ranging from the Sundarbans to Gujarat coast; while the 2nd variety is said to be distributed in the West coast, *i.e.*, Mumbai coast to Gujarat coastal mangals. The species *Avicennia alba* Blume. was not reported from the mangrove habitats of the Andaman & Nicobar Islands, while other large number of mangroves are reported from this Island mangals (Dagar, *et al.*, 1991).

Photo Plate 48. *Avicennia officinalis* L. A. Full grown tree on the river side - with stilt roots. B. Stilt roots developed on the trunk.C. Flowering branch of *A. officinalis* L. D. Close view of the flowering branch. E. Mature fruits of *A. officinalis* L. F. Crypto-viviparous germinated seeds of *Avicennia officinalis L.*

10. *Avicennia officinalis* L. Sp. Pl. 1: 110.1753.

Common Name : White Mangrove; **Local Name** (Bengali): Jat Ban, Jat Bani or Jat Baine

Very common in the inter tidal zones in Sundarbans; evergreen, medium to tall tree, attain the height between 15.0m - 20.0 m; bark smooth, whitish; much branched, deliquescent, irregularly spreading branches, gregariously grow in the inter-tidal forest; conspicuous, negatively geotropic aerial roots– pneumatophores or breathing roots numerous, erect pencil like, without any secondary growth; attain the height up to 0.30 cm and diam. 20.0 mm - 25.0 mm; pneumatophores developed in the inter-tidal flat ground, developed from the surface spreading underground horizontal roots, frequently bunches of stilt roots developed from the trunk base in stressed condition and when the tree bents on one side; old tree occasionally develops insignificant trunk buttresses in mature well developed soil. Leaves simple, opposite decussate, exstipulate, cauline, petiolate, petiole 1.5 cm long, green, glabrous, pulvinous; lamina elliptic obovate, broadly ovate or obovate - oblong, entire, obtuse or round apex, coriaceous, slightly fleshy, dorsiventral, dorsal side dark green, shining; lamina 8.0 - 12.0 cm × 5.5 cm, unicostate venation, leave turning black on drying. Inflorescence cyme compound spikes, peduncle up to 25.0 cm long; flower buds glabrous, round brown, flowers zygomorphic, complete, bisexual, regular, cyclic, rosaceous, sessile, dull yellow, 0.8 cm × 1.5 cm across, slightly foetid, with 10-12 flowers in each branch; sepals – 5, polysepalous; petals – 4, gamopetalous; stamens 4, epipetalous, filamen t- 0.2 cm long; ovary superior, 0.7 cm long; capsule fleshy, ovoid, opaque, green, densely hairy, 3.0 cm long and 2.8 cm broad, flattened with a short apical beak; seed simple, dark green, completely occupy the fruit cavity; seed germination epigeal, crypto- viviparous or incipient viviparous seedling germination. Flowering and fruiting during late summer to monsoon. Stem bark contain higher percentage of tannin, wood less used for carpentry works but largely used as fuel wood. Said to be used medicinally as poultice for boils; kernel bitter taste but occasionally edible as vegetable. Play dominant role as soil binder in the frequently inter-tidal silted up land and also add substantial bio-mass to the surrounding soil and tidal water; play important role by nutrient supply to the coastal area.

Photo Plate 49. *Avicennia alba* Blume. A. Full grown tree - with stilt roots and pneumatophores. B. Stilt roots developed on the trunk base.C. Branch with fruits & Close view of fruits. D.Trunk base with pneumatophores. E & F. Mature fruits with close view

11. *Avicennia alba* Blume, Bijdr. Fl. Ned. Ind. 14 : 821. 1826.

(*A. tomentosa* Wild. Sp.Pl. 3(1):395.1800; *A. marina* (Forsk.) Vierh. var. *alba* (Blume) Bakh. Bull. Jard. Bot. Buitenz, Series 3 (3): 205,207-219.1921).

Chromosome No. 2n = 66 (Raghaban & Arora, 1958)

Common Name : White Mangrove; **Local Name (Bengali)**: Kalo Ban, Kal Bani or Kal Baine

Very common in the inter-tidal zones; evergreen, medium to tall tree, attain the height between 15.0 m - 20.0 m; bark smooth, blackish; much branched, deliquescent, irregularly spreading, gregariously grow in the inter-tidal forest; large number of conspicuous negatively geotropic erect breathing roots, the pneumatophores without any secondary growth, attain the height up to 0.30 cm and diam. 15.0 mm - 25.0 mm in the inter-tidal flat ground; developed from the surface spreading underground horizontal roots; frequently bunches of stilt roots developed from the trunk surfaces in stressed condition and when the tree bent on one side. Leaves simple, opposite decussate, exstipulate, cauline, petiolate, petiole up to 2.0 cm long, green, glabrous, pulvinous; lamina lanceolate, 10.5 cm - 12.5 cm in length and 1.5 cm - 4.5 cm broad, entire, acute apex, cuneate base, whitish green glabrous or pale green dorsal side and bright silvery white or silvery grey with papilose in the ventral side, coriaceous; leaves dense in branches and terminal branches often pendulous. Inflorescence spicate raceme, long spikes 2.5 cm - 5.0 cm long, bracteate, bracteole 0.25 cm long, acropetally arranged in clustered, 8-12 flowers, dull yellow, flower sessile. Bracteate sessile, bisexual, regular, flower 0.5 cm – 0.8 cm long and 0.35 cm across, mild fragrant, cyclic, hypogynous, erect, small, one bract and two bracteoles; sepals – 5, polysepalous, ovate, entire, mucronate, imbricate, pubescent, 0.20 cm – 0.30 cm long and 0.1 cm – 0.15 cm broad, green, rosaceous, persistent in the fruit; petals - 4, gamopetalous, inferior, valvate, orange yellow, rosaceous, 4 lobed above and short tube below, petal ovate, entire, mucronate, 0.40 cm long × 0.20 cm broad; stamens 4, epipetalous, filaments attached with petals, opposite to sepals, inferior, anther dorsiventral, introrse, bilobed, fertile exserted; carpels : ovary 0.2 cm long, conspicuously hairy; capsule ellipsoid, 2.0 cm – 4.0 cm long, conical shape, extended into narrow terminal beak, green and yellow when ripe, pubescent. Germination epigeal, crypto -viviparous. Seedling leaves linear; seeds and seedling transmitted through tidal water. Flowering and fruiting season is during March to October. Stem bark contain higher percentage of tannin, wood less used for carpentry works but as fuel wood. Kernel bitter taste but occasionally edible.

Photo Plate 50. *Avicennia marina* (Forsk.) Vierh. A. Plant in the natural habitat. B.Twig with Flower buds. C. Pneumatophores. D. Branch with fruits. E. Germinated seeds with tender roots. F. Seedling plants in the earthen pots in the laboratory for growth study

12. *Avicennia marina* (Forsk.) Vierh., Denkschr. Akad. Wiss. Wien Math. Nat. 71.435.1907.

(*A. intermedia* Griff.Trans.Linn.Soc.Lond.Bot..20:1.1846; *A. mindanaense* Elmer, Leafl. Philipp. Bot. 20: pl.1.1915).

Chromosome No. 2n = 36 (Subramanian, 1980).

Common Name : Grey Mangrove; **Local Name (Bengali)**: Peyara Ban, Peyara Bani or Peyara Baine

Very common in the inter-tidal zones; evergreen, small to medium tree, attain the height between 3.0 m and 8.0 m, occasionally attain the height up to 10.0 m; bark smooth flaky, yellowish brown, bark peeling in patches like guava; much branched, deliquescent, irregularly spreading branches, gregarious in the inter-tidal forest; conspicuous negatively geotropic aerial roots, the pneumatophores or breathing roots numerous, erect pencil like, without any secondary growth; attain the height up to 0.30cm and diam. 80.0 mm, spreading on the underneath of the mature tree, developed from the surface spreading underground horizontal roots. Leaves simple, opposite decussate, exstipulate, cauline, petiole 1.0 cm long, green, glabrous, pulvinous; lamina lanceolate, elliptic - oblong or elliptic - obovate, 8.0 cm – 11.0 cm long and 2.5 cm – 3.5 cm broad, glabrous, entire, acute, cuneate or tapering towards the base, coriaceous, smooth, shining in dorsal side but whitish or silvery pubescent in ventral side. Inflorescence cyme globose, compound spike, penduncle long up to 10.0 cm, infrequently branched, glabrous, round, condensed terminal cyme; flower heads capitate; flower- 4 lobed, actinormorphic, condensed in terminal or axillary cyme, sessile, 0.5 cm – 0.9 cm long and 0.5 cm across in open condition, pale yellow or orange yellow, complete, bisexual, regular, cyclic, hypogynous, rosaceous; sepals–5, polysepalous; petals–4, gamopetalous; stamens 4, epipetalous, filament 0.2 cm long; ovary superior, 0.3 cm long; capsule fleshy, ovoid, opaque, green, densely hairy, 3.0 cm long and 2.8 cm broad, flattened with a short apical beak; seed simple, dark green, completely occupy the fruit cavity; seed germination epigeal, crypto-viviparous or incipient viviparous seedling germination. Flowering and fruiting during late winter to monsoon. Stem bark contain higher percentage of tannin, wood used for minor carpentry works and as fuel wood. Kernel bitter taste but occasionally edible as vegetable; this tree act as soil binder in the inter-tidal river slops and river flat lands; also help in the land accretion processes.

Family - 3. Sonneratiaceae

(Fire Fly Mangroves)

Sonneratiaceae is the 3rd important mangrove family of the Tropical and Sub - Tropical regions of the Old World and Australian coast. *Sonneratia* Linn. f. Suppl. Sp. Pl. 38.1781 is the Type Genus and true mangroves under the family Sonneratiaceae; this genus - *Sonneratia* Linn. f. was segregated from the family – Lythraceae for having different adaptations or habitats. The other non- mangrove genus under the family- Sonneratiaceae is *Duabunga* Ham. is reported from Andaman and Nicobar Islands. Tomlinson (1986) has described five distinct mangrove species under the family - Sonneratiaceae, *viz., Sonneratia apetala* Buch.-Ham., *S. griffithii* Kurz., *S. caseolaris* (L.) Engler, *S. alba* J. Smith and *S. ovata* Baker. In the Indian sub-continents the first four species are distributed widely in mangrove habitats (Baker & van Steenis, 1951). According to Banerjee, *et al.* (1989) *S. alba* J. Smith. is distributed in the Andaman & Nicobar Islands and west coast of India. However, the rest three species *viz., Sonneratia apetala* Buch.-Ham., *S. griffithii* Kurz., *S. caseolaris* (L.) Engler were collected from Sundarbans mangal,*viz.* Ganga delta. in West Bengal, India.

Table 36. Key characters of the species under *Sonneratia* Linn.f. (Based on P. B. Tomlinson,1986)

A	Leaves narrow - less that 5.0 cm wide and tapering toward the apex; calyx- 4 lobes but occasionally 6-lobes; ovary 5-8 celled, stigma expanded and calyx not exceeding 2.0 cm long; fruits 1.5 - 2.0 cm diam.	1. *S. apetala* Buch.-Ham.
AA	Leaves wider - usually 5.0 cm wide or more, round or oval apex; calyx usually 6-8 lobed; ovary 14-21 celled, stigma capitate but not expanded; fruits 3 0 -7 0 cm diam	See - B & BB
B	Calyx flat, extended horizontally, not enclosing the ripe fruits; twigs not pendulous; leaves obovate-suborbicular; petiole scarcely developed; petals absent; filament white.	2. *S. griffithii* Kurz.
BB	Calyx flat, extended horizontally, not enclosing ripe fruits; twigs pendulous, slender; leaves elliptic-oblong, oval-oblong, petiole short; petals usually present; filaments red.	3. *S. caseolaris* (L.) Engler..
C	Leaves ovate to oblong-ovate, with short thick petiole; apex of fruit not depressed at base of style; tube of the fruiting calyx smooth, lobes reflexed; petal present.	4. *S. alba* J. Smith.
CC	Leaves broadly ovate, as broad as long with narrow petiole; apex of fruit depressed at base of style; tube of fruiting calyx finely warted, lobes ascending; petal absent.	5. *S. ovata* Backer.

Most members under the genus- *Sonneratia* Linn. f. grow within the inter-tidal river flat lands; different species under this genus - *Sonneratia* Linn.f. grow in different saline habitat, *viz., S. caseolaris* (L.) Engler grow in the inner estuary, having less salt in soil and water phases and along with this the species often seen to grow in the polluted water and rapidly silt deposited places in Sundarbans and elsewhere; but the species, *S. griffithii* Kurz. is seen to grow very often in the mid- and outer-estuary in shelter inter-tidal river flats. On the other hand, *S. apetala* Buch.-Ham. is seen to grow almost throughout in the Indian Sundarbans mangal, but more vigorous growth of *S. apetala* Buch.– Ham. is observed in the North - Eastern parts of Indian Sundarbans. The species like, *S. alba* J. Smith is also reported from the west coastal mangals within the Indian territories. Fruits of these species is edible, either cooked or as green; its taste is sour. These medium trees are with dense canopy and very dark; as such, during night the fire-fly gather largely within the dark canopies of the species and hence these species are also known as *'fire fly mangroves'*. In Brunei, N. Borneo 3 species, *viz., S. ovata, A. alba* and *S. caseolaris* grow together along with the intermediate forms, suggested to be the hybrid between *S. alba X ovata* and *S. alba X caseolaris;* flowering and fruiting are poor in these hybrids (Tomlinson, 1986). Hybrids in Queensland between *S. alba* and *S. caseolaris* have been referred to as *Sonneratia X gulngai* by Duke (1984)

Table 37. The distinct characteristic features between the Members of the Genus - *Sonneratia* Linn. f.

Characters	***Sonneratia apetala***	***Sonneratia caseolaris***	***Sonneratia griffithi***
Distribution	Mid-estuarine species and elsewhere	Inner-estuarine, less salt resistant species.	Mid- and outer estuarine, high salt tolerant species.
Stem bark	Thin, light brown, irregularly fissured.	Pale brown, smooth or furrows at maturity.	Deep brown or pale brown; smooth in young and fissured at maturity.
Canopy of the trees	Leafy branches pendulous, leaves partially iso-bilateral.	Leafy branches not pendulous, globose, dark green.	Leafy branches not pendulous, globose, dark green.
Leaves	Obtuse, elongated narrow gradually, lamina 8.0 cm - 12.0 cm long X 1.5 cm - 3.0 cm width; obtuse, cuneate.	Ovate-elliptic, broad, not narrow gradually, lamina 8.0 cm -12.0 cm x 4.0 cm - 5.0 cm width.	Obovate, round apex and cuneate base, 5.0 cm - 10.0 cm x 3.0 cm - 6.0 cm width, glabrous.
Inflorescence	Terminal cyme from branch axis.	Solitary cyme or few flowers on outer pendulous twig.	Solitary cyme on terminal and lateral branches.
Flowers	Medium, 1.2 cm - 2.0 cm x 1.5 cm - 2.0 cm. Petal absent.	Large, 4.0 cm - 6.0 cm x 4.0 cm - 6.0 cm. Petals 6 polypetalous, purple to violate.	Large, 3.0 cm - 5.0 cm x 5.0 cm - 7.0 cm. Petals absent; stamens white.
Fruits	Capsule, 1.5 cm - 2.0 cm diam., depressed globose berry, with persistent 4 acute calyx teeth.	Berry, 5.0 cm - 8.0 cm diam., flattened globose with long style and persistent 6 calyx teeth	Globose berry, 5.0 cm -7.0 cm diam., with persistent-6 acute calyx.
Seeds	4-7 in each fruit, brown, triangular, ridged, 1.5 cm long	Numerous in fleshy pulp of the placenta, angular and irregular.	Seed numerous, angular, light brown.
Seed Germination	Epigeal, non-viviparous	Epical, non-viviparous	Epigeal, non-viviparous.

Photo Plate 51. A.& C. *Sonneratia alba* J.Smith & B. *S. caseolaris* (L.) Engler, collected from Goa, Mandovi estuary

Photo Plate 52. *Sonneratia apetala* Buch.– Ham. A. Tall full grown tree along with *Nypa fruticans* and *Avicennia* sp. on the river flat silted up soil. B. Tall full- grown tree on the river side silty soil. C. Flowering branch, close view. D. Branch with round fruits, E. Branch with fruits.

13. *Sonneratia apetala* Buch. - Ham. Syeme Embassy Ava 3 : 477.1800.

Chromosome No. 2n = 18 (Raghaban, 1959) and 2n = 24 (Sidhu, 1961)

Common Name : Fire Fly Mangrove; **Local Name** (Bengali): Keora

Common in the inter-tidal river flat land in the Sundarbans; evergreen, medium to tall tree, attain the height up to 20.0 m; deliquiscent branching; branches or twigs pendulous or drooping, from distance look almost similar to *Eucalyptus* sp., trunk base in an average full grown tree ranges between 20.0 cm – 50.0 cm diam., branches not dense but loosely arranged, pendulous downward; bark thin, light brown, irregularly fissured. Conspicuous negatively geotropic aerial roots - pneumatophores or breathing roots numerous, pneumatophores erect with secondary growth, attain the height up to 1.5 m and diam. 8.0 cm or more, developed from the surface spreading underground horizontal roots. Leaves simple, dorsiventral, opposite decussate, exstipulate, cauline, petiolate, petiole 0.5 cm long, pulvinous, glabrous, lamina narrow gradually, tapering towards the apex or oblong lanceolate, glabrous, shining, entire, obtuse, cuneate. Inflorescence terminal cyme; mostly 3-flowered or 7-flowered cyme, develop from the axils of branches; flowers ebracteate, pedicellate, pedicel 2.0 cm long, terete, glabrous, greenish, erect, 1.2 cm – 2.5 cm long and 1.5 cm – 3.5 cm across in case of open flower; without any petal; sepals - 4, gamosepalous, persistent, imbricate, cup-shaped, elliptic, entire, acute, 1.5 cm- 2.0 cm length and 0.8 cm – 1.0 cm broad, 4 lobed.; stamens - 4, epipetalous, filament - 0.2 cm long; ovary superior, 0.3 cm long; capsule fleshy, ovoid, opaque, green, densely hairy, 3.0 cm long and 2.8 cm broad, flattened with a short apical beak; seed simple, dark green, completely occupy the fruit cavity; seed germination epigeal. Flowering and fruiting during late summer to monsoon months. Stem bark contain higher percentage of tannin, wood used for carpentry works and fuel. Fruits occasionally cooked as condiment. This species is one of the pioneer species of mangrove succession in eastern part of Indian Sundarbans; but this species is comparatively less dominant in the western part of Indian Sundarbans.

Photo Plate- 53. *Sonneratia griffithii* Kurz. A. Tall full grown tree along the river flat silted up soil. B. Close view of branch with fruits. C. Flower buds in branch, close view. D. Branch with flower buds. E. Seeds of *S. griffithii*. F. Close view of fruits with persistent calyx teeth. G. Branch with mature flat fruit.

14 . *Sonneratia griffithii* Kurz., J. Asiat. Soc. Bengal 40 (2): 56.1871.

Common Name : Mangrove Apple; **Local Name (Bengali)**: Ora

Not much common in Indian Sundarbans; grow in the inter-tidal river flat land; evergreen, medium to tall tree, attain the height up to 15.0 m; deliquescent spreading branches, trunk base normally 30.0 cm diam.; bark deep brownish or pale brown, fissured. Conspicuous negatively geotropic aerial roots; pneumatophores or breathing roots numerous, erect with secondary growth, attain the height up to 1.5 m and diam. 8.0 cm or more, developed from the surface spreading underground horizontal roots. Leaves simple, opposite decussate, exstipulate, cauline, short petiolate, petiole 0.3 cm long, pulvinous, glabrous, lamina obovate, entire, round apex, cuneate base, fleshy, 5.0 cm – 10.0 cm long X 3.0 cm – 6.0 cm width, glabrous, shining, entire. Inflorescence solitary cyme, branches having terminal solitary flower. Flowers ebracteate, pedicellate, pedicel short, up to 1.0 cm long, terete, greenish, glabrous, solid, shining, situated at the basal small leaves, 3.0 cm – 5.0 cm long and 5.0 cm – 7.0 cm across, bisexual, incomplete, actinomorphic, whitish green; sepals 6- 8, gamosepalous, attached with globose ovary; petal absent; stamens numerous, free, filaments long, 2.0 cm – 3.6 cm long, white, glabrous, basifixed, exserted, showy, flowers white, anther bilobed, fertile, dehisced; carpels 6, syncarpous, ovary globose, 2.0 cm – 3.0 cm long, 4.0 cm - 5.0 cm diam; ovary multiloculed, axile placentation, ovary surface shining green, glabrous, style –1, terminal, 3.0 cm – 4.0 cm long, soft, terete, greenish, glabrous, persistent in fruit, stigma-1. Fruit globose berry, terminally into short style beak, 5.0 cm - 7.0 cm diam, partly enclosed by persistent green calyx at base; seed numerous, angular, light brown; seed germination epigeal but not viviparous. Flowering and fruiting during April - October. Stem bark contain higher percentage of tannin, wood used for carpentry works, fuel wood. Fruits occasionally cooked as condiment.

Photo Plate 54. ***Sonneratia caseolaris*** **(L.) Engler. A. Tree in natural habitat and close view of fruit. B. Full grown tree on the river flat and with pneumatophores. C. Close view of branch with fruits. D. Flower and flower buds in branch, close view. E. Branch with flower buds and fruits. F. Close view of fruits with persistent calyx teeth. G. Seedling.**

15. *Sonneratia caseolaris* (L.) Engler, Engler & Prantle Nachir. 261. 1897.

Common Name : Mangrove Apple; **Local Name** (Bengali): Chak Keora

Not much common in Sundarbans, grow in the inter tidal estuary on river flat lands; evergreen, medium to tall tree, attain the height up to 10.0 m; deliquescent spreading branches, twigs slender, slightly pendulous, trunk base generally 30.0 cm diam. for full grown tree; bark pale brown, smooth and furrows at maturity. Conspicuous negatively geotropic aerial roots, pneumatophores or breathing roots numerous, pneumatophores erect with secondary growth, attain the height up to 1.5 m and diam. 5.0 cm or more, developed from the surface spreading underground horizontal roots. Leaves simple, opposite decussate, exstipulate, simple cauline, short petiolate, petiole 0.5 cm - 1.0 cm long, glabrous, pale green or reddish, pulvinous, lamina ovate-elliptic, prominent reddish mid-rib, slightly angled, green glabrous, coriaceous, fleshy, dorsiventral, unicostate reticulate venation, entire, round apex, cuneate base, fleshy, 8.0 cm -12.0 cm long × 4.0 cm - 6.0 cm width, glabrous, shining, entire, apex become acute at young stage but occasionally degenerate at maturity. Inflorescence solitary cyme or few flowered on outer pendulous twig. Flowers ebracteate, pedicellate, pedicel short, up to 6.0 cm long, bisexual, regular, 4.0 cm – 6.0 cm long and 4.0 cm – 6.0 cm across, cyclic, perigynous, pendulous, large cup shaped, showy, opening in the late evening; calyx tube green with 6 valvate lobes, linear, entire, oblong-elliptic, acute, truncate, inferior, valvate green, fleshy, 2.5 cm - 4.0 cm length and 1.0 cm – 1.5 cm broad, glabrous, coriaceous, persistent in fruit, flattened; petals – 6, polypetalous, inferior, vestigial in nature, velvate, linear, entire, acuminate, truncate, up to 3.5 cm long and 0.5 cm broad, purple to violate red, exserted but not showy, short lived, glabrous; stamens indefinite, free, showy, fertile, filament 4.0 cm long, soft; carpels – 6, syncarpous, ovary cup-shaped, green, globose; fruit berry, green stalked, flattened, globose berry, persistent subtending calyx, at maturity flat with sharp pointed, long style; seed numerous; germination epigeal. Flowering and fruiting during March to October, often flower buds are seen throughout the year. Stem bark contain higher percentage of tannin, wood occasionally used for carpentry works and as fuel wood. Fruits occasionally cooked as condiment.

Family – 4. Combretaceae

Under this family - Combretaceae only two genera, *viz., Laguncularia* Gaertn.f and *Lumnitzera* Willd. are considered as mangroves. The species, *Laguncularia racemosa* (L.) Gaertn.f. is the New World mangrove species, reported from the West African coast and Carabbean Islands. The genus *Lumnitzera* Willd. has two true mangrove species, *viz., Lumnitzera racemosa* Willd. and *L. littorea* (Jack.) Voigt; along with these two species in the mangrove habitats there is one putative hybrid, between *Lumnitzera racemosa* Willd. X *L. littorea* (Jack.) Voigt, *viz., Lumnitzera rosea* (Gaud.) Presl.; this hybrid species occasionally present in the Old World Tropical mangal, in Hinchinbrook Island and Queensland (Tomlinson, 1986). *Lumnitzera racemosa* Willd. is the type species under this genus and described by Willd. in Neue Schr. Ges. Naturf. Fr. Berl. 4 : 187.1903 to ascertain the genus description. This species, *Lumnitzera racemosa* Willd. was also reported as *Bruguiera madagascariensis* DC. Prodr. 3 : 23. 1828 and as *Petalema alternifolia* Roxb. Fl. Indica 2 : 371.1832. These mangrove species can be distinguished for having inferior unilocular ovary with pendulous ovules, one seeded drupe like fruit without having any endosperm. Under the family - Combretaceae there is only one species, *viz., Lumnitzera racemosa* Willd found in the Indian Sundarbans. This is small to medium tree, up to 5.0 m high with deliquescent and diffused erect and much branched, dark green canopy; pneumatophores are looping from the lateral horizontal roots; but above the tidal zone no conspicuous aerial roots are formed; leaves cyclic, exstipulate, petiolate; inflorescence axillary raceme; flowers pedicellate, pedicel short, bracteate; fruits cylindrical with pointed apex, 1 -seeded drupe, with persistent sepals and style green, glabrous. Seed germination not viviparous.

This species is adapted best in the sand dominated outer estuary, coastal areas. *Lumnitzera racemosa* Willd. is the common mangrove species throughout the Indian mangals, while *L. littorea* (Jack.) Voigt was only collected and reported from the mangrove habitats of Andaman and Nicobar Islands, from the degraded mangrove areas, *viz.*, Mohanpur village side near Mayabandar of Andaman Island.

In Indian Sundarbans the species, *Lumnitzera racemosa* Willd. is distributed and grow gregariously in the tidal and inter-tidal zones of the coastal facing mangrove areas and western fringing mangals.

Table - 38. Key characters between two species of *Lumnitzera* Willd. from the Indian mangals

A.	Axillary raceme inflorescence attain a length up to 8.0 cm and flower is white and actinomorphic; reported from the main lands and mangrove habitats of India	1. *Lumnitzera racemosa* Willd.
AA.	Spike inflorescence attain length up to 3.0 cm and flower is reddish, slightly zygomorphic; reported from Andaman & Nicobar Islands	2. *Lumnitzera littorea* (Jack.) Voigt.

Photo Plate 55. *Lumnitzera racemosa* Wild. Branch with close view of the reddish fruits. Mohanpur-Mayabandar, Andaman

Photo Plate 56. *Lumnitzera racemosa* Willd. A. Close view of the plant on the river bank at Sundarbans & B. Inset Photograph of Herbarium specimen for *Lumnitzera racemosa* Willd.

Figure 16. *Lumnitzera racemosa* Willd. Neue Schr.Ges.Maturf.Fr. Berl. 4 : 187. 1803.
Common Name : Black Mangrove : **Local Name** (Bengali): Kripa or Kripal.

Evergreen small tree, height up to 10.0 m, deliquiscent and diffused, erect and much branched, dark green, much branched dense canopy; this species is also reported to attain up to 30.0 m in other S.E. Asian mangal habitats. Aerial roots looping from lateral roots and conspicuously developed, but above the tidal zones, no aerial roots are developed. Leaves cyclic, simple, exstipulate, petiolate, petiole 0.5 cm in length, round, flattened, green, glabrous, pulvinous, lamina subulate, entire or slightly wavy, obtuse or slightly notched in apex; lamina 7.0 cm - 0.5 cm long X 3.0 cm- 6.0 cm broad, dorsiventral, glabrous, coriaceous, fleshy, deep green. Inflorescence axillary raceme, peduncle 6.0 cm- 7.0 cm long green, round, soft, attain the length up to 8.0 cm; flower white, actinomorphic, pedicellate, pedicel short, bracteate; bracts 0.1 cm long and 0.1 cm broad, ovate, entire, acute, green, coriaceous, hermaphrodite, actinomorphic, complete, erect, pentamerous, epigynous; calyx – sepals 5, gamosepalous, forming a tube like structure, calyx tube 0.8 cm in length, deep green glabrous; corolla – petals 5, polypetalous, oblong lanceolate, entire, acute, white coriaceous, deciduous, longer than sepals, equal length, superior, imbricate, alternate sepal; androecium – stamens 10, free, arranged in two whorls, 5 stamens developed at the base of the petal base; filaments - 0.5 cm in length, basifixed, round, soft, white, anther bilobed, extrorse, exserted, longitudinally dehisced, superior; gynoecium – carpels 1, ovary inferior, 0.1 cm long, one chamber, 3-ovules, style 1; seed germination hypogeal but not viviparous. Flowering and fruiting during May to November. Stem bark contain higher percentage of tannin, wood used for carpentry works and fuel. Medicinally the fluid of the stem is said to be applied to itches and herpes.

Family – 5. Arecaceae (*nom. alt.* Palmae)

(Palm Family)

The common mangrove habitat palm genera are *Nypa* Steck., *Phoenix* L., *Calamus* L. and *Oncosperma* Bl.; from Indian mangrove habitats only two species are reported as halophytic or salt tolerant palm species, *i.e.*, *Nypa fruticans* (Thunb.) Wurmb. and *Phoenix paludosa* Roxb. The first species, *N. fruticans* (Water coconut) grows on the river banks and sides of newly silted up formative river flats; this plam looks very much similar to the sunken coconut palm; while, the other is *Phoenix paludosa* (Sea date), looks similar to the date palm; *P. paludosa* grow gregariously on the mature soil, in the climax stage on mangrove succession (Photo Plate – 60 & 61).

Table 39. Difference between the two mangrove habitat palm species of the Indian Mangals

A.	Unbranched, rhizomatous palm, looks very much similar to sunken coconut palm; several spathe, male inflorescence-catkin, carpels 3, aggregate, fruits medium up to 12.0 cm long.	1. *Nypa fruticans* (Thunb.) Wurmb.
AA.	Unbranched, caudex palm, look like Phoenix sylvestris - the date palm species; single spathe, male inflorescence not catkin, carpels - 3, united, fruit small, up to 1.5 cm long.	2. *Phoenix paludosa* Roxb.

Photo Plate 57. *Nypa fruticans* (Thunb.) Wurmb. A. Natural habitat in Sundarbns & B. Close View in Sundarbans

Photo Plate 58. *Nypa fruticans* (Thumb.) Wurmb. A. Natural habitat in the river slope land at Sundarbans. B. Natural habitat with dichotomous branching of stoloniferous rhizomatous stem. C. Planted in freshwater zone above the tidal limit. D. *Nypa fruticans* (Thunb.) Wurmb. in above tidal zone at Sundarbans - with inflorescence stalk

17. *Nypa fruticans* (Thumb.) Wurmb., Verh. Batav. Gensotsch 156. 1781.

(*Nipa fruticans* Thunberg Kongl. Vet. Acad. Nya Hanal 3 : 231.)

Common Name : Mangrove Palm or Water Coconut **Local Name** (Bengali): Gol -Pata

Dichotomously branched, rhizomatous monoecious palm, fibrous roots; no aerial roots present but rhizomatous roots exposed above ground due to soil surface erosion, it looks like roots; leaf-rachis or leaf base are spongy and perform the function of aerial roots for gaseous exchange or respiratory function in this tidal influenced soil; looks like sunken coconut palm. Leaves pinnately compound, spirally arranged on the top of the rhizomatous stem; leaves up to 5.0 m long, rachis stout, glabrous, sheathing base, up to 60.0 cm diam.; terete, shiny, green, glabrous, isobilateral, distinct mid-vein, unicostate parallel venation. Inflorescence – spadix on a long peduncle, up to 50.0 cm long, cylindrical; several spathe enclosed the flowers, peduncle erect, stout, glabrous, green; male flowers grow gregariously on each peduncle, covered by spathe; tepals – 6, polytepalous in two whorls, up to 0.8 cm long and 0.5 cm broad, ovate elliptic, entire, acute, tufted, glabrous, shiny; stamens – 3, united to form central column, anthers bilobed, up to 0.4 cm long, short; female flowers – enclosed by spathe, tepals-6 as male tepals, carpels-3, free, angular but funnel shaped, up to 1.0 cm high, unilocular; single ovule, basal placentation, style and stigma absent. Fruits drupe, glabrous surface, dark brown epicarp, up to 12.0 cm long and 25.0 cm periphery, ridged and angular, mesocarp fibrous, endosperm thick; seeds grooved adaxially; germination incipient viviparous. Flowering and fruiting period during November to July; but fruits are very often seen on the mother plant almost throughout the year. Commonly grow gregariously in patches on the inter-tidal canal and river sides; but in Andaman this plant is seen to grow above the tidal reaches. Its isothermic leaves (attap) are much used for thatching cottage; it is also reported that in Malaysia alcoholic drinks or mollases is prepared from fluid of young inflorescences.

Photo Plate 59. ***Nypa fruticans*** **(Thumb.) Wurmb. A. Natural habitat of *N. fruticans* at Port Blair (Photo of 1991). B. Seedling plants in the river falt land at Sundarbans. C. *Nypa fruticans* & *Cocos nucifera* grow side by side at Sipi Ghat, Port Blair((2001). D. Fruits at mature female inflorescence. E. Single fruit. F. Two halves of fruit, showing the cotyledon**

Photo Plate 60. *Phoenix paludosa* Roxb. A. Natural habitat of *P. paludosa* at Sundarbans. B. Natural habitat at Sundarbans showing the pneumatothods and C. Natural habitat at Sundarbans showing the fruit stalk at spadix.

18. *Phoenix paludosa* Roxb. Hort. Beng. 73. 1814 & Flora Ind. 3 : 789. 1832

Chromosome No. 2n = 36 (Venkatesubban, 1945 and Sarma & Sarkar, 1957)

Common Name : Sea Date **Local Name** (Bengali): Hental or Bokra

Unbranched perennial palm, attain the height up to 5.0 m or more, with foliages on top and sharp spines in the petiole and apex of leaflets. Fibrous roots not deeply sunken; generally without aerial roots but horizontal fibrous roots occasionally exposed to air due to soil erosion and the above ground aerial *'pneumatothodes'* also develop up to 25.0 cm high in the compacted soil, frequently inundated with tidal saline water. Stem slender, cylindrical, unbranched up to 10.5 cm diam., covered with dark fibrous sheath and long pointed base, distinct leaf scars, encircling the stem, rough surface, grey, solid, terete, woody; stem up to 5.0 m high or more; leaves pinnately compound, rachis spirally arranged on the top of the stems, base of the rachis sharp, spiny and stem base mostly covered with spines of fibrous leaf sheaths; rachis 1.0 m – 2.0 m long, sheathing at base. Spadix up to 45.0 cm long, peduncle branched, spathe up to 45.0 cm long, boat shaped, stalk 20.0.cm long, yellow, spongy, glabrous, deciduous, cover the inflorescence at young stages; gradually it ruptures into 2-3 halves; male flowers sessile, 0.65 cm long and 0.45 cm across, not showy; sepals – 3, gamosepalous, cup-shaped, 0.15 cm long and 0.3 cm broad; petals – 3, polypetalous, ovate-lanceolate, entire, acute, covered, thick, imbricate, 0.6 cm long and 0.2 cm broad, yellow, stout, stamens- 6, free, sessile, each two of 6 stamens on common ridged plate; anther bilobed, 0.3 cm long, yellow, longitudinally dehisced, twining, extrorse, exserted; female – sessile, erect – 0.5 cm long, 0.35 cm across, not showy, odourless; sepals – 3, gamosepalous, cup- shaped, 0.2cm long and 0.35 cm width, glabrous, valvate, stout; petals – 3, imbricate, ovate, 0.3 cm long and 0.45 cm width, curved, entire, acute, thick, glabrous; carpels – 3, syncarpy, ovary superior, 1-seeded, style - 0, stigma – 3, short. Fruits drupe, shining green, black when ripe; seed woody, hard on maturity; germination hypogeal. Flowering and fruiting period during March to June. Commonly grow gregariously in patches on the inter-tidal canal and river sides also grow above the frequent tidal reaches on mature soil, in Andaman this plant is also present. Its leaves are much used for thatching cottage and the stalk or stem is very hard, durable and largely used for cottages and other rural uses; occasionally ripe fruits are edible.

Notes : Tomlinson (1986) has reported other palm species under this family- Arecaceae, these are (i) *Calamus erinaceus* (Becc.) Dransfield, Kew Bull. 32 : 484.1978 and (ii) *Oncosperma tigillarium* (Jack.) Ridl. J.R.Asiat. Soc.Straits Brench 33. 173. 1900. These species, *C. erinaceus* (Becc.) Dransfield is liane palm with the spiny stem, diam. ranging 2.0 - 3.0 cm and attain the height upto 15.0 m; fruits globular, about 1.0 cm diam with numerous reflexed and regularly overlapping scales. *O. tigillarium* (Jack.) Ridl. is tall many stemmed palm, up to 25.0 m tall and about 10.0 cm diam with a crown pinnate leaves to 4.0 m long. The other palm species reported by Tomlinson from the occasional saline coastal habitats are *Phoenix reclinata* (a tropical African species), *Euterpe* sp., *Manicaria* sp., *Maturitia* (a tropical American palm) and *Raphia* sp. (central American and tropical African palms).

Family – 6. Meliaceae

The family- Meliaceae has two Tribes, *viz.*, Tribe–1. Xylocarpeae and Tribe–2. Aglaieae; under these two tribes there are three mangrove species, such as, 1.*Xylocarpus granatum* Koening, 2.*X. mekongensis* and 3. *Aglaia cucullata* (Roxb.) Pellegrin. These 3 species grow and naturalised in the Indo-West Pacific Islands of the tropical and sub-tropical mangals. In the Sundarbans mangal and mangals of the other Indian territory these members under the Genus – *Xylocarpus* Koening has two common species, *viz., Xylocarpus granatum* Koening and *X. mekongensis* Pierre; along with these the Genus – *Aglaia* Lour. has only one species, *viz., Aglaia cucullata* (Roxb.) Pellegrin. The other species, *Xylocarpus molluccensis* (Lamk.) Roem. is also reported from the 'back mangal communities' in the other Indian mangals. These above mentioned mangroves and back mangrove species are medium to large sized trees and timber of these species are fine grained, hard, durable and commonly used for the carpentry works. As such exploitation of these tree species are enormous in Sundarbans and elsewhere; big sized trees are now very scare in the Indian part of Sundarbans.

Table 40 : Key characters among the two mangrove genera under the Family - Meliaceae

A.	Tall tree with large globose or round fruits between 10.0cm to 20.0cm, diam., stem with buttresses and aerial horn like woody roots present, most common plants in Sundarbans	1. *Xylocarpus* Koen.
AA.	Small to medium tree with small globose or round fruits up to a diam. between 5.0 cm - 7.5 cm, stem without any buttresses and not with prominent aerial roots present, not common in Indian Sundarbans.	2. *Aglaia* Lour.

Xylocarpus spp. are partially or fully deciduous. Fruits of these species are 8.0 cm to 20.0 cm diam., with four to several seeds and hard woody fruit/seed coats; viviparous or incipient viviparous germination is absent, but their mature seeds can float on saline river water and remain viable for longer periods; germination of seeds start soon after it reaches the soil substratum. Their trunk bases are stout, with one to several stem buttresses; their erect aerial roots are hard, up to 1.0 m high, conical, woody, buffalo horn like, erect conical pneumatophores and conspicuous knee roots. Reddish black coloured wood of *Xylocarpus* spp. are valuable for carpentry works. On the other hand, *X. molluccensis* (Lamk) Roem. prefer the sandy and rocky areas, away from the frequent tidal inundation; as such, the species *X. molluccensis* (Lamk) Roem. is considered as the 'mangrove associated' species.

A	Medium tree; grow in the intertidal silty clay soil, as minor elements of mangroves; forming well developed buttress surrounding the trunk base, but aerial pneumatophores or root suckers generally absent; bark yellowish or whitish with thin flakes; mature fruits hang on the mother plants almost throughout the year; diam. of the fruits up to 20.0 cm or slightly more, round, yellowish brown woody fruit coat	1. *X. granatum* Koen.
AA	Medium or small tree; grow either in the silty clay soil or on the sand or rocky bay as 'minor elements of mangroves' or as 'back mangal'; buttresses generally absent but well developed aerial blunt pneumatophores or root suckers; bark reddish with thick flakes; fruits hang on the mother plant for certain months of the year; diam of the fruits about 10.0 cm - 12.0 cm, greenish and not completely round in shape, fruit coat not woody but scaly	See - B. & BB.
B.	Grow generally on the inter tidal silty but consolidated clay as true mangrove; buttresses occasionally formed but very inconspicuous, well developed root suckers or aerial woody pneumatophores present; bark peeling in narrow strips; inflorescence exceeding 8.0 cm	2. *X.mekongensis* Pierre.
BB.	Grow generally on the sandy; away from the frequent tidal inundation, buttresses absent and with no aerial root suckers or pneumatophores; bark longitudinally fissured; inflorescence less than 8.0 cm long	3. *X.molluccensis* Roem..

Table 42 : Differences between the two species of *Xylocarpus* Koen.

Characters	*Xylocarpus granatum* Koen	*Xylocarpus mekongensis* Pierre
Aerial Roots	Well developed buttresses surrounding the trunk base; pneumatophores woody, but less than *X. mekongensis.*	Buttresses occasionally form, inconspicuous, but well developed root suckers or aerial pneumatophores present.
Stem Barks	Bark yellowish or whitish with thin flakes.	Bark peeling in narrow strips.
Inflorescence	Upto 6.0 cm long.	Exceeding 8.0 cm.
Leaf fall	Partially leaf-fall occur during late Winter months	Fully deciduous during late winter months, i.e. beginning of February.
Fruits	Hanging large sized fruits, upped 20.0 cm diam are seen on the mother plants almost throughout the year.	Hanging medium sized fruits, upped 12.0 cm dam is seen on the mother plants.
Seed Germination	Non-viviparous	Non-viviparous

Photo Plate 61. *Xylocarpus* spp. at Sundarbans. A. Full grown tree of *X. granatum* B. Full grown tree of *X. mekongensis.* C. Hanging fruit of *X. granatum.* D. Two halves of the fruit of *X. granatum* with seeds. E. Single fruit of *X. mekongensis* with leaves

Photo Plate 62. *Xylocarpus granatum* Koenig. A.Tall tree in the natural mangrove forest, Sundarbans. B. Tree with fruits. C. Close view of twig with flowers. D. Branch with inflorescence and flowers. E. Close view of fruit, *X. granatum* Koen.

19. *Xylocarpus granatum* Koenig, Naturf. 20 : 2. 1784.

Common Name : Cannon Ball Mangrove **Local Name** (Bengali): Dhundul

Much branched, dark green, partially deciduous medium tree, attain the height up to 20.0 m, grow in the inter-tidal ridge forest and river bank; branches differentiated into nodes and internodes, deliquescent, nodes not ridges, internodes long, 8.0 cm – 12.0 cm long, terete, smooth surface, brownish or yellowish, peeling in patches, glabrous, erect, solid, woody; trunk base with well developed buttresses, but generally lacking aerial pneumatophores or root-suckers. Leaf compound, with leaflets, opposite, superimposed or alternate, cauline, exstipulate, leaflets – stalk 0.7 cm long, terete, corky rough surface, encircling strech found, pulvinous, lamina ovate-elliptic, 8.0 cm – 11.0 cm X 4.5 cm - 5.5 cm width, entire, apex round or blunt, occasionally acute apex, cup-shaped, green, dorsiventral, dorsal surface green, glabrous and ventral surface light green, whitish rough. Inflorescence mixed, up to 6.0 cm long, spreading irregularly branched, without well developed main axis; with ephemeral bracts, three flowers subtended by each such bract; bracteoles smaller; the terminal flower female and lateral two are male; flower irregular, tetramerous, erect, pedicellate, pedicel 0.25 cm long, terete, glabrous, slightly woody; epicalyx – 4, free, valvate, triangular, 0.5 cm long, basal 0.2 cm flat, entire, mucronate, green, coriaceous, glabrous, deciduous,; calyx – sepal 4, polysepalous, inferior, twisted, equal length, sepal elliptic oblong, 0.7 cm long and 0.25 cm broad, coriaceous, entire, acute, cup shaped; corolla – petals 8, gamopetalous, globose tube, basal periphery larger than apical periphery; androecium – stamens 8, short filament, white; gynoecium – carpels 8, syncarpous, ovary superior, globose, 8-chambered, two ovules in each chamber, axile placentation; style 1, terminal, 0.5 cm long, round. Fruits round spherical or globose with long woody stalk; fruit coat thick, woody, round pendulous and hanging on the branches; fruit diam. 12.0 cm - 15.0 cm, occasionally up to 20.0 cm. Flowering and fruiting times throughout the year; but the normal and profuse flowering time is April to August and the fruits are very often seen to hang on the mother plant almost throughout the year. Seeds after rupturing the fruit coat come out and can float for 2-3 months and remain viable. Timber is hard, durable and fine grained, as such exploited for carpentary works; seeds bear saponin. Distributed throughout the mangrove habitats in Indian coasts, deltas and Andaman & Nicobar Islands.

Photo Plate 63. *Xylocarpus mekongensis* Pierre. A.Tree with yellow leaves in the natural mangrove forest at Sundarbans. B. Tree with aerial breathing roots. C. Yellow leaves before shedding. D. Flowering branch. E. Mature fruits in hands, *X. mekongensis*

20. *Xylocarpus mekongensis* Pierre, Fl. For. Cochinch. Plate 359B, 1897.

Common Name : Cedar Mangrove **Local Name** (Bengali): Passur

Much branched, dark green, fully deciduous during middle of February, medium tree, attain the height up to 20.0 m, grow in the inter- tidal ridge forest and river bank; branches with nodes and internodes, deliquescent, nodes not ridges, internodes long, 8.0 cm – 12.0 cm long, terete, smooth surface, brownish or yellowish, peeling in patches, glabrous, erect, solid, woody; trunk base not with well developed buttresses, but with aerial pneumatophores or root-suckers. Leaf compound, new leaves appear with inflorescence during February-March after complete leaf shading, with usually 4 leaflets, opposite, superimposed or alternate, cauline, exstipulate, leaflet – stalk 0.7 cm long, terete, corky rough surface, encircling strech found, pulvinous, lamina ovate-elliptic, 8.0 cm – 11.0 cm X 4.5 cm - 5.5 cm width, entire, apex acute, dorsal surface green, glabrous and ventral surface tight green. Inflorescence united cymes, about 8.0 cm long, longer with distinct main axis; flowers irregular, tetramerous, erect, pedicellate, pedicel 0.4 cm long, terete, glabrous, woody in mature condition, brown but green in young condition, 0.7 cm long X 0.3 cm across, epicalyx – 4, free, valvate, ovate, 0.15 cm long; calyx- sepals 4; corolla - petal 8, white, thin, deciduous, shorter than calyx; androecium - stamens 8, epipetalous, short filament, white, anther bilobed, white 0.07 cm long; gynoecium – carpels 8, syncarpaous, ovary superior, globose, 0.2 cm. diam., 8-chambered, two ovules in each chamber, axile placentation, style – 1, terminal 0.1 cm long, white glabrous, stigma – 1, flat, 0.1 cm diam. Fruits smaller than *X. granatum* Koen., less than 12.0 cm diam, 10 – 16 seeded, seeds arranged compactly, individual seed can float in the tidal water for several days and germinate as soon as it reach the muddy soil substratum. Flowering and fruiting times dominant during March to June and mature fruits are seen to hang on the mother during July to November. Seeds after rupturing the fruit coat broadcasted on the water and surface soil and can float in saline water for long period. Timber is hard, durable and fine grained, blackish and exploited for carpentry works; seeds contain saponin. Distributed throughout the mangrove habitats in Indian coasts, deltas and Andaman & Nicobar Islands.

Photo Plate 64. ***Aglaia cucullata*** **(Roxb.) Pellegrin. A. Small tree in the natural habitat at Sundarbans. B. Close view of branch. C. Potted seedling in the Laboratory. D. Newly germinated seedling in the laboratory.**

21. *Aglaia cucullata* (Roxb.) Pellegrin

(Syn. = *Amoora cucullata* Roxb.).

Common Name : Latmi **Local Name** (Bengali): Amur or Latmi

Small to medium tree, occasionally grow in the river side dense swmp mangrove forest of the eastern part of the Indian Sundarbans, occasionally pneumatophores may form in the tidal inundated areas. Small tree with dense large sized partially pendulous or drooping foliages. Leaves - 30.0 cm to 40.0 cm long, leaflets - 15.0 cm - 20.0 cm X 10.0 cm, 2 - 4 pairs imparipinnate compound, oblong - elliptic, sub-falcate, sub-acute, glabrous on both surfaces. Male flowers in axillary branches, peduncles long, pedicels short; calyx 3 - lobed rounded, ciliolate; petals 3, longer than calyx; staminal tube scarcely shorter than the petals, ovoid; rudimentary ovary stalked, ovoid - oblong. Fruit globose or ovoid or round with rough surface and with strong petiole, 5.0 cm - 7.5 cm perifery. Occasionally grows in the mangrove forests in the northern part of the Sundarbans Tiger Reserve at Bagna Forest range and Marich Jhampi areas, within the frequently tidal inundated areas. Flowering and fruiting during March to August. Low quality timber and fuel wood.

This species can also grow in the above tidal zone and beyond the mangrove habitats or saline environment.

Family – 7. Euphorbiaceae

Among the good number plant species under the family – Euphorbiaceae, only three genera, *viz.*, *Excoecaria* L., *Glochidion* J.R. & G. Forst and *Hippomane* L. are considered as mangrove associated genera in the Globe. These first two genera, *viz.*, *Excoecaria* L., *Glochidion* J.R. & G. Forsst are distributed in the Old World Tropical mangals, while the species under *Hippomane* L. is distributed in the New World mangals. In the Indian mangals only the species, *Excoecaria agallocha* L. is present and distributed almost throughout in the inter - tidal and above tidal zones. The Genus - *Excoecaria* L. Syst. Nat. (Ed.10) 1288. 1759 has 35–40 species, but the species *Excoecaria agallocha* L. is the type species under the genus. Though this species is very common in the mangrove habitats, it is not considered as true mangrove but as 'mangrove associate' Tomlinson (1986).This is medium dioecious tree with dense dark green foliages; at maturity its leaves change to yellowish or slightly reddish; this is partially a deciduous tree. Presence of milky latex in it twigs and leaves is the characteristic features. This tree species has no aerial roots, but after washing and eroding the surface soil its spreading horizontal roots often exposed on the surface soil. For excretion of the excess salts and other excretory products, the old or mature trees very often form gall like or round ball-shaped perforated outgrowth on the trunk base. This plants take vital role for protecting soil erosion from frequent tidal washing in the inter-tidal areas; this species can also grow in non-saline zone.

Photo Plate 65. *Excoecaria agallocha* L. A. Full grown tree on the inter-tidal flat land. B. Horizontal roots of *Excoecaria agallocha* L. C. Natural habitat of the *Excoecaria agallocha* L. in Sundarbans within the inter-tidal zones.

Photo Plate 66. *Excoecaria agallocha* L. A. Natural habitat on the tidal river side. B. Male plant with catkin inflorescence. C.Female plant with tri-lobed fruits. D. Female plant with tri-lobed fruits. E. Close magnified view of fruits, *Excoecaria agallocha* L.

22. *Excoecaria agallocha* L. Syst, Nat. (Ed. 10) 1288. 1759.

Chromosome No. 2n = 130 (Sidhu, 1961) & 2n =140 (Datta, 1967)

Common Name : Milky Mangrove or Blinding Tree **Local Name** (Bengali): Geoa or Gneoa

Medium dioecious tree, 15.0 m– 20.0 m tall, deliquescent, woody, perennial, partially deciduous, much branched, very dense with dark green leaves and spreading branches; bark grey, fissured, with poisonous milky latex, lenticels prominent on young twigs and also present in mature trunk base; mature trunk base has gall like or bar like tumerous out-growths. Roots generally underground but spreading horizontal roots exposed on surface eroded soil. Stem with dense dark green foliage; leaves spirally arranged, somewhat cluster to the end of spirally arranged, erect short, simple, exstipulat, cauline, petiolate, lamina ovate-elliptic, 4.0 cm to 7.5 cm in length and 2.4 cm – 3.5 cm broad, young leaf dark green but turning yellowish to reddish on maturity, pulvinous, acute; mature leaves senescent during the winter or early summer and new leaves profusely grow during late summer or early monsoon months. Male inflorescence catkin, pendulous or erect, peduncle 4.0 cm – 7.0 cm in length, deciduous; male flowers bracteate, sessile, monochlamydous, dioecious, zygomorphic, cyclic; female inflorescence mixed cyme, female flower bracteate, bract glandular, sessile, with a pair of bracteoles, achlamydous, dioecious zygomorphic; gynoecium – carpels 3, syncarpous; fruit 3-lobed, schizocarpous, about 0.6 cm diam. Flowering and fruiting during April to August, active growth is during monsoon months. Timber is soft, used for packing boxes, fuel wood, paper pulp; latex acrid and poisonous. Distributed throughout the mangrove habitats in India.

Notes : Tomlinson (1986) reported another species,*viz*.. *Excoecaria dallachyana* (Baill.) Benth. Fl. Austr. 6 : 153. 1873, from South Queensland and New South Wales. Naskar & Mandal (1999) had also reported *Excoecaria indica* (Willd.) Muel. Arg (Syn. *Sapium indicum* Willd.), an occasional species from the Sundarbans. Both these species grow and reported from the above tidal zones of back mangrove areas. The species, *E. bicolar* Hassk is the rare shrub and planted as ornamental species in the ornamental garden.

Family - 8. Myrsinaceae

The family – Myrsinaceae has three mangrove associated genera, *i.e.*, *Aegiceras* Gaert., *Ardisia* Sw. and *Myrsine* L.; but no species were reported from India under the genera - *Ardisia* Sw. and *Myrsine*. Two species under the Genus - *Aegiceras* Gaertn., Fruct, et Sem. 1 : 216.1788, *viz.*, *A. corniculatum* (L.) Blanco and *A. floridum* Roem. & Schult. were reported from the Old World Tropical mangals. *A. corniculatum* (L.) Blanco is common in Indian and other Old World Mangals, while *A. floridum* Roem. & Schult. is only reported from N. Borneo, Java, Moluccas, Celebes and throughout the Philippines to Indochina (Tomlinson, 1986).

Photo Plate 67. ***Aegiceras corniculatum*** **(L.) Blanco. A. Natural habitat on the tidal river side/ river falt land. B. Branches with mature fruits. C. White scented flowers on the umbel inflorescence - honey-bees collecting nectar from flowers**

Photo Plate 68. *Aegiceras corniculatam* (L.) Blanco. A. Natural habitat on the tidal river flat land. B. Branches with mature fruits. C. Branches with white fragrant flowers. D. Close view of branches with white flowers. E.Fruits on branch.

23. *Aegiceras corniculatam* (L.) Blanco Fl. Filip. 79. 1837.

(*Rhizophora corniculatum* L. Sp. Pl. 448. 1753; *Aegiceras majus* Gaertn. Fruct, et Sem1 : 216. 1788 & *A. fragrans* Konig, Konig & Sims Ann. Bot. 1 : 131. 1805).

Chromosome No. 2n = 36 (Subramanian, 1980); 2n = 46 (Faure, 1968) & 2n = 18 (Sidhu, 1961).

Common Name : River Mangrove; **Local Name** (Bengali): Khalsi

Small tree or shrub, up to 4.0 m high; spreading diffused branches; grows patches on the river flats and river slopes, within frequent tidal inundated areas; trunk base swollen, without any aerial roots; stem with dense branching and foliages umbrella like spreading; leaves simple, alternate or spirally arranged, exstipulate, cauline, petiolate; petiole 0.5 cm - 1.0 cm long, glabrous, pulvinous; lamina ovate-lanceolate, oblong or ovate-oblong or elliptic, entire, apex round to emarginate, shiny, coriaceous, 6.0 cm – 10.0 cm long and 3.0 cm - 5.0 cm broad. Inflorescence simple umble in the shoot apex or on axillary branches, peduncle up to 0.5 cm long; flowers white, pentamerous, pointed end in bud condition, 1.5 cm – 2.0 cm long and 0.5 cm – 1.0 cm across, complete, bisexual, erect; sepals – 5, polysepalous, twisted, free, oblong; petals – 5, gamopetalous, twisted, slightly fused at base to form short tube; stamens – 5, opposite to petal, filament 0.3 cm long, base united to form a tube, basifixed anthers, sagittate, longitudinally dehisced, ovary 0.7 cm long, pyramidal, enlarged up to corolla mouth, ovary single loculed, several ovules in each chamber; style - 1, terminal, 0.3 cm long; fruit capsule, 2.0 cm - 9.0 cm long, completely curved with pointed apex, calyx persistent at base, capsule 1-seeded, 1.2 cm long; incipient viviparous germination and epigeal. Flowering and fruiting during January to September; active vegetative growth is during summer and monsoon months. Timber is not much used for carpentary works, but used for fuel wood; flower nectar sweet scented and best quality nectar/ honey collected by honey bees from this species. Distributed throughout the mangrove habitats in India.

Notes : For having non-capsular, non-fleshy, single seeded fruit, the genus- *Aegiceras* Gaertn. was treated under the family - Aegiceraceae by de Candole in Prodromus 141 - 143,1844.

Family – 9. Aegialitidaceae

(*nom alt*. Plumbaginaceae)

The genus – *Aegialitis* R. Br. (Prodr. Fl. Holl. 118. 426. 1810) has two species, *viz*., *A. annulata* R. Brown and *A. rotundifolia* Roxb.; the 1st species is reported from N. Australia and Malaysia, while the 2nd one is common in mangrove habitat and considered as the minor element of mangroves. The distinctive characteristic features between these two species are *A. annulata* has 9-10 corolla lobes, while *A. rotundifolia* has only 5 corolla lobes. Tomlinson (1986) has mentioned that these two species never mix with other.

Photo Plate 69. *Aegialitis rotundifolia* Roxb. A. Plants on the forest floor. B. Floating fruit gathered on the sandy soil. C. Natural seedling plants on the forest floor. D. Close view of the fruiting branches. E. Close view of the fruiting branches.

Photo Plate 70. *Aegialitis rotundifolia* Roxb. A. Shrub in the natural habitat. B. Branch with flowers and fruits. C. Close view of the flowering branch. D. Close view of the branch with fruits.

24. *Aegialitis rotundifolia* Roxb. Fl. Ind. 2 : 111. 1824.

Common Name : Club Mangrove **Local Name** (Bengali): Tora

Evergreen shrub, up to 3 m high, without any aerial roots, but trunk base broad due to basal fused upright roots; stem base swollen and fluted with spongy structure. Bark dark grey and fissured. Tap roots sunken. Stem straight, with inconspicuous leaf scars on shoot, branching only towards apex. Leaves simple, alternate, cauline, exstipulate, petiole 5.0 cm – 8.5 cm long, dilated at base, clasping the stem, tubular leaf-sheath, gradually narrow towards the lamina, terete, 0.8 cm diam. at base, green, lamina broadly ovate, 6.0 cm – 9.0 cm long and 6.0 cm – 9.0 cm broad, coriaceous, slightly flesh, dark green, dorsiventral, dorsal side shiny, entire, obtuse, cuneate at base. Inflorescence raceme, branched, peduncle up to 6.0 cm long, developed from sheathing petiole. Flower bracteate, pedicel 0.5 cm long, flower complete, pentamerous, bisexual, erect, 1.5 cm long X 1.0 cm across; sepals 5, gamosepalous; petals 5, 1.2 cm long X 0.15 cm width, white, obtuse; androecium – stamens 5, 1.6 cm long; carpels 5, syncarpy, ovary oblong, 0.4 cm long, 1-chambered, style-5, terminal, 0.5 cm long, white, stigma absent. Fruit capsule, up to 7.0 cm long, persistent calyx, yellowish at young stage but turn brown quickly after maturation; seed single in each fruit, 7.0 cm long, black; incipient viviparous seed germination, epigeal. Flowering and fruiting during April to October; active vegetative growth during monsoon . Distributed throughout the coastal areas of India.

Family – 10. Sterculiaceae

(Looking Glass Mangrove)

The family – Sterculiaceae has only one genus, *viz., Heritiera* Aiton Hort. Kew. 3 : 545.1789 along with its two species grow in the mangrove habitats or as back mangrove species in the Ganga delta, Andaman and Nicobar Islands and the Mahanadi delta of India. Banerjee, *et al.* (1989) has also reported 3rd species, *Heritiera kanikensis* from the mangrove habitat of Bhitarkanika, Orissa coast; this was reported for globose fruits and devoid of any ridge on the fruits. In the Indian mangrove forests or from the back mangal communities, two species, *viz., Heritiera fomes* Buch. - Ham. and *H. littoralis* Dryand are reported. *H. fomes* Buch.- Ham. is common in the Ganga delta, Sundarbans and Mahanadi delta, the Bhitarkanika; while the species, *H. littoralis* Dryand is collected and reported from the Port Blair and Maya Bandar of Andamans and also from the Bhitarkanika.

The local name of the species under *Heritiera* Aiton (in Bengal) is *Sundari* and it is assumed that the name Sundarbans of this lower Ganga delta was derived from the vernacular name of this tree species, *H. fomes* Buch Ham. Due to change of salinity regime, the natural habitats of *Heritiera* spp. has now become more salty and natural regeneration of these tree species is affected and it has now become rare in these lower Ganga delta, *i.e.*, in Sundarbans of India. In the eastern part of Indian Sundarbans the species *H. fomes* Buch Ham. is now also frequently present. This genus bear distinct leaf characters, like ventral surface of the leaves with scaly indumentum; fruits are samara and the wings reduced to form vestigial keels for getting additional help during seed dispersal through water current in the tidal river.

Table 43 : Key characters between the two common Indian species of *Heritiera* Aiton

A	Trunk base dominant with buttress, but buttress attain the height up to 1.0 m or less and trunk diameter up to 75 cm or slightly more; leaves lanceolate, 9.0 cm - 16.0 cm long X 3.0 cm- 5.0 cm width, acute; flowers showy, reddish; fruits diam. upto 4.0 cm almost round, with prominent circular ridged rib.	1. *Heritiera fomes* Buch.- Ham.
AA	Trunk base dominant with buttress, but buttress attain the height up to 2.0 m or more and trunk diameter up to 1.5 m or slightly more; leaves not lanceolate, 15.0cm - 26.0 cm long X 5.0 cm- 8.0 cm width, acute; flowers not showy, white or pinkish, fruits elongated, up to 7.0 cm long, nut like, with prominent longitudinal ridged rib.	2. *Heritiera littoralis* Dryan

The photographs of the tree trunks, leaves and fruits of the *Heritiera littoralis* Dryand, collected from the Andaman Islands are presented below for easy identification of the species from the *Heritiera fomes* Buch. – Ham, collected from the Indian Sundarbans.

Presently, this species is rare in the Indian Sundarbans due to less supply of perennial upstream freshwater; which has caused rapid and abrupt ecological changes here. Tall tree with stout, erect flattened pneumatophores, root suckers, knee like bending buttress conspicuous, above ground plank like horizontal roots present; with thick elongate to lanceolate foliage, white tomentum beneath.Inflorescence mixed, ebracteate, peduncle bend and pendulous, much branched; flowers regular, ebracteate, unisexual, achlamydous, odourless; fruits winged, nut like woody coat, fruits can float on saline river water and remain viable for several days; viviparous and pseudo - viviparous germination is absent. This plants prefer less saline or fresh water zone for better growth and regeneration. For gradual change of the physiognomy in the lower Ganga delta and rise of salinity along with large scale exploitation of *Sundri tree*, it become very rare and threatened species in the Indian Sundarbans (Photo Plate - 71). It was experimentally proved that *H. fomes* can grow and flourish in non-saline area and beyond the tidal habitat in southern parts of Bengal.

Photo Plate- 71. *Heritiera fomes* Buch. Ham. A. Tree in the natural habitat at Sundarbans. B. Tree showing the buttress. C. Branch with flowers. D. Mature fruits on branch. E. Germinated seed of *H. fomes* . F. Collection of specimen in Sundarbans.

25. *Heritiera fomes* Buch. Ham. Symes Kingdom Ava 3:319 – 320. 1800.

Chromosome No. 2n = 38 (Sidhu, 1968)

Common Name : Looking Glass Mangrove **Local Name** (Bengali): Sundri or Sundari

Evergreen tall tree, attain the height up to 25.0 m or more, but in the Indian Sundarbans the present ecological condition is not congenial for the luxuriant growth of this species; besides, the timber value of this species is very high and the degree of exploitation of this species is very high. In Indian Sundarbans average height of this tree species was noted about 10.0 m for over exploitation. Root much spreading, blunt end peg-like pneumatophores present and exposed plank like horizontal roots and with stem buttresses in mature tree; the aerial roots almost similar to *Xylocarpus* spp., with blind root suckers. Branches deliquescent, pale green, inconspicuous leaf scars present on apical shoots. Leaves alternate, simple, stipulate, stipules in pair in each node, petiolate; petiole 1.0 cm long, terete, grey, slightly woody; lamina elliptic-lanceolate, 8.0 cm – 15.0 cm long X 3.0 cm – 5.0 cm width, entire, acute, tapering, coriaceous, slightly hard, dorsiventral, dorsal surface pale green glabrous, ventral surface whitish grey, rough surface with scaly indumentum; reddish or orange yellow. Inflorescence mixed, ebracteate, peduncle bent and pendulous, much branched; peduncle up to 5.9 cm long, having both mature and young flowers. Flowers ebracteate, regular, unisexual, achlamydous, odourless, erect and pendulous, small, pedicellate, pedicel 0.65 cm long, terete, pubiscent, soft, easily dissociate from the thalamus, 1.0 cm long X 0.5 cm across. Sepals four-five, gamosepalous, globose, cup-shaped, tube beneath, upper 4-5 apical lobes, orange yellow outer surface but reddish orange towards base in inner surface, lobes ovate, entire, acute, pubescent; male flowers; stamens –5, fused to form cylinder dumble shaped, known as pistillode, anther encircling the apical portion of the pistillode and to form globose structure; anther distinctly bifurcated with apical white beak; female flower : carpels 4-5, apocarpous, loosely attached. Fruit capsule, sub-globose, 2.0 cm long, periphery up to 5.0 cm diam., pericarp brown, hard, woody divided in two halves by distinct ridged, adjoining into longitudinal circular keeled, 1-seeded. Flowering and fruiting period during March to October. Occasionally grow in the inter-tidal canal and river sides at eastern part of Indian Sundarbans. Timber is hard, durable and fine grained; seeds may be eaten as vegetable and fresh wood used as fuel in the villages at Indian Sundarbans.

Photo Plate 72. *Heritiera littoralis* Dryand. A. Full grown up tree at Sipi Ghat Farm, Port Blair. B. Seedling plant of *H. littoralis* at Wandoor Bay, Andaman. C. Branch with mature fruits. D. Showing single fruit. E. Two fruits hanging from the branch.

26. *Heritiera littoralis* Dryand, Aiton Hort. Kew. (ed.1). 3 : 546.1789.

Common Name : Looking Glass Mangrove **Local Name** (Bengali): Sundri or Sundari

Evergreen tall tree, attain the height up to 25.0 m or more; not collected from the Indian Sundarbans.The present collection was made from Port Blair. Root spreading, blunt end peg-like and exposed plank like horizontal roots and with stem buttresses in mature tree. Branches deliquescent, pale green, inconspicuous leaf scars present on apical shoots. Leaves spirally arranged, simple, coriaceous, oblong or ovate-elliptic, usually 12.0 cm - 20.0 cm X 5.0 cm - 10.0 cm; obtusely pointed apex and cordate base; coriaceous, slightly hard, dorsiventral, dorsal surface pale green glabrous, ventral surface whitish grey, rough surface with scaly indumentum, veins prominant below; dark green above, grayish white beneath; petiole 1.0 - 2.0 cm long and 1.0 cm wide, shortly bipulvinate; stipules in pairs at each node, short. Flowers 3.0 - 4.0 mm in diam and 4.0 - 5.0 mm long, with a slender short pedicel, male flower somewhat smaller; both sexes with a cup-shaped calyx. Male flowers more numerous, with an androgynophore arising from the centre of the disc and consisting of stamens. Carpels 4 - 5 (-6) sessile, united loosely. Fruit capsule, 5.0 cm long, periphery up to 7.0 cm diam., maturing in pendulous cluster, usually only 1 developing from a single carpel in each flower; always 1-seeded and with a raised dorsal keel up to 5.0 mm but sometime up to 2.0 cm; epicarp woody, mesocarp fibrous, endocarp hard. Embryo represented by fused cotyledons, radicle directed ventrally. Seedling germination hypogeal. Flowering and fruiting period during November to March. Occasionally grow in the inter-tidal sides. Timber is hard, durable and fine grained; seeds edible as vegetable occasionally.

Difference between the two species under the Genus *Heritiera* Aiton.

A.	Fruit knobby with a ventral ridge together with a transverse, circular ridge, 2.5 cm diam.; plant normally grow in the tidal swamps in Ganga delta	1. *Heritiera fomes* Ait.
AA.	Fruit smooth with rudderlike crest, but without a transverse circular ridge; plants with wide distribution in back mangal	2. *Heritiera littoralis* Dry.

Family – 11. Rubiaceae

From the mangrove habitat of the Indian Sundarbans, only two Genera, *viz.*, *Hydrophyllax* Linn. f. and *Scyphiphora* Gaertn. f. along with their two species, *viz.*, *H. maritima* L. and *S. hydrophyllacea* Gaertn. were collected and reported. *S. hydrophyllacea* Gaertn. is a shrub or small tree with white flowers, while *H. maritima* L is a herb with succulent stem and leaves, pinkish flowers and rare on the coastal sandy soil.

Scyphiphora hydrophyllacea Gaertn. f. and *Hydrophyllax maritima* L. of the family Rubiaceae are very recently reported from Indian Sundarbans (Mandal, *et al.*, 1995). The 1st species is a shrub, growing on the inter-tidal river slopes, while the 2nd species is a herb and was collected from the sea beach of the Bakkhali coast of Sundarbans. Aerial roots of *S. hydrophyllacea* is wavy on the above ground, but not much conspicuous; the underground horizontal roots may be exposed due to soil erosion for frequent tidal effect. Inflorescence condensed cyme, having 4–7 flowers in aggregate, ebracteate, reddish white. Fruits glabrous, up to 1.0 cm long, smooth with persistent calyx, fruit surface with narrow ridges; seeds ovoid; viviparous or crypto-viviparous seed germination absent (Photo Plate – 73 & 74)

Photo Plate 73. *Scyphiphora hydrophyllacea* Gaertn. f. ,View of the dense foliages with flowers and fruits

Photo Plate 74. ***Scyphiphora hydrophyllacea*** **Geartn. f. A. Shrub in the natural habitat at Sundarbans. B. Flowering branch. C. Close view of the flowering branch of *Scyphiphora hydrohyllacea* Geartn. f.**

27. *Scyphiphora hydrophyllacea* Geartn. f. Fruct. Sem. Pl. 3 : 91.1805.

Common Name : Wild Ixora **Local Name** (Bengali): Tagri Bani

Shrub or small tree, attain a height up to 4.0 m; much branched, evergreen dense foliages; in the inter-tidal river flats of outer estuary. Much spreading horizontal roots, under developed taproots and aerial roots wavy, not much conspicuous. Leaves simple, opposite decussate, inter stipulate, lamina ovate-elliptic or ovate-oblong, entire, blunt, connate at base, 6.0 cm - 10.0 cm long X 3.0 cm – 5.5 cm width. Inflorescence condensed cyme, having 4-7 flowers in aggregate, ebracteate, develop at node; flowers – 1.0 cm long X 0.5 cm across, tetramerous, complete, bisexual, erect, pedicellate, pedicel short, 0.15 cm long; sepals - 4, gamosepalous, persistent, 0.3 cm – 0.6 cm long, glabrous; petals – 4, gamopetalous, funnel shaped, 0.35 cm – 0.85 cm long, reddish white, 4 apical lobes; stamens - 4, inserted, epipetalous, filament short, 0.15 cm long, soft glabrous, anther bilobed, 0.15 cm – 0.2 cm long; ovary- inferior, short cylindrical, 0.3 cm long, glabrous, bilocular, 2-ovules in each locule; style – 1.0 cm long, terminal, slender, soft, glabrous, stigma cup-shaped, bilobed. Fruit glabrous, up to 1.0 cm long, smooth with persistent calyx; fruit surface with narrow ridge; seeds ovoid brown, endosperm persistent. Seed germination hypogeal. Flowering and fruiting times during monsoon and early winter months, *i.e.* during May to November. Distributed in Indian Sundarbans, Andhra Pradesh, Andaman & Nicobar Islands; throughout in the S.E. Asian countries. Medicinally the warm shoot extract is said to be useful in stomach-ache.

Photo Plate 75. *Hydrophyllax maritima* Linn. f. - flowering twigs in hand.

28. *Hydrophyllax maritima* **Linn.** f. Suppl. 126.1781.

Prostrate much branched herb, grow in the coastal sand beach area; creeping or twining . Tap root system and adventitious roots developed from the nodes; these nodal profuse roots anchor the herb on the sandy substratum. Internodes slightly ridged, easily differentiated, 1.2 cm long, fleshy, hollow, terete-flattened, glabrous, thick, succulent, dark green, poorly axile branches develop. Leaves simple, opposite decussate, interstipulate; sessile, short, spathe cover the node, dark green, glabrous; lamina ovate elliptic, succulent, glabrous, dark green, 2.5 cm long and 1.7 cm width, acute, entire, main vein not distinct, isobilateral, unicostate, reticulate venation inconspicously found; solitary flower developed at the leaf base, *i.e.*, axile solitary flower; ebracteate, complete, bisexual, regular, erect, odourless, sessile, 3.4 cm long, hollow tube-shaped, tetramerous; sepals – 4, gamosepalous, 2.5 cm long × 1.0 cm width, superior, 4 - lobes above, lobes 0.8 cm long, thick, fleshy, green, glabrous, valvate, alternate to petal, persistent; petals – 4, gamopetalous, hollow tube, 4 - lobes above, lobes purple colour but tube white, herbaceous, superior, valvate, glabrous, deciduous, 1.9 cm long; stamens – 4, epipetalous, filament attached to the petal but anther remain free, filament 0.9 cm long, white, terete, glabrous, soft, anther bilobed, 0.25 cm long, dorsiventral, violet, longitudinally dehisced, introse, exserted, shorter than gynoecium; carpels – 2, syncarpous, ovary inferior, 1.5 cm long, ovary single chambered, basal placentation, one ovule, style – 1, terminal, 2.5 cm long, white, soft, terete, terminating into a tapering end, stigma absent. Fruit achene, 2.5 cm long and 1.0 cm width with 4 - persistent calyx lobes. Germination hypogeal. Flowering and fruiting period generally during July to September. Distributed in the coastal areas of India, Sri Lanka and other S.E. Asian countries.

Family – 12. Acanthaceae

Acanthus L. Sp. P. 639.1753 is the Type Genus of the family- Acanthaceae and there are three mangrove associated species, *viz., Acanthus ilicifolius* L., *A. volubilis* Wall. and *A. ebracteatus* Vahl. These first two species were collected and reported from the Sundarbans mangals but the last species was collected and reported from Port Blair, Andaman Island.

Table 44 : Key characters of Indian species under the Genus - *Acanthus* L. (Based on Tomlinson, 1986)

A	Plant, twining in nature, stem and leaves without any vines or arms; flowers white	1. *A. volubilis* Wall
AA	Plant erect shrub and much erect or slender branches and with sharp vines and arms on the branches and leave margin; flowers generally violet	See - B & BB
B	Common erect shrub, grow throughout the mangrove habitats, much common	2. *A. ilicifolius* L
BB	Rare, erect shrub, grow in the mangrove habitats at Port Blair, Andaman Islands; not much common	3. *A. ebracteatus.* Vahl

Table 45 : The distinct characteristic features of three mangrove species of *Acanthus* L.

Characters	Acanthus ilicifolius L.	Acanthus volubilis Wall.	Acanthus ebracteatus Vahl.
Habitats	Viny shrub, erect, upto 1.5 cm long, scarcely woody, bushy with very dense growth.	Vineless; twining herb or shrub, climb on mangrove thickets.	Viny shrub, erect, upto 1.5 m long, scarcely woody, bushy with very dense growth.
Rooting patterns	Shallow tap roots, occasionally stilt roots	Shallow tap roots, occasionally with stilt roots.	Shallow tap roots, occasionally stilt roots
Leaves	Leaves margin and the mid - vein with sharp poisonous spines; lamina elliptic oblong.	Leaves without any spine, thick, fleshy, oblong lanceolate	Leaves margin and the mid - vein with sharp poisonous spines; lamina elliptic oblong.
Flowers	Violet colour	White colour.	White colour
Occurrences	Very common in the Sundarbans and almost in every saline habitats.	Rare in the Indian Sundarbans and less reported from the other locality in Indian mangals.	Common in Sipy Ghat/ Port Blair, Andaman Island

Photo Plate 76. A. *Acanthus ilicifolius* L. B. *A. volubilis* Wall. and C. *A. ebracteatus* Vahl

Photo Plate 77. *Acanthus ilicifolius* L. A. Plants in the Natural habitat in Sundarbans. B.Twig with flowers and fruits. C. Twig with flowers. D. Flowering plant in the natural habitat. E. Inflorescence with flowers & fruits. F. Inflorescence with fruits

29 *Acanthus ilicifolius* L. Sp. Pl. 639.1753.

Common Name : Holly Mangroves **Local Name** (Bengali): Horkoch Kanta or Horgoja

Vine shrub, up to 2.5 cm high, scarcely woody, bushy, with very dense growth. Shallow tap roots, but occasionally stilt roots are conspicuous; form the tap roots, fibrous nutritious roots develop. Stem differentiated into distinct nodes and internodes. Leaves simple, opposite decussate, cauline, exstipulate, petiole short, 0.3 cm long × 0.4 cm broad, flattened, glabrous, pulvinous to sheathing base, solid, lamina oblong, elliptic-oblong, spiny to very spiny margin and wavy, apex acute, basal angled, dorsiventral, green, unicostate reticulate venation, 15.0 cm long × 6.0 cm broad. Inflorescence racemose, spike, up to 16 .0 cm long peduncled; flowers- 0.8 cm long × 0.5 cm broad, curved, ovoid, entire, acute; flowers up to 5.5 cm long × 3.8 cm across, bisexual, typically zygomorphic, complete, showy, erect, sessile, hypogynous; calyx – sepals 5, polysepalous, twisted, one cover completely, outer one large, while inner one small, 1.7 cm – 2.0 cm long × 0.5 cm – 0.8 cm broad, ovate, margin folded, acute, basal curved, coriaceous, persistent; corolla – petals large, showy, violate, 3.5 cm – 4.0 cm long × 3.4 cm – 3.6 cm broad, ovate, entire, obtuse, coriaceous, slightly fleshy, slightly curved, margin outer folded, deciduous; androecium - stamens 4, free filament, but anther adhere loosely, filament 1.5 cm long, fleshy, solid, white, glabrous, 0.3 cm broad, flattened, erect, exerted, introrse, anther bilobed, 1.0 cm long, densely white, pubescent, basifixed longitudinally dehisced, 4 filaments base assemble into same place; gynoecium – carpels 2, syncarpy, ovary dumbell-shaped, 0.4 cm long, glabrous, terete, ovary superior, two chambered, two ovules in each chamber, axile placentation, style 1, terminal, 3.0 cm long, persistent, terete, glabrous, white, stigma - 2, bifid, short; fruit 2.5 cm – 3.0 cm long, 1.0 cm diam., capsule, glabrous, 4 seeded, kidney shaped, grey, 0.5 cm - 1.0 cm long; germination epigeal, non-viviparous. Flowering and fruiting during May to August. Common both in the inter-tidal and above tidal regions with high salinity and also in low saline soil. Distributed throughout the mangrove habitats in the Old World Tropics. Medicinally the leaves is said to be useful as fomentation in neuralgea and rheumatism, also as expectorant; roots are said to be useful for asthma, paralysis, leucorrhoea and debility.

Phot Plate 78. *Acanthus volubilis* Wall. Flowering twig in hand, collected from Indian Sundarbans

30. *Acanthus volubilis* Wall. Pl. As. 2, 56. T. 172.1881.

Common Name : Holly Mangroves **Local Name** (Bengali): Lata Horgoja

Twining herb or shrub, attain the length, up to 4.0 m, much branched. Tap root system develop and sometime stilt roots are found. Leaves simple, opposite decussate, exstipulate, petiolate; petiole up to 1.0 cm long, slightly flattened, glabrous, solid, pulvinous, green, lamina oblong-lanceolate, entire, blunt, angled, 8.0 cm long × 5.6 cm width, dorsiventral, succelent, green, glabrous, without spines. Inflorescence raceme, ebracteate, peduncle, up to 9.8 cm long, erect; flowers, sessile, complete, bisexual, irregular, erect, up to 2.5 cm long; sepals - 4, polysepalous, mixed size, 0.7cm-1.1 cm long × 0.3 cm – 0.6 cm width, green, glabrous, leathery, curved, entire, acute, superior; petals – 4, gamopetalous, fused to form single plate, base fused to form hollow short tube, while apex flattened to form 5 short lobes, 2.5 cm long and 1.6 cm width in flattened portion, white, fleshy, glabrous, entire; stamens - 4, free, equal size, filament – introrse, inserted, basifixed, 1.0 cm long and 0.15 cm width, anther bilobed, brownish, 0.5 cm long × 0.2 cm width, gregarious hairs arising from ventral sides; carpels - syncarpous, ovary superior, two chambers, one ovules in each chamber, axile placentation, style – 1, terminal, white, glabrous, terete, 1.3 cm long, stigma absent. Fruit capsule, up to 2.0 cm long, seed with black seed coat. Flowering and fruiting during June to September.

Photo Plate 79. ***Acanthus ebracteatus*** **Vahl. A. Plants in the natural habitat at Sipighat, Port Blair, Andaman. B. Flowering branches, close view. C. Flowering branches, close view. D. Flowering branches, close view.**

31. *Acanthus ebracteatus* Vahl. Pl. As. 2, 56. T. 172.1881.

Common Name : Holly Mangroves **Local Name** (Bengali): Horgoja

Twining herb or shrub, attain the length, up to 3.0 m, much branched. Tap root system develop and sometime stilt roots are found. Leaves simple, opposite decussate, exstipulate, petiolate; petiole up to 1.0 cm long, slightly flattened, glabrous, solid, pulvinous, green, lamina oblong-lanceolate, entire, blunt, angled, 8.0 cm long × 5.6 cm width, dorsiventral, succelent, green, glabrous. Inflorescence raceme, ebracteate, peduncle, up to 9.8 cm long, erect; flowers, sessile, complete, bisexual, irregular, erect, up to 2.5 cm long; sepals - 4, polysepalous, mixed sized, 0.7cm -1.1 cm long × 0.3 cm – 0.6 cm width, green, glabrous, leathery, curved,entire,acute,superior; petals- 4, gamopetalous, fused to form single plate, base fused to form hollow short tube, while apex flattened to form 5 short lobes, 2.5 cm long and 1.6 cm width in flattened portion, white, fleshy, glabrous, entire; stamens - 4, free, equal size, filament – introrse, inserted, basifixed,1.0 cm long and 0.15 cm width, anther bilobed, brownish, 0.5 cm long × 0.2 cm width, gregarious hairs arising from ventral sides; carpels- syncarpous, ovary superior, two chambers, one ovules in each chamber, axile placentation, style – 1, terminal, white, glabrous, terete, 1.3 cm long, stigma absent. Fruit capsule, up to 2.0 cm long, persistent calyx, seed with black seed coat. Flowering and fruiting during January to September. Specimens collected from the inter-tidal canal sides of Sipighat, Port Blair. From the mainland of Indian territories this species, *Acanthus ebracteatus* Vahl. was not reported, till date.

Family – 13. Apocynaceae

The family-Apocynaceae has only 2 species reported from the back-mangrove communities; among these the Genus - *Cerbera* L. is reported from the mangrove habitats of the Old World, while the species *Rhabdenia biflora* (Jacq.) Muell.-Arg. is reported from the tropical America in the Caribbean region both in fresh water and salt water swamp. Among the Genus - *Cerbera* L. there are three species, *viz., C. odollam* Gaertn., *C. manghas* L. and *C. floribunda* Schum. The species *Cerbera manghas* L. is only collected and reported from the Indian Sundarbans and other mangrove habitats in India, including Andaman & Nicobar Islands.

Photo Plate 80. *Cerbera manghas* L. A. Flowering branch of *Cerbera manghas* L. B. Two mature fruits hang from the branch. C. Seedling plant in the earthen pot at laboratory condition.

32. *Cerbera manghas* L. Sp. Pl. 208. 1753.

(*C. lactaria* Hamilt. *ex* DC. Prodr. 8 : 353. 1844. *C.odollum* Bl. Bidjr, Fl. Ned. Ind. 1032. 1826)

Local Name : Dabur

Occasional on the sandy soil of coastal areas at Bagmara Block of STR, beyond the frequent tidal inundated zone. Small tree with dense foliages. Stem green in the young stage but turning brownish on maturity; branching dichotomous, nodes and internodes prominant and with milky latex; leaves dorsiventral, petiolate, glabrous, glossy, 14.0cm × 4.5 cm, obovate to acute tip, petiole 2.5 cm, tufted at apical part, alternate, leaf scar present on the mature branches, apical branches soft, woody, prominant mid veins unicasted, reticulate veination. Inflorescence termibal with yellow to white flowers, bracts and bracteoles present, bisexual, pedicellate, pedicel long, 1.6 cm - 2.9 cm, regular, complete, sweet with yellow colour, 5 corona like extension with woody appearance, mouth of the corolla tube with fibrous structure; sepals - 5, gamosepalous, imbricate, lanceolate, outer part green and inner part pale green, deciduous; stamens - 5, free, anthers - 5, epipetalous, hidden within the corolla tube, bilobed, anther tip united, sagittate, alternate to petal; carpels 2, syncarpous, differentiated into ovary. Fruit with long petiole and usually two from each stalk, with smooth dark green surface oval, looks green mango like, up to 8.0 cm long 5.0 cm diam. Flowering and fruiting normally during February to August; though the mature fruits hang in the mother plants almost throughout the year. Barks and nuts or fruits are said to have purgative properties, narcotic, poisonous and useful in hydrophobia.

Family – 14. Asclepiadaceae

The family -Asclepiadaceae has five genus in the Sundarbans mangrove forest area, *viz. Sarcolobus* R. Brown, *Pentatropis* W. & A., *Tylophora* R. Brown, *Finlaysonia* Wall. and *Hoya* R. Brown. Out of these, only two species under *Sarcolobus* R. Brown, *i.e., S. globosus* Wall. and *S. carinatus* Wall. grow in the inter-tidal river flats and river slopes as halophytic shrubs, while *Pentatropis capensis* (Linn.f.) Bullock, *Tylophora tenuis* Bl. and *Finlaysonia obovata* Wall. grow in the mangrove cleared areas; these species occasionally grow in the tidal inundated areas. But, *Hoya parasitica* Wall. grow as parasitic herb, especially on the *Xylocarpus* tree. All these species are not halophytic in nature, though they grow along with the mangroves, as mangrove associates or back mangals.

Table 46 : Key to the Genera under the family - Asclepiadaceae in Sundarbans mangal

A.	Halophytic shrubs grow in the frequently inundated river flat land	1. *Sarcolobus R.* Brown.
AA.	Not true halophytic herbs, grow above frequently tidal reaches	See - B & BB.
B.	Parasitic shrub or herbaceous, leaves broadly lanceolate and very thick, succulent; generally grow on the old Xylocarpus sp. tree	2. *Hoya R.* Brown.
BB.	Climbing shrubs, common on the mangrove reclaimed areas	See - C & CC.
C.	Twining herb, leaves small upto 2.5 cm long × 1.8 cm broad, pedicel weak and pendulous; flower with corona, fruit long	3. *Pentatropis* W. & A.
CC.	Twining herb or shrub	See - D & DD.
DD.	Leaves linear lanceolate, puberulous above, acute; cyme lateral, dense branching, peduncles about 2.0 cm long, corolla pink, globose	4. *Tylophora R.* Brown.
D.	Parasitic with succulant, lanceolate leaves	5. *Finlaysonia* Wall.

Photo Plate 81. *Sarcolobus carinatus* Wall. in the natural habitat at Sundarbans

Photo Plate 82. *Pentatropis capensis* (Linn.) Bullock - twining twigs on the *Phoenix* sp. hedge in natural condition.

33. *Pentatropis capensis* (Linn.) Bullock, Kew Bull. 1955. 284.1955.

Common Name : Ambarvel (Hindi). **Local Name** (Bengali): Dudhi Lata

Climbing herb with milky latex; simple, opposite decussate, exstipulate, petiolate leaves; lamina ovate-elliptic, entire, mucronate, green, fleshy, glabrous, 2.5 cm long X 1.8 cm width. Cyme pedunculate; flowers bracteate, bisexual, regular, pedicellate, yellowish, pedicel upto 2.6 cm long, terete; sepals - 5, polysepalous; petals - 5, gamopetalous, inferior, upper surface brownish white and glabrous, imbricate; stamens - 5, gynostegium, corona -5; carpels -2. Fruit follicle, 3-4 cm long, cylindrical, gradually tapering end, rigid; seeds numerous. Flowering and fruiting during November to January. Medicinally useful as alternative and refrigerant.

Table 47 : Key to the Species under the Genus - *Sarcolobus R.* Brown

A.	Leaves thick and coriaceous, upto 8.5 cm long and 4.0 cm width, flowers ebracteate, rosaceous, petal white-pink coloured having no brown spots, inflorescence branched cyme	1. *Sarcolobus globosus* Wall.
AA.	Leaves coriaceous upto 6.5 cm long and 2.7 cm width, flower bracteate, slightly cup-shaped with (below) hollow tube, petal white-pink, scattered brown spots on petal, inflorescence corymb	2. *Sarcolobus carinatus* Wall.

Photo Plate 83. ***Sarcolobus carinatus*** **Wall. A. Flowering twig. B. Twig with single flower - close view. C. Natural habitat in Sundarbans mangal. D. Mature winged fruits - close view.**

34. *Sarcolobus carinatus* Wall. As.Res.Xii: 570. t. 4.1816.

Local Name (Bengali): Baole Lata

Prostrate or climbing herb or shrub on the river banks, river slope, river-flat away from the strong tidal current; climbing branches weak but prostrate branches thick, stout and sheathing with secondary thickening; nodes slightly ridged, but not jointed, internodes up to 12.0 cm long, solid, glabrous, terete, green grey, 0.15 cm - 0.25 cm diam. Corymbose, peduncle unbranched, 2.0 cm long × 0.15 cm width, glabrous, solid, arise at the nodal axis, bract scars present on the peduncle; bracteate, bract short, alternately arranged, ovate, entire, blunt; flowers - 0.5 cm long and 0.8 cm across, complete, bisexual, regular, pedicellate, pentamerous, hypogynous, cup-shaped; sepals - 5, polysepalous, inferior, imbricate, ovate, entire, acute, 0.25 cm long and 0.20 cm width, tufted, glabrous, slightly thick, deciduous; petals - 5, gamopetalous, inferior, ovate, entire, acute, white-pink, scattered brown dot like spots on outer surface of petals; stamens - 5, gamostegium attached to the stigma, corona absent, filament short, not conspicously round, white, globose 0.1 cm long, introrse, whitish; carpels - 2, syncarpous, superior, ovary pyramidal, 0.1 cm long and 0.1 cm width, glabrous, whitish green, two chambered, axile placentation, style -1, terminal, 0.15 cm long, whitish, soft, glabrous, stigma-2, globose, bipartite, short, 0.08 cm long. Fruits winged follicle, brown, large as long about 5.0 cm and 7.5 cm periphery; 2-3 flattened ribs on fruit coat, fruit coat terminated into a short beak, yellowish green, seeds numerous. Flowering and fruiting generally during July to October. Fruits occasionally cooked as vegetable.

Photo Plate 84. *Sarcolobus globosus* Wall. - A. Twig with flowers. B. Twig with fruits.

35. *Sarcolobus globosus* Wall. As. Res. XII. 568.t.5.1816

Local Name (Bengali): Caw Phal

Prostrate or climbing shrubs on the river bank, river slope, river-flat away from the strong tidal current; climbing branches weak but prostrate branches thick, stout and sheathing with secondary thickening; nodes slightly ridged, but not jointed, internodes up to 8.0 cm long, solid, glabrous, semi-terete. Branched cyme, peduncle first trifurcate, then bifurcate; flowers whitish-green, 0.2 cm thick, ebracteate, complete, bisexual, regualr, erect, medium, pedicellate; pedicel 0.45 cm long, cylindrical, 0.5 cm long and 1.0 cm across, pentamerous, hypogynous, rosaceous; sepals- 5, polysepalous, 0.15 cm long and 0.15 cm width at base, imbricate, ovate, entire, acute, shiny, alternate to petal, persistent; petals - 5, gamopetalous, inferior, ovate, entire, acute, pink-white, deep pink, 8.5 cm long and 0.3 cm width, twisted, distinct hairy on both the surfaces; stamens -5, gynostegium, attached to the stigma, corona absent, filament short, 0.1 cm long, white; carpels 2, syncarpous, ovary superior, pyramid shaped. Fruits follicle, globose. brown, large as long as about 6.0 cm and 8.5 cm periphery or more; seeds numerous, brown, flattened. Flowering and fruiting generally during October to January.

36. *Finlaysonia obovata* Wall. Pl.As. Rar.2 : 48, t. 162.

Local Name (Bengali): Kali Lata

Large evergreen climber with stout branches; leaves obovate - oblong or oblanceolate, obtuse or apiculate. Flowers foetid. Occasional on the mangrove cleared / degraded above tidal zones in Sundarbans.Flowering and fruiting during Novemver to April.

Photo Plate 85. *Finlaysonia obovata* Wall. - Plant in hand, with flowers and fruits

Photo Plate 86. *Hoya parasitica* Wall.- parasite on the *Xylocarpus mekongensis* tree at Sundarbans

37. *Hoya parasitica* Wall.

Local Name (Bengali) : Pargachha

Parasitic shrub or herbaceous, leaves broadly lanceolate and very thick, succulent; generally grow on the old *Xylocarpus* sp. tree. Flowers small.

38. *Tylophora tenuis* Bl. Bijdr. 1062. 1826

Local Name : Antomul

Slender twining herb; leaves linear-lanceolate, puberulous above, acute; cyme lateral, dense branching, peduncles about 2.0 cm long, corolla pink, globose. Flowering and fruiting during November to March. Occasionally grow on the river bank and side of the creeks but above the tidal reaches. Plants are used for excessive perspiration, bilious swelling and small pox.

Family – 15. Bignoniaceae

Only one species, *i.e., Dolichandrone spathaceae* (Linn.f.) K. Schum. occasionally grow in the back- mangrove communities of the Indian Sundarbans and elsewhere.

Photo Plate 87. *Dolichandrone spathaceae* (Linn.f.) K. Schum. A. Showing the canopy structure. B. Branch with the mature fruits in the branch. C. Branch with the mature fruits in the branch.

39. *Dolichandrone spathaceae* (Linn.f.) K. Schum. Fl. Kais. Wilh. Land. 123.1889.

Local Name (Bengali): Gorshingiah

Medium tree, occasionally grows in the mangrove reclaimed river banks and village side; leaves pinnately compound, leaflets 5.0 - 7.5 cm long; flowers are showy; capsule 25.0 - 35.0 cm long, seeds rectangular with corky wings. Flowering and fruiting during November to March. Seed powder is said to be antiseptic and used in spasmodic affections.

Family – 16. Malvaceae

The family – Malvaceae has about 80 genera and 1500 species, distributed throughout the tropical and temperate region; but only two genera, *viz., Hibiscus* L and *Thespesia* Solander ex Correa are occasionally present in the coastal areas as back mangals or as mangrove associates. *Thespesia populnea* (L.) Solaner ex Correa, *T. populneoides* and *T. lampus* (Cav.) Dalz. & Gibs., *Hibiscus tiliaceous* L. and *H. tortuosus* Roxb. are collected and reported from the Indian Sundarbans as mangrove associated or as back mangroves (Naskar & Mandal,1999)

Table 48 : Mangrove Genera under the Family - Malvaceae

A	Epicalyx present, calyx conspicuously 5-toothed, leaves purely cordate with usually 9-11 main veins; capsule dehisced into 5 separate valves.	1. Hibiscus L.
AA	Epicalyx absent, calyx fused to form cup-shaped structure, with 4 inconspicuously minute teeth; leaves cordate-lanceolate, with 7 main veins; capsule dehisced into 5 longitudinal sutures	2. Thespesia Soland ex Correa.

The Genus- *Hibiscus* L. Gen. Ed. 1: 207. 1737 has several non mangrove, wild and cultivated species; only two species *viz., H. tortuosus* Roxb. and *H. tiliaceous* L. are collected from the Sundarbans as back mangrove species or as mangroves associates.

Table 49. Key to the Mangrove Associated or Back Mangrove species under the Genera - *Hibiscus* L.

A	Petal yellow, longitudinal red veins upward parallel, deep red spot found in the base at inner side of flowers, stigma blackish - red	1. H. tortuosus Roxb.
AA	Petal light yellow, yellow-white, longitudinal red veins upward parallel, crimson eye spot in base at inner side of flowers, stigma pink yellow	2. H. tiliaceous L.

Table 50 : Key to the Mangrove Associated or Back Mangrove species under the Genera - *Thespesia* Soland.

A	Medium tree, up to 15.0 m tall; flower solitary, raceme, 5.0 cm diam.	1. *T. populnea* (L.) Soland
AA	Shrub or small tree, up to 7.0 m tall; flower solitary, 7.0 cm diam.	2. *T. populneoides* Kostel

Photo Plate 88. *Hibiscus tiliaceous* L. A. Long view of the branches. B. Close view of the branch with flowers

Photo Plate 89. ***Thespesia populnea*** **(L.) Soland ex Correa - flowering branch**

40. *Thespesia populnea* (L.) Soland ex Correa, Ann. Mus. Par.IX : 290.1807.

Common Name : Umbrella Tree or Indian Tulip Tree **Local Name** (Bengali): Paras or Habal

Medium tree, up to 15.0 m high, deliquiscent branched, evergreen, latex mucilagenous. Leaves simple, alternate, stipule present at both the sides of petiolar base, lanceolate, entire, acute, green, glabrous, coriaceous; petiole, long, pulvinous, gregarious indistinct hairy, green, solid, terete, 10.5 cm long and 2.0 cm width, slightly twining, lamina cordate, 15.0 cm long X 11.7 cm width, entire, acute, coriaceous, dorsiventral, dorsal green with smooth surface, ventral side whitish green with rough surface, unicostate reticulate venation, whitish yellow mid vein distinct. Flower- solitary raceme, mature flower at base, ebracteate, pedicel – 5.0 cm – 12.0 cm long X 0.3 cm width, solid, green, terete, bisexual, regular, complete 5.5 cm long and 4.5 cm across, showy, large, slightly fragrant; sepals – 5, gamosepalous, cup-shaped, inferior, valvate, 5 minute lobes, green, rough surface; petals - 5, polypetalous, twisted, inferior, slightly touched at base, obovate, herbaceous, wrinkled, 5.0 cm long X 4.0 cm width at terminal part, while 0.6 cm width at base, dark red spot at basal interior side, showy yellow but turned into pink through gradual maturation, deciduous; androecium – stamens indefinite, monoadelphous, filaments fused to form staminal column surrounding the style; carpels - 5, syncarpy, superior, cup shaped, 0.5 cm long and 0.8 cm diam., white yellow, glabrous, fleshy, 5-chambered, two ovules in each chambers, axile placentation, style – 1, terminal, soft, white-yellow, glabrous, fleshy 2.5 cm long, terete, stigma-5, 0.5 cm long, 5 parts, each part longitudinally flattened, glabrous, white yellow, distinct. Fruit capsule, 2.5 cm long and 2.5 cm width, cup-shaped with partly covered by persistent calyx, 5 halves inconspicously found, longitudinally dehisced, brown after maturation; seeds flattened, black; germination hypogeal. Flowering and fruiting during November to March, but occasionally throughout the year. Timber producing plant, distributed throughout the coastal states in India and also tropical S. E. Asian countries. Medicinally - barks, leaves, flowers are used in cutaneus affections; seed oil is said to be used in skin troubles; barks, roots and fruits are astringent and said to be used in dysentery and hemorrhoids; seeds are purgative.

Other species under this genus- *Thespesia* Soland ex Correa, Ann. Mus. IX : 290.t.8,f. 2.1807 in the coastal habitats are *T. populneoides* (Roxb.) Kostel and *T. lampus* (Cav.) Dalz. & Gibs.

Photo Plate 90. *Hibiscus tiliaceous* L. A. Branch with flower buds. B. Branch with flower buds.

41. *Hibiscus tiliaceous* L. Sp. Pl. 694.1753

Common Name : Yellow Mallow Tree **Local Name** (Bengali): Bhola

Tall tree, up to 20.0 m high, deliquiscent branched, evergreen plant with mucilagenous latex. Leaves simple, alternate, stipules ovate, entire, acute, basal curved, green, pubescent, coriaceous, 1.0 cm long X 0.5 cm width; petiolate, terete, green, pulvinous, lamina cordate, 17.5 cm long × 16.5 width, coriaceous, entire or slightly wavy, acuminate, dorsiventral, dorsal side green glabrous, ventral whitish-green, rough, unicostate reticulate venation. Cyme two flowers on each peduncle, one mature, while other is young; bracteate, paired, bracts ovate, entire, acute, 1.0 cm long and 0.5 cm width, coriaceous, pubescent, green, pedicellate; bisexual, complete, regular, solid, soft, 4.0 cm long and 5.0 cm across, pentamerous, showy; epicalyx fused, 7-10 lobes, cup-shaped, persistent, lanceolate, entire, acute, outer surface green, inner surface whitish green, pubescent, 0.4 cm long and 0.3 cm width, valvate, inferior; sepals – 5, gamosepalous, 5 long lobes, inferior, rosaceous, valvate, each lobe 1.2 cm long and 0.7 cm width, entire, acute, pubescent, outer whitish green with rough surface, inner shiny with smooth surface persistent, alternate to corolla; petals - 5, polypetalous, twisted, slightly fused at base, 3.8 cm long and 2.5 cm width, obovate, wrinkled, curved, basal thick, whitish-yellow, herbaceous, crimson eye spot fused in base at inner side, glabrous, longitudinal, 10-12 yellow, white veins upward parallely, deciduous; androecium – stamens indefinite, monadelphous, filaments fused to form staminal column surrounding style, 1.8 cm long and 0.3 cm width, filamentous outgrowth, 0.25 cm long, terete, white, soft, glabrous, anther unilobed, kidney shaped, yellow, extrorse, basifixed; carpels – 5, syncarpous, ovary superior, globose, 0.5 cm long and 0.4 cm width, whitish green, glabrous, two locules, many ovules in each locule by series, axile placentation, style – 1, terminal, white, soft, glabrous, 2.0 cm long, 0.05 cm width, terete, stigma – 5, short, cylindrical, 0.5 cm long, pink yellow, gregarious, minute glands on stigma. Fruits capsule, 2.5 cm long and 1.5 cm width, globose, with 5-persistent calyx and epicalyx, dehisced into 5 parts, through 5 longitudinal sutures. Seeds numerous, black, flattened; seed germination is hypogeal. Flowering and fruiting mainly during November to March, but occasionally throughout the year. Distributed throughout the back mangals. Medicinally, the leaves are laxative and resolvent; infusion used as a vulnerary; mucilage of the bark is said to be useful in dysentery; roots are used for the preparation of embrocation for the treatment of rheumatism.

Photo Plate 91. *Hibiscus tortuosus* Roxb. - single branch with flowers

42. *Hibiscus tortuosus* Roxb. Hort. Beng. 51. 1814 & Fl. Ind. Iii: 92.1832.

Common Name : Coast Cotton Tree **Local Name** (Bengali): Bhola

Medium tree, up to 15.0 m high, deliquiscent branches; leaves simple, alternate, ovate, acute, green, pubescent, coriaceous, 1.5 cm long X 1.2 cm width, petiolate, petiole 5.5 cm long; cyme, pendunde having generally two flowers, one mature while other young, bracteate, bracts – 2 on both side, 1.5 cm long × 1.2 cm width, ovate, entire, acute, leather, brownish green, bracteate, pedicellate; petals 5, polypetalous, twisted, inferior; seeds black, flattened. Flowering and fruiting mainly during November to March, but occasionally throughout the year. Distributed throughout the coastal areas as back mangrove species.

43. *Thespesia populneoides* (Roxb.) Kostel, Allg. Med. Pharm. Fl. 5. 1861.

Common Name : Portiatree **Local Name** (Bengali): Paras

Shrub or small tree, grow naturally in the coastal areas; up to 7.0 m tall, much-branched, leaves ovate, entire, acuminate, petiole 6.0 - 10.0 cm long; flower solitary, axillary, 6.0 - 7.0 cm in diam., yellow, peduncle 8.0 - 10.0 cm long, capsule pyriform. Flowering and fruiting almost throughout the year.

Family – 17. Tiliaceae

Under the family - Tiliaceae, the only species *Brownlowia lanceolata* (L.) Kosterm. is collected and identified from the Indian Sundarbans. An uncommon or rare shrub, twining habit, attain up to the height 3.0 m, grow above tidal zones of Sundarbans mangal. Its leaves are looked very much similar to the leaves of *Heritiera fomes* Buch. - Ham.; for that locally it is known as 'Lata Sundri'. Aerial air roots and characteristic viviparous germination is absent,but this species can grow and survive in inter tidal mangrove habitats. This species is considered as the mangrove associate or back mangal in the Indian Sundarbans. The genus, *Brownlowia* Roxb. Pl. Corom.3 : 61, t. 265.1819 has 3 species, occasionally grow either in the swampy mangrove habitats or as terrestrial species. While the species *Brownlowia lanceolata* (L.) Kosterm. is collected from the Sundarbans mangal, the other species, *i.e., B. argentata* Kurz. and *B. tersa* Koestern. are reported from the mangals of S. E. Asian countries..

Photo Plate 92. *Brownlowia lanceolata* (L.) Kosterm. A, B & C. Branches with flowers and wrinkle fruits

44. *Brownlowia lanceolata* (L.) Kosterm

Local Name (Bengali): Lata Sundari

Twining, much branched shrub, attain the height up to 3.0 m and the twining twigs seem to be brownish; tap root system developed. Bark brownish black, smooth, leaf simple, alternate, exstipulate, petiolate; petiole 0.5 cm – 2.0 cm long, brown, terete, not pulvinous, rough surface, solid, lamina 10.0 – 14.0 cm long and 2.5 cm – 4.0 cm width, lanceolate, entire, acute, round, coriaceous, dorsiventral, dorsal surface green, glabrous, shiny, ventral surface pale green, leathery, unicostate reticulate venation. Mixed cyme, pendulous, pedunculate branched, ebracteate; flower-ebracteate, bisexual, complete, regular, pendulous, pedicellate; pedicel short - 0.15 cm long, terete, rough surface, brownish, flower 0.45 cm long and 0.4 cm across, not showy; sepals – 4, polysepalous, ovate, entire, blunt, curved, outer surface rough having dark brown spots, inner surface glabrous, coriaceous, 0.3 cm long and 0.2 cm width at base, brownish; petals – 6, polypetalous, obovate, entire, curved, angled, narrow venation towards apex, herbaceous, yellow, 0.45 cm long and 0.2 cm width at apex; stamens – indefinite, free, filament exserted, extrorse, glabrous, soft, terete, yellowish, basifixed, 0.2 cm long, anther – bilobed, 0.05 cm long, dark yellow, basifixed; carpels –3, apocarpous, ovary superior, one chambered, two ovules in each chamber, axile placentation, 0.05 cm long, style – terminal, 0.2 cm long, glabrous, terete, greenish stigma absent. Fruit capsule, small. Flowering and fruiting during June to September. Not much common, both in the inter-tidal and above tidal regions at Sundarbans mangrove forest, including TR.

Family – 18. Rutaceae

Under this family only one genus- *Atalantia* Corr. (= Syn. *Merope* Roem. Syn. Monogr. 1 : 44. 1846) occasionally grow in the tidal inundated mangrove habitats. This monotypic genus has the only species, *viz., Atalantia correa* M. Roem. [Syn.= *Merope angulata* (Willd.) Swingle], is occasionally found from the mangrove habitats in Sundarbans, as back mangrove species. This species grows in the inter-tidal zones but in very threatened condition. It looks very much similar to citrus.

Photo Plate 93. *Atalantia correa* M. Roem. A. Branch with flowers. B. Branch with fruit. C. Plant in the natural habitat. D. Three branches. E. Branch of *Atalantia correa* M. Roem.

45. *Atalantia correa* M. Roem. (= Merope angulata (Willd.) Swingle Bull.Wash.Bio. Soc.2: 22. 1915.

Common Name : Mangrove Lime **Local Name** (Bengali): Ban Lebu

Shrub, 2.5 m – 3.0 m tall, evergreen, much branched, stout swollen stem base; tap roots well developed, no aerial root, but trunk or stem base swollen, flat; bark blackish, longitudinally fissured; stem glabrous, shiny, green, round, solid, two spines at each node; lemon like scented; nodes not ridged. Leaves simple, alternate, stipulate (stipule as two spines), petiolate, petiole 0.3 cm long and 0.2 cm broad, green, glabrous, pulvinous, lamina lanceolate, oblong, entire, acute, dorsiventral, green, ventral side pale green, glabrous, coriaceous, 7.0 cm long and 3.0 cm - 4.0 width, aromatic, unicostate reticulate venation. Inflorescence solitary, flower- bisexual, complete, actinomorphic, medium , 1.8 cm long × 0.5 cm across, not showy, erect, pentamerous; sepals – 5, gamosepalous, green, 5-teeth above, cup shaped, 0.2 cm long and 0.25 cm across, valvate; petals –5, polypetalous, each petal lanceolate, white, fleshy, acute, entire, 0.9 cm long and 0.4 cm width, not showy, twisted; stamens – 10, in single whorl, filament long, upto 0.25 cm and 0.08 cm width, fleshy, white, one third of the total stamen, anther introse, inserted, longitudinal dehisced, white, 0.4 cm long and 0.1 cm width; carpels – syncarpy, ovary superior, triangular, green, 0.4 cm long, many ovules in each chamber, axile placentation; style – 1, terminal, solid, green, glabrous, 0.3 cm long, cylindrical, stigma absent. Fruit triangular, green, globose; seeds many; germination hypogeal. Flowering and fruiting October to March. This species grows occasionally in the back mangrove areas; occasionally inundated with the high tidal water. Essential oil from fruits said to be useful for the treatment of rheumatism. Distributed throughout the mangals of the S. E. Asian countries.

Family -19. Poaceae

(*nom. alt.* Gramineae)

In the Indian Sundarbans mangrove habitat there are several species under the family - Poaceae; out of these the most common halophytic grass species growing in the inter-tidal habitats are *Porteresia coarctata* (Roxb.) Takeoka, *Myriostachya wightiana* (Nees ex Steud) Hook. f., *Phragmites kakra* (Retz.) Trib. Ex Steud., and *Aeluropus lagopoides (L.)* Trin. Ex Thw.

Porteresia coarctata (Roxb.) Takeoka is the pioneer grass species in mangrove forest of Indian Sundarbans; this species looks very much similar to the paddy plant, *Oryza sativa* L.; so the synonym of this species is *Oryza coarctata* Roxb. Its caryopsis fruits and seeds looks similar to paddy and often eaten by the poor villagers of the rural Sundarbans. The other grass species in the Sundarbans mangrove habitats under the family Poaceae are important as natural fodder grass for the wild deer; the deer species, *Cervis axis* is common and play important role in the ecosystem. All these species under the family Poaceae also play important role as soil binder.

Table 51 : Key to the Species under the family - Poaceae, collected from the mangrove habitats in Sundarban

A	Common grass on newly silted up land; culms standing erect from a stout creeping rhizomes or root-stock branched polished, node swollen	1. *Porteresia coarctata* Takeoka
AA	Occasionally grow at coastal sandy soil, looks similar to *Cynadon* sp.	2.*Aeluropus lagopoides* Trin. ExThw.
AAA	Occasional grass on the newly silted up soil	See B & BB
B	Culms densely tufted, erect from a stout sheathed, sometime floating with lomg branching roots	3. *Myriostachya wigtiana* Hook. f.
BB	Culms densely tufted, creeping rootstock and with extravaginal young shoots	4. *Phragmites kakra* Trib. Ex Steud.

Photo Plate 94. *Porteresia coarctata* (Roxb.) Takeoka, Natural habitat in the river flat land with inflorescence.

Photo Plate - 95. *Porteresia coarctata* (Roxb.) Takeoka. A. Vegetative growth in the river flat at Sundarbans B. Enlarged view of single leaf with serrated margin. C. Single stalk of panicle inflorescence. D. Single stalk of panicle inflorescence, close view. E. Grass with mature inflorescence. F. Germinated seeds of *Porteresia coarctata* (Roxb.) Takeoka.

46. *Porteresia coarctata* (Roxb.) Takeoka, Bull. Nat. Sci. Mus. Tokyo 8 : 406. 1965.

Common Name : Wild Rice. **Local Name** (Bengali): Dhani Ghash

The first or the pioneer species in the mangrove succession of the inter-tidal river flat lands. Culm standing, erect from a stout creeping rhizome or root stock, internode 8.0 cm - 14.0 cm long, terete, greenish yellow, glabrous, nodes swollen. Rhizome or root stock much branched and polished, pseudo-tap roots formed in the loosely silted up soil layer, roots up to 1.5 cm long penetrate up to the deep consolidated soil; leaf-blade margin unequal, mid-vein obscure, coriaceous, linear acuminate, serrate or dented, leaf sheath long, undulately reticulate, polished, ligules very short, lamina up to 30.0 cm long, 1.5 cm broad, parallel venation, isobilateral. Spikelets 15.0 - 20.0 cm long, arranged on stem, stalked, each spikelet consisting of special bracts; the 1st and 2nd glumes do not subtend flower and 0.35 cm and 0.20 cm long, respectively, 3rd glume having flower, known as lemma, 1.3 cm long and 0.4 cm width, teeth, acuminate, hard, entire, rough outer surface, while inner surface glabrous, shiny, folded, slight longitudinal ridged row on outer surface; overlap the edges of the opposite bracteoles, known as palea; 1.0 cm long and 0.3 cm width, bifolded; upper rough surface, inner surface glabrous, shiny, hard, entire, acute; flower - hermaphrodite, zygomorphic, complete, erect, 1.3 cm long and 0.4 cm width; tepals are known as lodicules - 2, free, herbaceous, glabrous, white, soft, 0.15 cm long and 0.15 cm width, ovoid, entire, obtuse; stamens-6, free, filament 0.5 cm long, white, glabrous, soft, versatile, exerted, anther bi-lobed, 0.45 cm long, extrorse, yellow, longitudinal dehisced; carpel-1, ovary superior, conical flax shaped, glabrous, shiny, 1-chambered, white, 0.15 cm long, 1-ovuled, style 2, terminal white, soft, 0.1 cm long, stigma 2, 0.3 cm long, white, feather. Fruit caryopsis, 1.4 cm long and 0.45 cm width; seed brown coloured, thin 1.0 cm long. Flowering and fruiting period is from July to December. Ripe seeds are cooked like rice in the rural areas of the Indian Sundarbans and by the poor people.

Photo Plate 96. *Myriostachya wightiana* (Nees ex Steud) Hook. f. A. Single inflorescence and foliage. B. Tuft of the Grass, *Myriostachya wightiana* (Nees ex Steud) Hook. f.

47. *Aeluropus lagopoides* (L.) Trin. Ex Thw. Enum. Pl. Zwyl. 374.1864.

Common Name : Saline Gran **Local Name** (Bengali): Nona Durba

Prostrate, creeping with erect branches; frequently found in the sandy areas of coastal Sundarbans; stem solid, leafblades linear; panicles pinkish. Flowering and fruiting during March to July.

48. *Phragmites kakra* (Retz.) Trin. Ex Steud. Nom. (ed. 2) 2 : 324. 1814.

Common Name : Reed Gran **Local Name** (Bengali): Nalor Dharma

Culms erect, stout, creeping, found in the marshy places; leafblades flat, lanceolate, ligule narrow; inflorescence panicles, silky hair, spikelets bi-sexual, 3-4 fid, pedicel and glumes glabrous, fruit grain free, loosely enclosed by the lemna and palea, oblong semiterete, hilum oblong short. Flowering and fruiting during December to March.

49. *Myriostachya wightiana* (Nees ex Steud)Hook. f., Fl. Brit. India 7: 327. 1896.

Common Name : Saline Gran **Local Name** (Bengali): Nalai

Culms tufted, erect, stout, sheathed, sometimes floating with long branched, frequently found in the riverbeds of Sundarbans mangal mixed with *P. coarctata;* roots smooth, simple, leafblade broad, serrated, finely acuminate, sheath very long, inflorescence panicles, rachis smooth, crowded, fascicled and whorled flowering, spikelets 4 - 8 flowered, compressed, pedicels short; fruit grains obliquely ovoid, sub-terete. Flowering and fruiting during June to December.

Family – 20. Verbenaceae

The family - Verbenaceae has two mangrove habitat genera and under these two genera there are three species, collected from the Indian Sundarbans mangal. These are *Clerodendron inerme* (L.) Gaertn., *Clerodendron neriifolium* var. *macroparpa* and *Premna corymbosa* (Burm. f.) Rottl. & Wild. All these three species grow within the inter-tidal zones, though these species have no specific morphological features likewise mangroves. The species, *Clerodendron inerme* (L.) Gaertn. is a common straggling much branched shrub, up to 3.0 m high and terminal branches very often slender twiggy, dark evergreen shrub; flowers white with long styles and stigma; it forms dense bush on the river banks and river side thickets; grows in dense patches, occasionally inundated during high tide. This species very often planted as hedge plant in garden periphery. Other species, *Clerodendron neriifolium* var. *macrocarpa* is also collected and identified for the first time from the Indian Sundarbans.

Table 52 : Key to the Genera under the family-Verbenaceae of the inter-tidal habitats in Sundarbans

A	Drupe with one 4-celled pyrene; in the river side thickets	1. *Premna* L
AA	Drupe with four 1-seeded pyrenes; leaves simple, lyx campanulate or tubular	2. *Clerodendron* L.

Genus - *Clerodendron* L.

The Genus - *Clerodendron* L. has two species in the inter-tidal mangrove habitat at Indian Sundarbans; though these species grow and adapt best in these mangrove habitats, these are not considered as mangrove species, but as mangrove associates. These species can be identified seperately with the following characters:

Table 53 : Key to the Species under the Genus-*Clerodendron* L. from the inter-tidal habitats in Indian Sundarbans

A	Common on the river bank; flowers white; fruits upto 1.5 cm diam; green with 4 seeds	1. *C. inerme* (L.) Gaertn.
AA	Very rare and occasional, reported for the 1st time from the Indian Sundarbans mangal; fruits 2.5 cm diam or more, green with 4 seeds	2. *C. neriifolium* var. *macrocarpum*

Photo Plate 97. *Clerodendron neriifolium* var. *macrocarpum* L. A. Plant in the natural habitat at the mangrove forest of Sundarbans Tiger Reserve. B. Close view of the branches. C. Close view of the fruits of *Clerodendron neriifolium* var. *macrocarpum* L.

50. *Clerodendron neriifolium* var. *macrocarpum* L.

(Syn. = *C. macrocarpum* Kurz.)

Perennial woody shrub, occasionally in the inter-tidal zone at Sundarbans mangal; attain 2.0-3.0 m high. Stem tetrangular, solid, slightly swollen at nodes, glabrous; barks greyish-white. Leaves mostly opposite, simple, lanceolate, rare often opposite, simple, lanceolate to oblong - lanceolate, often elliptic, variable, 7.0-10.0 cm long, 1.5-2.0 cm broad, acute to acuminate, entire, cuneate at base, thick-coriaceous, vein reticulate with 5-7 pairs of alternate or opposite secondaries, glabrous, deep-green above, dull-green below, petiolate; petiole slender, 1.0 - 1.2 cm long, glabrous. Flowers 5-12 in axillary or terminal dichasial cymes, peduncles tetrangular, 5.0-10.0 cm long, glabrous, bracteate, pedicellate. Calyx cupular-campanulate, 5.0-8.0 mm long, glabrous, persistant; lobes triangular-ovate, variable in size; corolla tubular, lobes - 5; stamens 4, long filaments. Fruits oval, 1.5 cm - 2.0 cm, lobed, green.

Photo Plate 98. *Clerodendron inerme* (L.) Gaertn. A. Plant in the natural habitat. B. Close view of the flowers on twig. C. Flowers on the branches. D. Foliages and flowers in natural habitat.

51. *Clerodendron inerme* (L.) Gaertn., Fruct. 1 : 271. 1788.

Common Name : Hedge plant **Local Name** (Bengali): Ban-Jnui or Batraj or Ban-Jhampi

Straggling much branched shrub, up to 3 m long and terminal branch very often twining, slender, twiggy, dark evergreen and form dense bush in the river banks and river slopes, occasionally inundated with the saline tidal water. Stem - internodes variable in length, 2.0 cm - 8.0 cm long, terete, glabrous, yellowish green, soild, up to 0.3 cm diam., nodes slightly ridged; leaves simple, opposite decussate, cauline, exstipulate, petiole - 0.6 cm - 1.0 cm long, terrete, glabrous, solid, soft, brownish green, pulvinous, lamina ovate, elliptic, entire, acute, angled, dorsiventral, green, fibrous, herbaceous, 3.5 cm - 4.5 cm long and 1.5 cm - 2.0 cm broad, unicostate reticulate veination. Inflorescence cymes, axillary pedunculate; flower - 5.6 cm long X 1.0 cm across, ebracteate, pedicellate, pedicel up to 1.0 cm long, terete, soft, greenish white, solid, glabrous; complete, bisexual, zygomorphic, erect, odourless, medium hypogynous; sepals - 5, gamosepalous, 5 short lobes above, 0.15 cm long, entire, beneath cup shaped tube, 0.5 cm long, green, glabrous, thin, persistent, valvate; corolla: petals-5, gamopetalous; 5 petal lobes above, equal length, imbricate, each lobe 0.5 cm long and 0.35 cm broad, elliptic, entire, acute, herbaceous, white, glabrous, beneath cylindrical, long tube attaining 2.0 cm height and 0.2 cm diam., glabrous, white; androecium - stamens - 4, epipetalous, anthers remain free, filament long, 3.5 cm long, white, soft, terete, glabrous, exserted, extrorse, dorsiventral, anther bilobed, longitudinal dehisced; gynoecium - carpels 2, syncarpy, ovary 0.2 cm long, cylindrical glabrous, greenish, one chambered. Fruit pyriform drupe, woody, 1.5 cm - 2.0 cm long; seeds four in each fruit. Flowering and fruiting almost throughout the year. Medicinally, the leaves are used as febrifuge; it contain an amorphous bitter principle.

Family – 21. Tamaricaceae

The family - Tamaricaceae has 4 genera and about 120 species, distributed throughout the tropical, sub-tropical and temperate zones of the Globe. The Genus – *Tamarix* L. Sp. Pl. 279. 1753 has about 90 species and only three species are distributed in the coastal, above tidal zones; these species are *T. troupii* Hole (Syn = *T. gallica* L.), *T. dioica* Roxb.and *T. aphylla* (L.) Kanza. In the Sundarbans coastal areas these species, viz., *T. troupii* Hole and *T. dioica* are most common.

These species were collected and reported by Naskar & Guha Bakshi (1987), Naskar (1993) and Naskar & Mandal (1999). *Tamarix* spp. grow on the river bank flat land in the saline soil; though, these species are not mangroves; these are considered as mangrove associates, mangrove weeds or as back mangals. Members under the genus-*Tamarix* L. are mostly shrub to small tree with dense leaves and wide spreading pendulous branches, dark green and showy.

Table 54 : Key to the Species under the Genus - *Tamarix* L.

A	Flowers monoecious in short close, cylindric spikes	1. *T. dioica* Roxb
AA	Flowers bi-sexual	See - B & B
B	Branchlets articulate at base of sheath, slender, cylindric, jointed; leaves reduced to small sheath with minute tooth; inflorescence in long slender spikes, sometime in loose panicles	2. *T. aphylla* (L.) Kanz
BB	Branches pendulous, slender articulated; leaves minute or scale like, acute, subulate with triangular semiamplexicaul base; panicled, speciform raceme	3. *T. troupii* Hole

52. *Tamarix dioica* Roxb. Hort. Beng. 22.1814 *nom. nud.* & Fl. Ind. 2 : 101.1824.

Common Name : French tamarisk **Local Name** (Bengali): Lal-Jhau

Shrub to small tree, 3.0 - 5.0 m tall with drooping branches; leaves minute, adpressed, sheathing; sheath tubular, apex acuminate. Flowers monoecious in short close, compact cylindric, loosely panicled spikes, purple or pink. Capsules oblong, tapering. Flowering and fruiting during the month July to October. Frequently found on the river banks and the edges of marshes, well adapted to the sandy and salty areas. Bark is bitter and astringent and has tonic properties; occasionally employed to cure skin diseases.

53. *Tamarix aphylla* (L.) Kanza, Boll. Orte. Bot. Palermo 8 : 82.1909.

Common Name : Athel tamarisk **Local Name** (Bengali): Lal-Jhau

Medium tree, well adapted in sandy and salty areas on the river banks; branches articulated at base of sheath, often greyish, fastigate, slender, cylindric, jointed - 5.0 cm long or more. Leaves reduced to small sheath with a minute tooth. Inflorescence in long slender spikes, sometime in loose panicles; flowers pink, 0.25 cm diam., bisexual or monoecious, sessile, scattered on long slender spikes, sometime in loose panicles at the ends of branches; stamens 5, inserted in alternate notches of 10-lobed disk; capsule 0.25 cm long. Flowering and fruiting during the month of July to October. Bark bitter and astringent and has tonic properties, useful for skin diseases.

Photo Plate 99. *Tamarix troupii* Hole. A. & B. Branches with white cat-kin inflorescence.

54. *Tamarix troupii* Hole, Ind. For. 45: 248.1919.

(Syn. *Tamarix gallica* L. Sp. Pl. 270.1753).

Common Name : French tamarisk; **Local Name** : (Bengali): Nona Jhau

Shrub with deliquiscent branching, a halophytic shrub, distributed throughout the mangrove reclaimed areas, in the river banks, river flat lands, more dominant in sandy areas; occasionally small tree, up to 8 m high but sometime bushy appearance; nodes and internodes not distinctly distinguished, presence of leaf scars; older trunks black brown, longitudinally fissured, while in young - twiggy shoot covered with leaves or leaf scars; woody, solid, terete, pubescent, green. Leaves simple, exstipulate, opposite, decussate, sessile, lamina, triangular, 0.15 cm long and 0.15 cm width at base, green, fleshy, thick, entire, tapering end, covered. Catkins, peduncle pendulous, 12.0 cm long, branches, bracteate, bracts acicular, 0.15 cm long, tapering long end; flower - pink violet, complete, bisexual, regular, pedicellate; pedicel short, 0.05 cm long, terete, soft, pink violet, pubescent, erect, 0.10 cm long and 0.05 cm across, hypogynous; sepals - 5, gamosepalous, cup-shaped inferior, valvate, 0.05 cm long X 0.10 cm across; petals - 5, apical acicular lobe, coriaceous, glabrous, pink red, persistent, valvate, inferior, 0.05 cm long, herbaceous, thin white ovate elliptic, entire, blunt, equal length in one whorl, deciduous, not showy; stamens -5, free, filament white, 0.05 cm long, soft, terete, glabrous, basifixed, inserted, anther bilobed, sagittate, introrse, 0.04 cm long, white greenish, inferior; carpels - 3, syncarpous, ovary superior, 0.05 cm long,1 chambered, basal placentation, style short, stigma short, globose, glabrous. Capsule gradually tapering, 0.35 cm long with persistent calyx, red violet coloured, dehiscent into 3 parts. Gregarious growth of new seedling plants were seen on the inter-tidal mud-flat lands at the coastal phases, where sand is dominant in the soil substrate. These species are not the mangroves, but grow as mangrove weed in mangrove reclaimed land. Flowering and fruiting are generally from November to March. Wood and twigs are much used by the villagers as fuel wood. Stem and leaves have higher percentage of tannin; useful in skin diseases, diarrhoea and dysentery.

Family - 22. Solanaceae

The type genus - *Solanum* L. Sp. Pl. 187.1753 has two species, *viz.*, *Solanum trilobatum* L. and *S. surattense* Burm.f.; these species were collected from the above tidal zones and from the slightly raised mangrove thickets. *S. trilobatum* is the twining shrub in the mangrove thickets; while *S. surattense* is erect shrub, normally grow on the sandy beaches in coastal areas. These species are not the mangroves or the mangrove associates, but commonly grow in the mangrove zones along with other mangroves and non-mangrove coastal species.

Table 55 : Key to the Species under the Genus - *Solanum* L.

A	Weak stemmed, diffused herb, grow on the sandy beach area	1. *Solanum trilobatum* L.
AA	Scandent or trailling shrub on the thickets on back mangal	2. *Solanum surattense* Burm.f.

55. *Solanum surattense* Burm. f. Fl. Ind. 57, 1768.

Common Name : Yellow-berried; **Local Name** (Bengali): Kanti Kari

Perennial shrub, 1.0 m high, on the coastal sandy soil; branches numerous, densely clothed with stellate tomentum; prickles 1.0 - 1.2 cm, straight, yellow; leaves up to 10.0 × 6.0 cm, ovate or elliptic, sinuate or sub-pinnatifid, obtuse or sub-acute; petioles 2.0 - 3.5 cm long; berry 1.2 - 1.6 cm diam., glabrous; seeds up to 2.0 cm diam., glabrous. Flowering and fruiting during the months May to August.

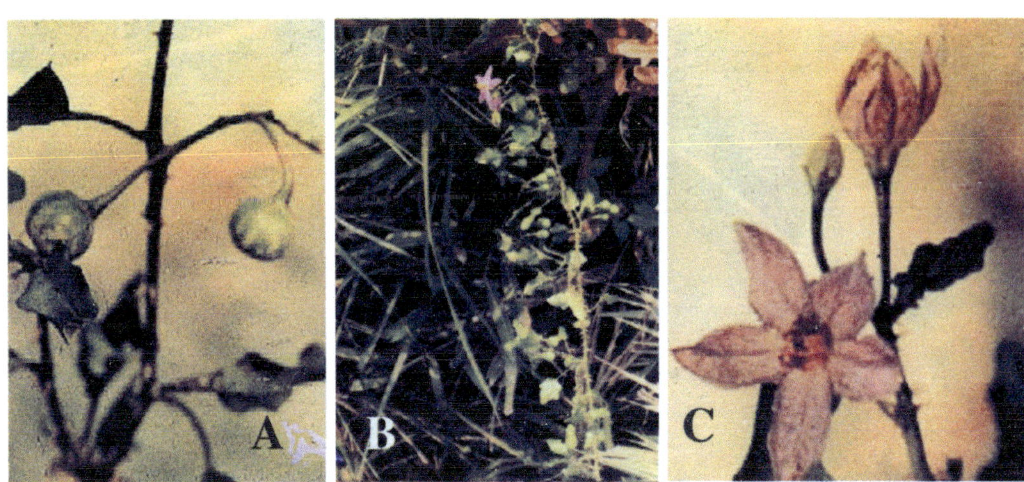

Photo Plate 100. *Solanum trilobatum* L. A. Twig with fruits. B. Natural habitat in the thickets of the river bank. C. Close view of flowers.

56. *Solanum trilobatum* L. Sp. Pl. 188.1753.

Common Name : Twining Brinjal; **Local Name** (Bengali): Lata Gurbegun

Trailing shrub with hooked prickles, weak branches, grow on bushy dry soil; nodes and internodes rigid; internodes covered by infrequently flattened prickles, gradually tapering towards end, yellowish internodes terete, dark violet, green, glabrous, solid, one side green and other side dark violet-green, woody; leaves simple, exstipulate, alternate, petiolate; petiole 1.0 cm long and 0.1 cm broad, terete, dark violet, glabrous, soft, solid, pulvinous, covered infrequently by prickles, lamina trilobed but not distinctly serrate, coriaceous, 3.5 cm long × 1.0 cm - 2.0 cm broad, entire, round, dorsiventral, dorsal surface dark violet, ventral surface green, unicostate, reticulate venation conspicuous. Corymbose raceme; flower - ebracteate, bisexual, regular, complete, pedicellate, pedicel 0.5 cm - 1.5 cm long, terete, solid, soft, pubescent, violet green, erect, medium, 2.5 cm long and 1.5 cm across, showy, hypogynous; sepals - 5, polysepalous, inferior, ovate, entire, acute, fleshy, pubescent outer surface, inner surface shiny green, 0.2 cm long and 0.08 cm broad, persistent, rosaceous, imbricate; petals - 5, gamopetalous, inferior, bell-shaped, ovate, entire, acute, thin, herbaceous, violet, distinct venation on petals, slightly adhere at base, imbricate, showy, 1.0 cm long and 0.5 cm broad; stamens - 5, epipetalous, filament 0.1 cm long, white, soft, glabrous, basifixed, anther 0.4 cm long and 0.2 cm broad, yellow, extrorse, inserted, globose, inferior; carpels - 2, syncarpous, superior, short cylindrical, 0.2 cm long and 0.1 cm diam., white, glabrous, fleshy, 4 lobes, axile placentation, style -1, terminal, terete, soft, glabrous, 1.0 cm long, stigma absent. Fruits berry, globose, 1.0 cm diam, with persistent calyx, seeds numerous, flat, yellowish, albuminous. Flowering and fruiting during October to March.

Family -23. Sapotaceae

The family- Sapotaceae has only one species, *viz.* *Manilkara hexandra* (Roxb) Dub. grow in some restricted area at Netidhopani within S.T.R. and other above tidal zones in Indian Sundarbans. This species is not mangrove and not considered as mangrove associate or backmangal, but its presence and naturaL growth beyond this inter tidal area is a big question about its existence and distribution. It may be introduced in this area by the old civilised people or by natural way it grow in this remote mangrove forest zone. A tall to medium sized evergreen tree with dense leafy crown.

Photo Plate 101. *Manilkara hexandra* (Roxb.) Dub. - Close view of the branch with foliage and flower stock.

57. *Manilkara hexandra* (Roxb.) Dub. Ann. Mus. Col. Marseille (ser. 3) 3 : 9.1915.

Common Name : Milk-tree; **Local Name** (Bengali): Bilati Bakul

Tall evergreen tree with dense leafy crown, up to 25.0 m high, beyond the frequent tidal inundated zones in the core areas at Sundarbans Tiger Reserve, *viz.* Netidhopani Forest Block. This species is purely non mangrove plant of the north Indian evergreen forest, but its existence in the mangrove forest zones of the STR is interesting. It is assumed that this species was introduced in the early period or may be transmitted by the natural ways by the birds or other animals. Flowering and fruiting November to March. Bark is febrifuge; seed produce edible oil. Wood may be used for carpentary works and for poles.

Family - 24. Sapindaceae

The species, *Dodonaea viscosa* (L.) Jacq. under the family-Sapindaceae occasionally grows on the sandy beach area or coastal zone at Sundarbans Tiger Reserve. This species was not reported earlier.

Photo Plate 102. *Dodonaea viscosa* (L). Jack. A. & B. Twigs/ branches with flowers and fruits

58. *Dodonaea viscosa* (L.) Jacq. Enum. Syst. Pl. Ca. 10. 1760.

(Syn. = *Ptelea viscosa* (Jacq.) L. Sp. Pl. (ed. 2) : 173. 1962, *non* Linn. 1753).

Common Name : Jamaica switch sorrel; **Local Name** (Bengali): Bilati Mehendi

Evergreen woody shrub or small tree with very dense branching of about 4.0 m - 5.0 m high; grow on the supra-tidal sea beach or coastal areas, occasionally inundated during high tides; stem woody, round, back yellowish to light brown colour; leaves elliptic to elliptic lanceolate, oblanceolate, base cuneate, attenuate, exstipulate, simple, petiolate, dorsiventral, glabrous, length 8.0 cm - 10.0 cm, breadth 2.0 cm - 3.0 cm, alternate, petiole very short, 2.0 mm - 3.0 mm long, margin entire, tip acute green. Inflorescence terminal with many flowers; flowers - greenish yellow, in short terminal panicles with raceme branches, ebracteate, 10.0 mm long, pedicellate, pedicel long, around 7.5 mm, perianth, epigynous; numbers of lobes varies from 3 - 5, green sometime 2-3 lobes united with each other to form a large lobe. Fruit capsule with thin wing, wing not the extension of perianth, yellowish brown, seed black. Flowering and fruiting during November to March. Leaves said to be useful for the treatment of wounds, swelling and burns.

Notes : Prain (1903) has reported this species from (Tirhut) Tripura, Chhota Nagpur, N. Bengal and Chittangong; A. K. Mukherjee has collected this taxon from the Dhablat/ Sagar Island, Sundarbans area on 01.05.1967 (CAL-5944). But, till date no mention was made for its distribution from the Sundarbans Tiger Reserve, facing the coastal areas. Field collections of this species was also made by the present author from the Radhanagar Beach (Havlok Island) of the Andamans on 10th February, 2001. It is very interesting to note that this species is distributed in the saline habitats of the coastal areas and also from the hilly regions of North Bengal, Tripura, *etc*.

Family – 25. Amaryllidaceae

The Genus - *Crinum* L. Gen. Ed. I : 97.1757, belongs to the family Amaryllidaceae, which have about 110 species; but only one species, *viz., C. defixum* Ker - Gawler grow within the inter-tidal zones, specially in the less saline habitat in the Sundarbans. *Crinum defixum* Ker-Gawler is not the mangrove or mangrove associate species, but in the inter-tidal zone of Ghusighata area of Sundarbans this species grow on the river flats and inter- tidal zone. It is the stout, bulbous herb, attain the height up to 1.5 m, roots fibrous and without any aerial out growth; rhizomatous bulb, ovoid or globose; leaves simple, lamina up to 1.0 cm long; inflorescence umbel, scape axillary, erect cylindrical, up to 1.0 m long and 0.3 cm broad; flowers white, ebracteate.

Photo Photo 103. *Crinum defixum* Ker-Gawler – mixed with *Cryptocoryne ciliata* (Roxb.) Fischy. Ex Wydler. A. Dense growth on the tidal river-flat . B. On the tidal river-flat also showing the flowers, at Ghusighata areas of Sundarbans, India.

Photo Plate 104. *Crinum defixum* Kar - Gawler. A. Natural habitat along with close view of single plant. B. Close view of flowering plant. C. Natural habitat in the Kulti River bed, Sundarbans.

59. *Crinum defixum* Kar - Gawler, Quart J. Sci. 3 : 105.1817.

Common Name : Mangrove Lili **Local Name** (Bengali): Sukha Darsan

Stout bulbous herb, attain the height up to 1.0 m; fibrous roots, without any aerial out growth; rhizome bulb, ovoid or globose, caudex; lamina simple rosette, up to 1.0 m long and 12.0 cm breadth, slightly sheathing at base, linear, entire, acute, adaxially convex, dark green, fleshy, glabrous, iso-bilateral, unicostate. Inflorescence - umbel, scape axillary, erect, cylindrical, up to 1.0 m long and 3.0 cm broad, tri-angular, fleshy, solid, spongy, glabrous, green, 5-10 flowers in each peduncle; flower ebracteate, bisexual, regular, erect, epigynous, up to 5.5 cm long and 5.0 cm across, showy; tepals - 5, gamotepalous, 5 lobes above, linear, mucronate, herbaceous, imbricate, 3.0 cm long and 0.3 cm broad, white, convex, slightly folded, glabrous, deciduous, beneath fuse to form hollow tube, fleshy, spongy, up to 1.8 cm long and 4.0 cm width; stamens - 6, free, filaments jointed at base by white thin layer, filament up to 2.2 cm long and 0.1 cm width, terete soft, glabrous, white mid position curved, anther bilobed, 0.15 cm width and 1.8 cm long, dark yellow with mid-narrow evergreen linear line, exserted, extrorse; carpels - 2, syncarpy, ovary inferior, globose, 0.5 cm long, 4-chambered, 1-ovule in each chamber, axile placentation; style -1, terminal, up to 4.5 cm long, terete, soft, glabrous, white below, navy green upper, stigma-1, globose. Fruit sub-globose; seed 1-2, rugose. Flowering and fruiting during June to September. Grow on the silted up sedimented soil in the inter-tidal river flats and river banks at low saline zone. Medicinally used as diaphoretic, emollient and used for treatment of burns and curbuncles.

Family -26. Araceae

Single species, *viz., Cryptocoryne ciliata* (Roxb.) Fischy. Ex Wydler was collected from the inter-tidal river flat lands in the inner estuary of the Sundarbans, where the salinity in soil and water phases are less.

Photo Plate 105. *Cryptocoryne ciliata* (Roxb.) Fischy. Ex Wydler- in the natural habitat of inter-tidal river slope.

60. *Cryptocoryne ciliata* (Roxb.) Fischy. Ex Wydler, Linnaea 5 : 428. 1830.

Local Name (Bengali): Kerali

Erect herb with terranean stolons on the inter-tidal river-flats or canal sides, in the marshy or moist localities. Leaves linear - oblong or linear - lanceolate; spathe axillary, cuspidate with purple fimbriate margins; ovary 5 - 7; fruits syncarpous, 5 - 7 loculed with 6 - 8 seeds in each locules, peduncle stout. Flowering and fruiting during May to September. Occasionally grow on the water-side on the river flat tidal inundated zones; collected from the river side at Kulti, Sundarbans.

Family – 27. Aizoaceae

(*nom. alt.* Ficoidae Benth & Hook.f.)

The family - Aizoaceae has about 130 genera, including 1,200 species. Out of these the Genus - *Sesuvium* L. Syst. Ex. X. 1058. 1759 is the saline habitat and under this the species *Sesuvium portulacastrum* L. grow on the occasionally tidal inundated coastal sandy areas. It is a prostrate or creeping herb, creeping on the muddy, bare flat land of saline coasts, frequently on the saline river flat lands.

Photo Plate 106. *Sesuvium portulacastrum* L. A. Natural habitat in dry coastal area - on dry river bed. B. Close view of foliages and fruit. C. Close view of a flower and twigs

61. *Sesuvium portulacastrun* L. Syst.(ed. 10) 1058. 1759.

Common Name : Salt wort **Local Name** (Bengali): Gada Bani.

Decumbent or ascending, much branched and often ceptitose herb, 10 - 35 cm high; stem creeping, rooting from the nodes, terete, succulent, red shining glabrous. Leaves opposite, sessile, linear - spathulate, apex narrowed obtuse or acute, entire, glabrous, dark green, thick fleshy. Flower rather inconspicous, solitary regular, glabrous, bisexual; perianth 5 - partite. thick, yellowish green. Fruits dry, scarious, seeds numerous, black. Flowering and fruiting during October to April. Common on the muddy river banks of the Sundarbans area. Occasionally edible.

Family - 28. Barringtoniaceae

(nom. alt. Lecythidaceae)

Two species under this family are *Barringtonia asiatica* (L.) Kurz. and *B. racemosa* (L.) Blume, reported from the coastal areas in the tropical Asian regions, as back mangrove species. These two species can be differentiated by the following key characters (Tomlinson, 1986) :

Table 56 : Key to the Species under the Genus - *Barringtonia* L.

A	Leaves with entire margin; erect terminal raceme with few flowers, pedicels 5.0 cm - 9.0 cm and flowers white, petals 6.0 cm - 8.0 cm filaments with white, red tipped; fruit cubical 10.0 cm in diameter	1. *B. asiatica* (L.) Kurz.
AA	Leaves with toothed margin; racemes from the axile of fallen leaves or older wood, pendulous long; many flowered, pedicels at most 5.0 cm long; flowers pink, petals at most 3.0 cm long, filament red, white or yellow tipped; fruit narrow, longer than wide, 5.0 cm - 7.0 cm X 3.0 cm	2. *B. racemosa* (L.) Blume

62. *Barringtonia asiatica* (L.) Kurz. J. Asiat. Soc. Bengal 45 : 131. 1876.

Common Name : Freshwater Mangrove **Local Name** (Bengali): Hijal

Medium sized, deliquiscent, much branched; grow in the river banks of the Sundarbans and also grow throughout the coastal areas; with buttress and dense crown; bark rough, thick in mature tree; leaves dense in terminal part of branches, lamina oblong-elliptic, obovate, fleshy petiole 30.0 cm - 40.0 cm × 12.0 cm 30.0 cm, apex rounded, base slightly auriculate, petiole short, stout. Inflorescence racemes with pink coloured or cream coloured flower, pendulous, flowers in short erect terminal raceme, showy, with 4 white petals, up to 10.0 cm diam; fruits dumble shaped, quadrangular with 2 persistent calyx lobes. Flowering and fruiting during May to September. Root bark is important as fish poison and seed powder is said to be used as an emetic and expectorant.

63. *Barringtonia racemosa* (L.) Spreng. Syst. Veg. 3 : 127. 1826.

Common Name : Freshwater Mangrove **Local Name** (Bengali): Sumudra

Medium sized tree, up to 10.0 m tall, grow occasionally on the river banks in Sundarbans; bark grey, leaves obovate-lanceolate, 20.0 - 25.0 cm long × 6.0 - 8.0 cm width, elliptic - oblong, glabrous, membranous, main costa 10-20 pairs. Flowers on slender pendulous spikes 40.0 cm - 50.0 cm long; flowers up to 2.0 cm long with a slender filiform style, cream coloured with pendulous terminal raceme; stamens numerous, filaments exerted, fruits rounded, narrow, 4 - angled or grooved, seeds solitary. Flowering and fruiting during April to October. Seed powder contain saponin and useful as fish toxicant. Distributed in the Indo-Malaya to Polynesia - occurs in the coastal areas, edges of the mangrove swamps. Root bark is important as fish poison and seed powder is said to be used as an emetic.

Family – 29. Boraginaceae

The family - Boraginaceae has only one halophytic species, *viz., Heliotropium curassavicum* L. and this herbaceous species is common in the Indian Sundarbans. Under the Genus- *Heliotropium* L. Sp. Pl. 130. 1753 there are several other mesophytic habitats species. The species *Heliotropium curassavicum* L., is a prostrate, much branched and spreading delicate week stemmed semi-fleshy herb; frequently grow on the salt crusted and high saline zone of the above tidal mangrove reclaimed areas and river bank or river flats. Its tap-root is much branched and deep rooted, about 0.3m depth.

Photo Photo 107. A. & B. *Heliotropium curassavicum* L. A. Natural habitat. B. Close view of flowering twig.

64. *Heliotropium curassavicum* L. Sp. Pl. 130. 1753.

Local Name (Bengali): Nona-Hatisnur

(Chromosome Number : 2n = 24 (Subramanuam & Kamble, 1967); 2n = 26 (Heiser & Whiteker, 1948; Britton, 1951 and Sarkar *et al.*, 1973) and 2n = 28 (Pal, 1957).

Prostrate, much branched and spreading succulent herb, on maturity form sub-erect branches; branches and foliages are dark-bluish, on the salt crusted / high saline clay-silty soil, beyond the frequent tidal inundated zone. Tap root much branched, deep rooted, about 0.3 m depth. Nodes swollen profusely branched and slightly swollen or flattened internodes or round, fleshy, soft, solid, pubescent, whitish green stem; length of internodes depends on the nutrient availability, moisture content and saline nature of the substratum; leaves simple, exstipulate, opposite decussate, sessile, 2.0 cm long, 0.5 cm broad and 0.1 cm - 0.3 cm thick, lanceolate, entire, blunt, succulent, glabrous, isobilateral, dark sky colour or blunt, unicostate reticulate venation inconspicuously. Inflorescence unipodial helicoid cyme, peduncle terminally bifurcate, up to 20.0 cm long; flower - ebracteate, bisexual, regular, complete, sessile, erect, small, 0.3 cm long and 0.25 cm across; sepals - 5, polysepalous, inferior; petals - 5, gamopetalous, rosaceous, short hollow cylindrical tube below; stamens - 5, epipetalous, inferior, filament very short, anther bilobed, introrse, sagittate; carpels - 2, syncarpous, ovary superior, shortly cylindrical; style - 1, terminal, 0.06 cm long, gradually tapering end, stigma absent. Fruits capsules, 0.1 cm long and 0.1 cm broad, green with persistent calyx. Flowering and fruiting during September to December. Medicinally, the root powder is said to be important for sores and wounds.

Family – 30. Fabaceae

(*nom. alt.* Leguminosae)

Under the family-Leguminosae, two sub- families, *viz.*, Papilionaceae and Caesalpiniceae are present in the mangrove habitats of Indian Sundarbans. Among these large number of genera and their species, only few species grow naturally in the halophytic condition as back mangrove species or mangrove associates. In the Sundarbans mangal the following species under this family are *Cynometra ramiflora* L., *Dalbergia spinosa* Roxb., *Derris scandens* Benth., *D. trifoliata* Lour., *D. indica* (Lamk.) Bennet, *Caesalpinia crista* L., *C. bonduc* (L.) Roxb.

Table 57 : Key to the Different Genera under the Family Fabaceae

A	Medium tree of the inter-tidal forest, leaves unijugate or bijugate, small bud scales, 4-calyx segments, 5 petals, stamens 10, 1-seeded fruit	*Cynometra* L.
AA	Climbing shrub, small to medium tree; leaves bipinnate or unipinnate leaves	See - B & BB
B	Climbing shrubs; leaves bipinnate; flower caesalpinous, hermaphrodite or unisexual, absence of vaxillary aestivation; pods flattened, ovoid-elliptic, terminated into long spine like beak	*Caesalpinia* L.
BB	Leaves unipinnate, flowers papilionaceous, hermaphrodite, presence of vaxillary aestivation; pods flattened orbicular, not terminated into beaks	See - C & CC
C	No stipules or spines; leaves up to 12.0 cm long and 7.5 cm width	*Derris* Lour.
CC	Nodes having spine like stipules; leaflets 1.2 and 0.4 cm width	*Dalbergia* Linn.f.

Photo Plate 108. *Cynometra ramiflora* L. A. Flowering branch. B. Flowering branch. C. Close view of branch.

65. *Cynometra ramiflora* L. Sp. Pl. 382. 1753.

Common Name : Wrinkle Mangrove **Local Name** (Bengali): Shinger

Small tree, attain the height up to 5.0 m – 7.0 m, deliquiscent branched, evergreen perennial. Leaves uni-jugate, compound, exstipulate, paripinnate, glabrous; leaflets 2.5 cm – 4.0 cm long and 1.0 cm – 1.5 cm width, lower pairs of leaflets comparatively small, mid costa obliquely placed, elliptic, entire, obtuse, angular, coriaceous, dorsiventral, glabrous, green, unicostate reticulate venation. Flowers 3.5 cm diam, in axillary racemes, reddish or purple; pods - 5–10 in a raceme, 2.5 cm long, elliptic-oblique, wrinkled and 1-seeded; seed flat, smooth. Flowering and fruiting time is October to March. Distributed in the mangrove forests of STR. This species is considered as back mangrove; commonly grow in the ridge forests of the inter-tidal zones. Wood of this full grown tree has timber value; root decoction is purgative; seed oil is said to be used for leprosy, scabies and other cutaneous diseases.

Photo Plate 109. ***Dalbergia spinosa*** **Roxb. A. Plant in the Natural habitat. B. Branches showing the flowers and pod**

66. *Dalbergia spinosa* Roxb. Fl. Ind. 3 : 223.1832.

Common Name : Nona-Sisoo **Local Name** (Bengali): Chulia Kanta

Much branched shrub or small tree with twining habitats; attain a height up to 8.0 m; tap root deep sunken but without any aerial roots; occasionally grow in the mangrove forest within the frequently inundated zones; but this species is mostly considered as back mangroves. Leaves compound, alternate stipulate, paripinnate, leaf axis up to 25.0 cm long, leaflets up to 8.0 cm long, each with gregarious spines, axis glabrous, terete, solid; dark green, axis glabrous, short petiolate; petiole 0.1 cm long, 0.05 cm width, glabrous, pulvinous. Raceme lateral or axillary, up to 20.0 cm long, peduncle, solid, green; flower pedicellate, uni-sexual, irregular, hypogynous, showy, bracteate; male flower 1.0 cm long X 1.8 cm across; sepals – 5, polysepalous; ovate- lanceolate, entire, pubescent, blunt, tufted, curved, up to 1.0 cm long and 0.5 cm width; petals – 5, polypetalous, imbricate, dark yellow, oblong, 1.5 cm long and 0.6 cm width, alternate to sepals; stamens 10, free, filaments hairy, terete, soft, solid, up to 0.8 cm long, dorsifixed, exerted, anther bilobed, extrorse; female flower by abortion (0) - carpel 2, bisexual flower syncarpy, superior, ovary dumbell shaped, 0.5 cm long and 1.1 cm diam., pubescent, greenish, single chambered, two ovules, parietal placentation, style – 1, 0.4 cm long, short terminal, stigma – 1, capitate glabrous. Fruit legume, flattened, epicarp woody, having gregarious spines, 5.0 cm long and 3.5 cm width, immature pod green and mature pod brown. Seeds round, shiny woody, testa glabrous, grey, 1.0 cm long, 1.0 cm diam., with short hard placenta, 0.2 cm long, glabrous. Flowering and fruiting during July to February. Medicinally fruits are tonic and antipyretic; seed oil is said to be used as a cosmetic and for discharge from the ear; leaves and stem bark is febrifuge, emmenagogue and anthelmintic.

Genus- *Derris* Lour. Fl. Cochinch. 432. 1790.

From Sundarbans mangrove habitats and mangrove reclaimed areas, three species under *Derris* Lour. were collected; these are *D. scandens* Benth., *D. trifoliata* Lour. and *D. indica* Bennet.

Table 58 : Key to the Species under the Genus - *Derris* Lour.

A.	Climbing shrub in the inter-tidal river banks; leaves elliptic oblong with acute tips; fruits flat with thin coat	See B. & BB.
AA.	Straight or bending medium tree in the mangrove cleared lands, grow as back-mangals, occasionally planted as avenue tree; leaves broadly ovate, with acute tip; fruit flattened but thick with blunt tip and hard woody coat	1. *D. indica* (Lamk.) Bennet
B.	Leaflets elliptic, acuminate; number 3-7, fruit sub-orbicular, thin margin	2. *D. trifoliata* Lour.
BB.	Leaves elliptic oblong or lanceolate with acute tip; fruits shuttle with both narrow end, one margin thicker than other	3..*D. scandens* Benth.

Photo Plate 110. *Derris trifoliata* Lour. A. Twining shrub in the natural habitat. B.Twig with flowers C.Twig with flowers.

67. *Derris trifoliata* Lour. Fl. Cochinnch. 433. 1790.

Local Name (Bengali): Pan Lata

Climbing or twining evergreen shrub. Compound leaves, leaf axis up to 15.0 cm long, pulvinous, terete, green, alternate, imparipinnate, leaflets up to 8.0 cm long and 3.1 cm broad, 3 - 7 numbers, petiolate; petiole 0.5 cm long, terete, glabrous, green, leaflets elliptic - oblong, entire, acuminate, cuneate, dark - green, coriaceous, dorsiventral, unicostate reticulate venation. Raceme on peduncle, up to 8.0 cm; ebracteate, pedicellate, pedicel 1.0 cm long, terete, green; flower - zygomorphic, bisexual, complete, medium, 1.0 cm long, pendulous, odourless; sepals - 5, gamosepalous, 5 short lobes above, beneath cup - shaped tube, 0.3 cm long; petals - 5, polypetalous, vaxillary, 1 standard, white - pinkish; stamens - 10 (9+1), diadelphous, anthers remain free, filament 0.8 cm long, white, anther - 0.2 cm long, bilobed, basifixed, introrse, inserted; carpels - 2, syncarpous, ovary superior, conical, 0.6 cm long and 0.2 cm width, flattened, glabrous, green, single chambered, marginal placentation, style -1. Fruit - pod or legume, sub-orbicular, 3.2 cm long and 2.5 cm width, seed one. Flowering and fruiting during April to August. Medicinally this plant is said to be useful as a stimulant, antispasmodic and counter irritant.

Photo Plate 111. *Derris scandens* Benth. A. Plant in the natural habitat. B. Close view of twig with fruits (Pods). C. Branch with fruits (Pods).

68. *Derris scandens* Benth. J. Linn. Soc, 4 : 103 (Suppl.) 1860.

Local Name (Bengali): Noalata

Climbing or twining evergreen shrub. Compound leaves, exstipulate, leaf axis up to 19.0 cm long, pulvinous, terete, green, alternate, imparipinnate, leaflets up to 6.3 cm long and 2.5 cm broad, more than 7 numbers or more, petiolate; petiole 0.5 cm long, terete, glabrous, green, leaflets elliptic - lanceolate, entire, blunt, cuneate, dark - green, coriaceous, dorsiventral, unicostate, reticulate venation. Raceme, peduncle 20.0 cm long, ebracteate, pedicellate, pedicel - 0.8 cm long; terete, green; flowers - zygomorphic, up to 1.0 cm long, bisexual, complete, medium, pendulous, odourless; sepals-5, gamosepalous, above short lobes, beneath cup-shaped tube, 0.25 cm long, green, glabrous, valvate, slightly upper level oblique; petals - 5, polypetalous, vaxillary, one standard, 0.8 cm long, 0.75 cm width, ovate, apical grooved or notched, with two edges, curved, two wings, 0.8 cm long and 0.35 cm width, white; stamens - 10 (9+1), diadelphous, anthers - 0.1 cm long, remain free, filament - 0.65 cm long white, anther bilobed, basifixed, introrse, inserted; carpels - 2, syncarpous, ovary superior, conical, 0.5 cm long basal flattened, 0.18 cm width, glabrous, single chambered, marginal placentation, style - 1, terminal, long 0.25 cm, glabrous, terete, stigma absent. Fruit- pod or legume, sub-orbicular, 4.0 cm long and 1.0 cm width, shuttle shaped, flattened, 1 - margin thicker than other, green, glabrous, two seeds in each pod, round with short grooved, 0.5 cm long and 0.7 cm diam, flattened, brown. Flowering and fruiting during July to December. Grow in the mangrove thickets and riverside jungles at Sundarbans. The root powder is said to be useful as fish toxicant or as fish poison.

Photo Plate 112. *Derris indica* (Lamk.) Bennet. A. Branch with pods. B. Flowering branch. C. Branch with pods(fruits).

69. *Derris indica* (Lamk.) Bennet, J. Bombay nat. Hist. Soc.68 (1)303.1973.

(Syn. *Galedupa indica* Lamk.; *Pongamia glabra* Vent.; *Pongamia pinnata* (L.) Pierre).

Common Name : Indian Beech **Local Name** (Bengali): Karanja

A back mangrove species, but frequently grow in the fresh water zones and also planted as avenue tree. Medium tree, attain up to 20.0 m high, evergreen, deliquiscent branching. Leaves compound, alternate, exstipulate, rachis upto 25.0 cm long, imparipinnate, green, pulvinous, 0.1 cm width, leaflets petiolate; petiole 1.0 cm long, 0.1 cm width, ovate, entire, acuminate, angled, green, dorsiventral, glabrous, unicostate reticulate venation conspicously, coriaceous, 10.0 cm long and 5.0 cm width. Raceme peduncle, up to 18.0 cm long; ebracteate, pedicellate; pedicel 1.0 cm long, terete, green; flower zygomorphic, 1.2 cm long, bisexual, complete, pendulous, odourless; sepals - 5, gamosepalous, cup-shaped, green, glabrous, valvate, inferior, persistent, 0.6 cm across; petals - 5, polypetalous, vaxillary, 1 chambered, 1 cm long and 0.9 cm width, apical notched, white pinkish, herbaceous, basal short appendages, curved, two wings, 1.2 cm long and 0.35 cm width, white, herbaceous and two keels, 1.0 cm long, 0.3 cm width, white, glabrous, deciduous; stamens - 10(9+1), diadelphous, staminal column up to 1.0 cm long and 0.2 cm width, anthers remain free, anther bilobed, basifixed, introrse, inserted; carpels - 2, syncarpous, conical, ovary superior, 1.3 cm long and 0.15 cm width, mid-region flattened, 2 ovules in each chamber, marginal placentation, style - 1, terminal, 0.25 cm long, glabrous, stigma absent. Fruit - pods, nut like, 5.8 cm long and 3.5 cm broad with persistent calyx, elliptic, glabrous, green, but black brown on maturity and woody on drying, thick up to 5.0 cm. Seeds 2, more or less square, brown seed coat. Flowering and fruiting during April to August, but fruits/pods hang upto December in the mother tree. Seed oil is said to be useful in leather dressing and also for the preparation of candle and soap; medicinally, it is said to be useful in scabies, herpes, leucoderma and other cutaneous diseases. It is useful in dyspepsia with stuggish liver. The active principle of this species is *'Karanjin''*.

Family- 31. Caesalpiniaceae

The Genus – *Caesalpinia* L. Sp. Pl.380.1753. has two species in the above tidal scrub jungles at Sundarbans mangrove forest, i.e., *Caesalpinia crista* L. and *C. bonduc* (L.) Roxb.

Table 59 : Key to the Species under the Genus -*Caesalpinia* L.

A	Shoot apex is not covered by gregarious spines; flowers bisexual; fruit coat smooth, seeds flat	1. *C. crista* L.
AA	Shoot apex covered by gregarious spines; flowers unisexual by abortion; fruit coat with dense spines; seeds globose	2. *C. bonduc* (L.) Roxb.

Photo Plate 113. *Caesalpinia crista* L. A. Branch with fruits. B. Branch with flowers

70. *Caesalpinia crista* L. Sp. Pl. 380. 1753.

Common Name : Molucca bean; **Local Name** (Bengali): Shingri Lata

Climbing, much branched shrub on the thickets and hedges on sandy coastal areas and occasionally on the ridge forests or mangrove cleared scrub jungles, form dense bush. Nodes not ridged, internode green, solid, woody, frequent brown spines. Leaves compound, alternate exstipulate, paripinnate, leaf axis up to 25.0 cm long, leaflets up to 6.0 cm long, with basal spines, dark green, axis glabrous, short petiolate; petiole 0.2 cm long, terete, pulvinous. Raceme, peduncle branched, up to 15.0 cm long, flower pedicellate, bisexual, irregular, hypogynous, showy; sepals – 5, polysepalous, imbricate, inferior -1.0 cm long × 0.5 cm width, basal fleshy and thick, curved as boat, inner concave, deep yellow, glabrous, entire, acute, base flattened; petals – 5, polypetalous, inferior, imbricate, ovate elliptic, entire, acute, 0.6 cm long X 0.4 cm width, dark yellow; stamens 10, monadelphous, anthers free with short filament stalk, carpels 2, syncarpous, ovary 1- chambered, marginal placentation. Pods 6.5 cm long X 3.5 cm width, with hard gradual pointed apex, hard epicarp, green, glabrous, without spines; seeds ovoid, flattened, 3.0 cm long and 3.5 cm width, shiny testa, grey black. Flowering and fruiting during November to June. Medicinally used as diuretic and lithontriptic and also the leaves powder is used for uterine tonic after parturition.

Photo Plate 114. *Caesalpinia bonduc* (L.) Roxb. A. Spreading branches with Inflorescence. B. Branch with fruits. C. Branch with Inflorescence

71. *Caesalpinia bonduc* (L.) Roxb. Fl. Ind. 2 : 362.1824.

Common Name : Nut or Fever Nut. **Local Name** (Bengali): Nata

Climbing, much branched shrub in the thickets, bush and hedges on sandy coastal areas and occasionally on the ridge forests or mangrove cleared scrub jungles, form dense bush with gregarious spines present on the plant. Nodes not ridged, internode with dense spines green, solid, woody, frequent brown branches. Leaves compound, alternate stipulate, paripinnate, leaf axis up to 25.0 cm long, leaflets up to 8.0 cm long, each with gregarious spines, axis glabrous, terete, solid; dark green, short petiolate; petiole 0.1 cm long, 0.05 cm width, glabrous, pulvinous. Raceme lateral or axillary, up to 20.0 cm long, peduncle, solid, green; flower pedicellate, unisexual, irregular, hypogynous, showy, bracteate; male flower 1.0 cm long × 1.8 cm across; sepals- 5, polysepalous; ovate - lanceolate, entire, pubescent, blunt, tufted, curved, up to 1.0 cm long and 0.5 cm width; petals – 5, polypetalous, imbricate, dark yellow, oblong, 1.5 cm long and 0.6 cm width, alternate to sepals; stamens 10, free, filaments hairy, terete, soft, solid, up to 0.8 cm long, dorsifixed, exerted, anther bilobed, extrorse; female flower by abortion (0) - carpel 2, bisexual flowers syncarpy, superior, ovary dumbell shaped, 0.5 cm long and 1.1 cm diam., pubescent, greenish, single chambered, two ovules, parietal placentation, style – 1, 0.4 cm long, short terminal, stigma – 1, capitate glabrous. Fruit legume, flattened, epicarp woody, having gregarious spines, 5.0 cm long and 3.5 cm width, immature pod green and mature pod brown. Seeds round, shiny woody, testa glabrous, grey, 1.0 cm long, 1.0 cm diam., with short hard placenta, 0.2 cm long, glabrous. Flowering and fruiting during July to February. Medicinally fruits are tonic and antipyretic; seed oil is said to be used as a cosmetic and for treatment of discharge from the ear; leaves and stem bark is febrifuge, emmenagogue and anthelmintic.

Family – 32. Chenopodiaceae

The family - Chenopodiaceae has two halophytic genera, *viz. Suaeda* Forsk and *Salicornia* L. in the Indian Sundarbans and the other halophytic habitats in India. Among them, the Genus- *Suaeda* Forsk. has two species in the Indian part of Sundarbans, *i.e., S. maritima* Dumort and *S. nudiflora* Roxb.; the 3rd species under this genus, *i.e., S. monoica* Forsk. ex. Gmel is also reported from Tamil Nadu and Andhra Pradesh coasts. The two species under this genus *Suaeda* Forsk the Indian part of Sundarbans, can be differentiated by the following characters:

Table 60 : Key to the Species under the Genus – *Suaeda* Forsk. ex Scop. Fl. Aegypt. Arab.69.1775.

A	Leaves broad, green, styles-2, seeds-2, seeds usually horizontal	1. *S. maritima*Dumort.
AA.	Leaves semi-terete, green, reddish after maturation, styles-3,seeds erect	2. *S. nudiflora*Roxb.

Photo Plate - 115. A. & B. Twigs of *Salicornia brachiata* Roxb.

72. *Salicornia brachiata* Roxb. Fl. Ind. 1 : 82. 1820.

Local Name: Nona Palang

Stem fleshy, nodes ridged, easily separated; leaves simple, fleshy, elliptic, green; flower bisexual, petals yellow, showy; stamens : 1 - 2; seeds with testa containing hooked hairs. Flowering and fruiting during July to November. Occasionally grow on the river flat land in the coastal mangals.

Photo Plate 116. A. & B. Natural habitats of *Suaeda maritima* Dumort

73. *Suaeda maritima* Dumort, Fl. Belg. 22. 1827

Common Name : Common Indian Saltwort; **Local Name** (Bengali): Giria Sak/ Gire Sak

Erect much branched herb on the river flat open land, grow gregariously, 0.30 m - 0.56 m tall; internodes not rigid, solid; leaves simple, alternate, exstipulate, sessile, 1.0 cm long and 0.4 cm broad, succulent, glabrous, 0.05 cm thick, green, ovate-oblong, entire, round, isobilateral; spikes at leaf axis; ebracteate, small flowered; flower - 0.1cm long and 0.15 cm across; bisexual, complete, regular, sessile; tepals-5, polytepalous; stamens-5, free, filament white; carpels - 3, syncarpous; fruit drupe, 0.15 cm width and 0.10 cm long with persistent perianths. Flowering and fruiting during October to December. Rural people occasionally cook the vegetative part as vegetable; a famine food.

Photo Plate 117. ***Suaeda nudiflora*** **Roxb. A. Natural habitat in the inter-tidal river flat land. B. New growth on river flat. C. Twig in hand. D. Close view of the twig. E. Branched twig.**

74. *Suaeda nudiflora* Roxb. Moq. Ann. Sci. Nat. 23. 1831.

Common Name : Saltwort; **Local Name** (Bengali): Giria Sak/ Gire Sak

Prostrate much branched herb on the river flat open land, having erect branches, grow gregariously, 0.30 m - 0.56 m tall; internodes not rigid, solid; leaves simple, alternate, exstipulate, sessile, 3.0 cm long and 0.3 cm width, succulent, glabrous, 0.05 cm thick, green and turns dark red to reddish on maturity, oblong-lanceolate, entire, round, isobilateral; spikes at leaf axis; ebracteate, small flowered; flower - 0.1 cm long and 0.15 cm across; bisexual, complete, regular, sessile; tepals - 5, polytepalous; stamens - 5, free, filament white; carpels - 3, syncarpous; fruit drupe, 0.15 cm width and 0.15 cm long with persistent perianths. Flowering and fruiting during November to January.

Family-33. Clusiaceae

(*nom. alt* . Guttiferae)

Members under the family-Clusiaceae are not mangroves, but occasionally considered as the mangrove associate; the species, *Calophyllum inophyllum* L. Sp. Pl. 1: 513.1753 is the only mangrove associated species collected from the Sundarbans Tiger Reserve, especially from the beach areas, beyond the frequent tidal action.

Photo Plate 118. *Calophyllum innophyllum* L. A. & B. Branches with dark green leaves.

75. *Calophyllum innophyllum* L. Sp. Pl. 1 : 513. 1753.

Common Name : Indian laurel; **Local Name** (Bengali): Kath Champa or Sultan Champa

Handsome, moderate sized evergreen tree, without any buttress or modified roots, glabrous, bark smooth, grey or blackish brown, latex yellow to whitish, sticky; leaves glabrous, simple, opposite with a short petiole, blade elliptic to oblong about 15.0 cm × 8.0 cm, margin entire, parallel distinct venation, leathery, broadly elliptic, shining on both surface; flower buds with minute rusty hairs, flowers stalked, white, fragrant in lax, few flowered axillary racemes; fruit 1- seeded globose drupe, 2.0 cm - 4.0 cm in diam., drupe round, green with long stalk, pulp scanty. Flowering and fruiting during November to March. A littoral species, on the sandy soil, sometime grow along with the mangrove trees in the transitional habitat; occasionally planted in the ornamental garden as ornamental tree for its dark dense beautiful coriaceous foliages. A native of the coastal areas in South East Asia, East African Islands, Australia and Polynesia. Wood is coarse grained but valuable and used occasionally for cabinet making; an aromatic yellowish green resinous gum exudes from the bark and volatile oil is distilled from its gum; leaves also have some medicinal properties.

Family – 34. Convolvulaceae

The family- Convolvulaceae consists of about 50 genera and 1,200 species; out of these the species *Ipomoea pes-caprae* (L.) R. Brown is common in the coastal sandy soil of Indian Sundarbans. Banerjee, *et al.* (1989) have also reported *Ipomoea tuba* (Sch.) G. Don from the hinterland of other coastal areas. *Ipomoea pes-caprae* (L.) R. Brown (Syn. *Convolvulus pes-caprae* L..) is actually an important species in sand dune areas and acts as sand binder; common in the beach area, beyond the normal tidal reaches. A prostrate, creeping herb; much branched, branches slightly erect and the main stalk remain horizontal with the ground. Tap roots and several adventitious roots develop from the nodes. Leaf apex bifurcated and lamina fleshy, green.

Photo Plate 119. ***Ipomoea pes-caprae*** **(L.) R. Brown - Natural habitat in the sand, dune along with flowers.**

76. *Ipomoea pes-caprae* (L.) R. Brown, Tuckery, Narr. Exped. Zaire, 1816,(Mart.1818).

Local Name (Bengali): Chhagal Knuri

Prostrate, creeping herb on the sandy coastal areas; much branched, branching with fast growth during monsoon months; branches slightly erect, while main stalk horizontal to the ground. Tap roots and several adventitious roots develop from the nodes; weak stem, internodes up to 11.0 cm long, terete, green, solid, slightly fleshy, 0.3 cm diam., glabrous. Leaves alternate, cauline, upright, exstipulate, simple, petiolate; petiole up to 12.0 cm long, terete, glabrous, base brownish green, fleshy, solid, basal portion conspicuously twisted, pulvinous; lamina apically bilobed, slightly folded, each fold broadly ovate, entire, straight, common notch of both folded lamina about 0.3 cm deep, succulent, dorsiventral, glabrous, unicostate reticulate venation. Solitary cyme, peduncle long, about 7.0 cm long, solid, glabrous, green, terete, erect, peduncle branched, long stalk bear large flower, while dwarf stalk having bud; flower - 5.5 cm long X 5.0 cm across, ebracteate, bisexual, complete, regular, erect, pedicel - 2.5 cm long, solid, terete, green, erect, odourless, pentamerous, perigynous; sepals- 5, polysepalous, 3 inner large and 2 outer small; petals - 5, gamopetalous, twisted, funnel shaped, 5 apical pointed beaks present; stamens - 5, epipetalous, filament base slightly attached to the corolla base; carpels - 2, syncarpous, ovary superior, conical, short, 0.2 cm long, ovary 4-chambered, 1 ovule in each chamber, axile placentation; fruit capsule with apical beak, 1.0 cm long and 1.3 cm periphery, with persistent large calyx, reddish at margin. Flowering and fruiting during October to February. Leaves are useful in rheumatism, stomachic tonic and the extract of leaves have the astringent, diauretic, laxative properties. Acted as the sand binder in the coastal areas.

Family -35. Cyperaceae

Several species under the sedge family- Cyperaceae were reported from the Indian Sundarbans (Majumdar, 1962 & Naskar, 1993); these are *Carex* spp., *Cyperus* spp., *Eleocharis* spp., *Fimbristylis* spp., *Bulbostylis* spp., *Fuirena* sp. and *Scirpus* spp. These species grow both in the dry habitats and also in the marshy places in the brackish-water zones. Among these the genera *Cyperus* and *Scirpus* are most dominant in the marshy places and also in the brackish-water zones of the Sundarbans; but the other species are not scarce.

Table 61 : Key to Some Common Genera Under the Family- Cyperaceae of the Sundarbans Mangals

A	Stem triquetrous, compressed, solid; flowering glumes distichous; spikelets many flowered	1. *Cyperus* L.
AA	Stem cylindric, spongy or hollow; spikes few to many in the stem at different height	2. *Scirpus* L.

Photo Plate - 120. *Cyperus* sp. in the marshy areas at Goa

77. *Cyperus rotundus* L. Sp. Pl. 1 : 45. 1753.

Common Name : Sedge; **Local Name** (Bengali): Mutha Ghash

Perennial herb; stolons slender, elongated, bearing elliptic or sub-globose, small to large tubers, stem erect, triquetrous; leaves flat, 1-nerved, acuminate; umble simple or compound; bracts usually 3; spikelets straw coloured to reddish brown, 10-50 flowered, compressed; rachilla with hyaline wings; glumes densely imbricate, usually 3 or 5 nerved, obtuse, minutely apiculate at apex, margin hyaline. Nut greenish black, ellipsoid or elliptic-obovoid, trigonous, minutely apiculate. Flowering and fruiting during June to October. Common sedge in the marshy places of the Sundarbans areas. Root/rhizome is said to be useful in diarrhoea.

78. *Scirpus littoralis* Schrad., Fl. Germ. 1 : 142. T.5. 1806.

Common Name : Sedge; **Local Name** (Bengali): Chenchco

Frequently found in the river bank and river flats and also found in the reclaimed areas of Indian Sundarbans. Root stock stout, hard, with fibrous roots, leaves membranous, acuminate, sheath soft, fleshy, long, inflorescence at the top of the stem, spikelets pale brown, oblong, terete, pedicellate, nuts broad, obovoid, glabrous with long apiculation, dull-brown. Flowering and fruiting during June to November.

Family - 36. Loranthaceae

Under this family - Loranthaceae, there are three genera in Indian Sundarbans, *viz. Viscum* L., *Dendrophthoe* Mart. and *Macrosolen* (Bl.) Reichenb; members of these genera grow as semi-parasitic or as epiphytic shrubs or under-shrubs on the mangrove trees. These species are not mangroves or halophytes, but as they grow in these mangrove forests of Indian Sundarbans, these are mentioned in this treaties. These species are *Dendrophthoe falcata* (Linn.f.) Etting., *Viscum orientale* Willd., *V. monoicum* Roxb. and *Macrosolen cochinchinensis* (Lour.) van Teig. (Syn. *Loranthus cochinchinsis* Lour.). *V. orientale* Willd. is the semi-parasitic or epiphytic evergreen shrub; common on the branches of the *Excoecaria agallocha* L., while *V. monoicum* Roxb. is rare in the Indian Sundarbans. *Dendrophthoe falcata* (Linn.f.) Etting. is the common epiphytic species on the mangrove trees, particularly on the branches of *Excoecaria agallocha* L. , *Xylocarpus mekongensis*, *Avicennia officinalis* and others. The other species, *Macrosolen cochinchinensis* (Lour.) van Teig. is also found to grow on the branches of other old mangrove trees. Other plants, *viz., Dendrophthoe falcata* (Linn.f.) Etting. and Hoya parasitica Wall. (Family- Asclepiadaceae are common epiphytes on the tree trunk of *Xylocarpus mekongensis* in Sundarbans.

Table 62 : Key to the Genera under the Family - Loranthaceae

A	Grow as epiphytic shrub on mangrove trees; flowers unisexual	1. *Viscum* L
AA	Grow as epiphytic shrub on mangrove tree; flowers bi-sexual	See B & BB
B	Petals 5-clefts, tubular, styles not articulated	2. *Dendrophthoe* Martius
BB	Petals 6-clefts, tubular, styles articulated above	3. *Macrosolen* Blume

Photo Plate 121. *Viscum orientale* Willd. A. Natural habitat on Excoecaria sp. and B. Collected branch in hand

79. *Viscum orientale* Willd. Sp. Pl. 4(2) : 737. 1806.

Common Name : Mistletoe; **Local Name** (Bengali) : Manda

Much branched, glabrous shrub, common on the branches of Excoecaria sp., Avicennia spp. and semi-parasitic, also grow on the branches of the non-mangrove trees outside the mangrove habitats; not developed normal roots, but for its parasitic or epiphytic nature it form haustorium and anchor the branches of the host trees; branches terete, grooved; leaves falcate, unequal, ovate elliptic or elliptic- oblong with round or obtuse apex, entire, 2.5 - 5.0 cm long and 1.0 - 2.0 cm broad, petiole 3.0 mm long. Inflorescence cyme, 0.4 cm long X 0.4 cm across, flowers 1.3 cymes together in leaf-axils; female flowers in the middle and the male in lateral position; peduncles 1.0 - 3.0 mm long; male flowers smaller, compressed; female flowers oblong; gynoecium : carpel - 1, ovary inferior, rod shapped, 0.1 cm long, one chambered, one ovule in each chamber, basal placentation, style absent, stigma - 1, spherical, globose, minute; fruit 5.0 mm long, oval or sub-globose, smooth, green, fleshy. Flowering and fruiting during February to June.

Photo Plate 122. ***Dendrophthoe falcata*** **(Linn. f.) Ettingsh, flowering branches**

80. *Dendrophthoe falcata* (Linn. f.) Ettingsh, Denk. Akad. Wiss. Math-Nat. 32: 52. 1872.

Common Name : Mistletoe; **Local Name** (Bengali): Bara Manda

Semi-parasitic or epiphytic shrub; normal roots absent but haustoria formed to attach on the host trees. Stem well differentiated into nodes and internodes, branched, deliquiscent, nodes ridged. Leaves opposite, decussate, exstipulate, simple, petiole 0.5 cm long and 0.4cm broad, flattened, glabrous. Peduncle unbranched; flowers - bisexual, complete, regular, pendulous, showy, deep red, 5.7 cm long, pedicellate, pedicel 0.6 cm long, glabrous, terete, green, pulvinous, slightly curved soft, sweet scented, pentamerous; calyx : sepals - 5, gamosepalous, cup-shaped, grooved, oblique in upper surface, glabrous, green, not apical lobe or beak; corolla : petals -5, gamopetalous, forming a hollow tube, 3.5 cm long, basal portion larger than the upper portion, apical 5 lobes of equal length, lobes 1.5 cm long and 0.1 cm broad, thick, glabrous, acicular, acute, entire, red, pendulous, terminally bent, total length of the corolla is about 5.0 cm, superior, imbricate; androecium - stamens 5, epipetalous, filament attached to the apex of the corolla tube from where apical lobes furcated, filament 0.5 cm long, round, soft, glabrous, white, basifixed, anther bilobed, globose, 0.8 cm long, longitudinal dehisced, red coloured, inserted, extrorse, same level surface to stigma, anthers equal length and upper equal levelled, superior; gynoecium - carpels 2, syncarpous, ovary inferior, cup-shaped, 0.4 cm long and 0.25 cm diam., 1 - chambered, one ovule in each chamber, basal placentation; style terminal, 4.7 cm long, gradually tapering apex, terete, soft, glabrous, white, stigma -1, minute, globose, apical slightly grooved. Flowering and fruiting during October to April.

81. *Macrosolen cochinchinensis* (Lour.) van Tiegh., Bull. Soc. Bot. Fr. 41 : 122.1894.

Local Name (Bengali): Bara Manda

Perennial undershrub, parasitic on the *Avicennia* spp. and mangrove trees; stem twisted, much branched; leaves up to 12.0 X 6.0 cm, elliptic lanceolate, with a narrow base, acute, entire, petioles 0.3 - 1.0 cm long. Inflorescence axillary racemose, solitary or sometimes up to 3, 4 -10 flowered; peduncles 2.0 - 6.0 mm long; pedicels up to 0.7 mm long; fruits up to 6.0 mm long, crowded with conical style base and calyx limb; seeds ellipsoid. Flowering and fruiting during March to July.

Family - 37. Opuntiaceae

(*nom. alt.* Cactaceae)

82. *Opuntia dillenii* (Ker. Gawler) Haw., Suppl. Pl. Succ. 79. 1819. Page - 527.

Common Name : Cactus **Local Name** (Bengali): Fani Mansa

Purely xerophytic shrub, jointed and phylloclades stem, flowers yellow, berries, purple, pear shaped, tubercled. Flowering and fruiting during March to August. Fruits and phylloclades are said to be edible.This is the only natural cactus in the lower Bengal; grow in the coastal sandy areas and also in the dry habitats in road sides and village thickets or scrub jungles. Its large flowers looked very beautiful when it is fully bloomed.

Photo Plate 123. *Opuntia dillenii* (Ker. Gawler) Haw. A. Natural habitat in the sandy coastal area of Sundarbans. B. Close view of single phylloclade with yellow flower. C. Close view of the phylloclades, showing the spines after leaf modification.

Family - 38. Pandanaceae

83. *Pandanus odoratissimus* Soland ex Parkinson, J. Voy. H.M.S. Endeav. 46. 1773. 525.

Common Name : Screw-pine; **Local Name** (Bengali): Keya

Small palm like with dichotomously branched, up to 5.0 m tall, sub-erect, supported by aerial prop roots; leaves up to 2.0 m long; serrated margin, sword shaped, acuminate, spadix long; inflorescence sweet scented, carpels confluents, stigma short, staminodes absent, fruits oblong, drupes numerous. Flowering and fruiting during March to November. Leaves are used for making cordage and obtain strong white fibres; also used for the remedy of small pox, leprosy, *etc.* The tribal population in Andaman and Nicobar Islands often eat the fruit as their staple food.

Photo Plate 124. *Pandanus odoratissimus* Soland ex Parkinson - collected from Wandoor Bay, Andaman

Family - 39. Pteridaceae

The Cryptogamic family - Pteridaceae is the back mangrove habitat genus, *Acrostichum* L. and this genus has three mangrove habitat species, *viz. A. aureum* L., *A. speciosum* Willd. and *A. danaeifolium* Langsd. & Fisch. Among these, the species *A. aureum* L is common in the Indian Sundarbans and both these species *A. aureum* L. and *A. speciosum* Willd. were collected and reported from Port Blair of Andamans. *Acrostichum aureum* Mett. is the common mangrove habitat ferm in the Indian Sundarbans mangal; this species is frequently found to grow in the mangrove degraded or mangrove reclaimed areas (Photo Plate – 125). Other species, *viz., Acrostichum speciosum* Willd. and *A. aureum* L. are also common in the Andaman & Nicobar Islands. These are the erect rhizomatous ferns, attaining upto 2.0 m high and with bushy appearance; frond simple, semi-pinnatifid, up to 1.0 m long and 4.0 cm broad, iso-bilateral, distinct mid-vein and reticulate vein pairs in their frond.

Photo Plate 125. *Acrostichum sp.* A. *Acrostichum aureum* L. - grow in the mangrove degraded area at Sundarbans. B. *Acrostichum aureum* L. - collected from the Indian Sundarbans.

84. *Acrostichum aureum* L. Sp. Pl. 2 : 1062.1753.

Common Name : Mangrove Fern; **Local Name** (Bengali): Hudo

Erect fern, rhizomatous fern attain the height up to 2.0 m; having small bushy appearance. Fibrous, frequently stilted, aerial fern like rhizomatous roots. Rhizome globose, horizontal, caudex, circinate vernation, young rachis covered by gregarious hairs like structure, known as ramenta. Frond simple, semi-pinnatifid, up to 1.0 m long and 4.0 cm width, isobilateral, distinct mid-vein and reticulate vein present, entire, blunt end, green and yellow on maturity with sorus, glabrous, coriaceous, unicostate reticulate venation. Mature fronds become sporophyllous, diffused sporangia at abaxial surface, mixed sporangia on both side of mid-vein, brown sporangia stalked, upper globose; sporangia formation is generally during winter months, *i.e.*, November to February. Tender fronds are edible and cooked as vegetable by the poor people; medicinally the paste of rhizome are said to be useful in boils.

Family - 40. Ruppiaceae

85. *Ruppia maritima* L. (Syn. *R. rosetelle* Koch)

Common Name : Sea Gran; **Local Name** (Bengali): Nona Jhangi

Ruppia maritima L. under the family- Ruppiaceae (*nom. alt.* Najadaceae) is the submerged, delicate, soft, green, much branched herb in the brackish water fisheries or stagnant brackish water pool; common in the Indian Sundarbans. It is the nutritious food for brackish water fish and prawn species; sometime grow profusely and choked the water bodies in brackish water areas.This submerged plant is considered as the preferred food for shrimps and mullets, but luxuriant growth sometime chock the water bodies and affect movement of shrimps and fish species.

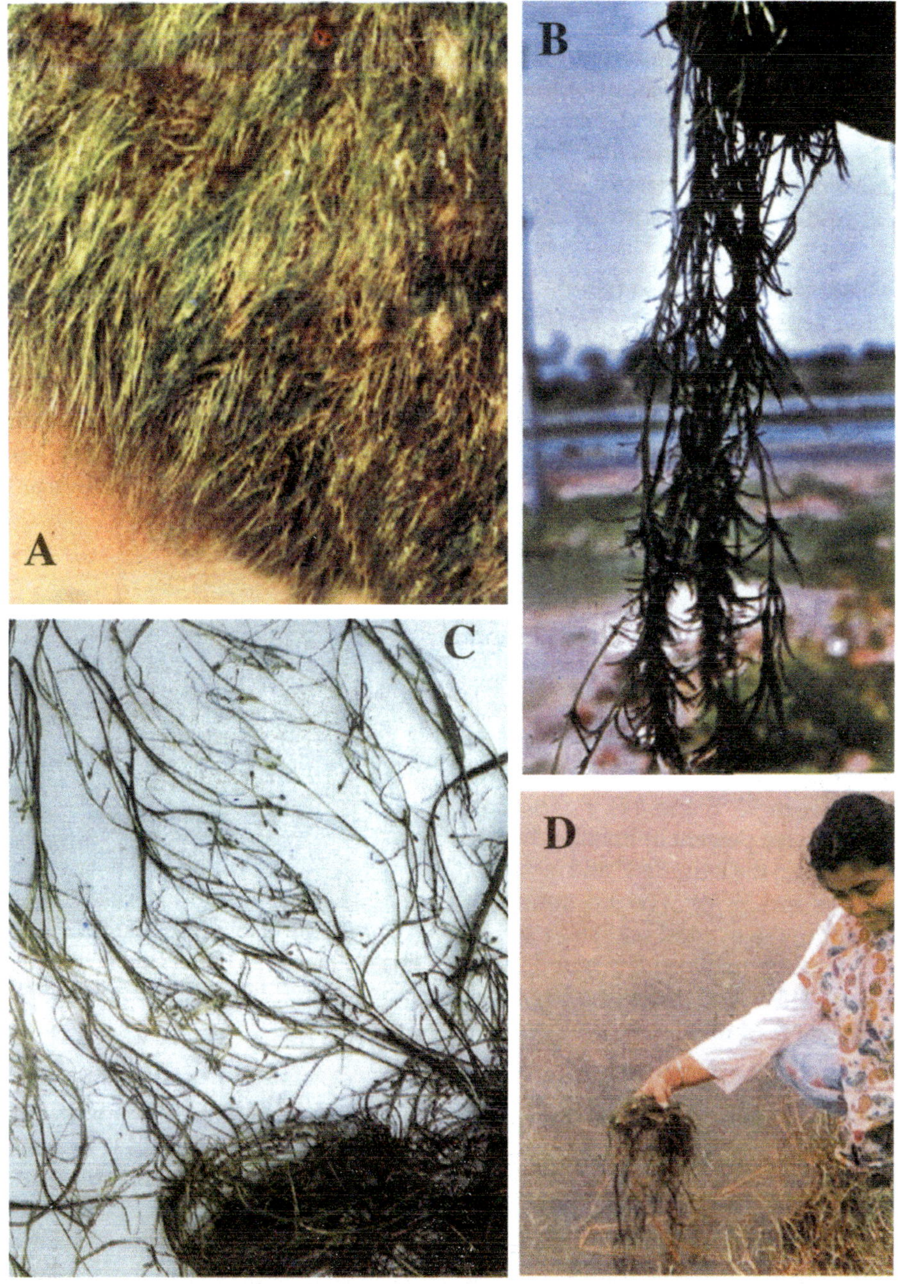

Photo Plate 126. A. Luxuriant growth of *Ruppia maritima* L. in brackishwtar fisheries. B. Collected specimen in hand, close view. C. Submerged condition in brackishwater fisheries at Sundarbans. D. Collection of *Ruppia maritima* L.- samples for studies

8

POSSIBILITIES & CONSERVATION NEEDS

Sundarbans Mangals - the Largest Single Mangrove chunk of the Globe

Finlayson and Moser (1991) have reported that the total mangrove areas of the world was about 14 million hectares. Out of these, the Old World Tropical mangroves, *i.e.*, the Indo-West Pacific Tropical zones and Tropical Australia have the most dominant mangrove areas and mangrove species abundance along with its unique succession pattern.

Finlayson and Moser (1991) have also reported that about 1.48–1.73 million acres or 6,00,000 – 7,00,000 ha mangrove cover area was present in the southern part of the four coastal districts of Bangladesh, *viz.*, Khulna, Barisal, Noakhali and Chittaganj and two coastal districts of India, *viz.*, 24 Parganas South and 24 Parganas North. These coastal zones popularly known as Sundarbans, where good number of mangrove species gregariously grow. Mangrove species of the Sundarbans are estimated about 80% of total world mangrove species. This undivided Sundarbans mangal of Bangladesh and India spread over within the latitude 21°00′N and 22° 31′N and the longitude between 88°10′ E and 92°15′E.

In an recent estimate it was shown that the total highest mangrove cover area is in Indonesia, which is about 42,510 km², situated within the latitude 6°N and 11°S and longitude between 95°E and 141°E. The 2nd highest mangrove cover area present in Brazil, which is about 25,000 km² and within the latitude 0° and 30°S and longitude between 30°W and 52°W. The 3rd dominant mangrove areas present in the tropical and sub-tropical Australia, within the latitude between 8°S and 38°45′S and longitude between 115°E and 155°E, which cover about 11,500 km². (Naskar & Mandal, 1999).

The total coverage under the mangroves in Indonesia is spread over in several countries and thousands of islands, *viz.*, Java, Sumatra, Sulawesi, Irian Jaya, Kalimantan and several other small to large islands; these Indonesian mangroves also spread over both in the Indian and Pacific Oceans. Mangrove areas in the New World zones at Brazil and the Old World zones of Australia are more than Sundarbans, but that zones are scattered and mangrove species diversity is less than the Sundarbans mangals.

This undivided Sundarbans mangroves in Bangladesh and India is the single largest mangrove chunk of the world, where the mangrove species diversity is much more and important for harboring several endemic and threatened plants and animal species.

The mangrove areas in the undivided Sundarbans, of the Ganga–Brahamaputra deltas, *i.e.*, the Hugly-Padma-Meghna mouths, grow very compactly in hundreds of deltaic islands; lies within 21°00′N and 22° 31′N latitude and the longitude between 88°10′E and 92°15′ E, which cover about 10,000 km². These are very much dense and the species diversity there is higher than the other world mangals. Several endemic, threatened, endangered and rare flora and fauna are present there with pristine glory. As such, IUCN has included these Sundarbans mangrove forest as the *'World Heritage Site'* since 1984 and very recently it is proposed to include as the *'Ramsar Site'* for effective conservation.

Photo Plate 127. Tide water current in the mangrove forest of Indian Sundarbans, West Bengal.

Sundarbans Mangals Act as the Sink - for Metropolitan Pollutants

Quite a substantial amount, an average of 396,939 kg/ day of pollution load is discharged by the Calcutta Metropolitan city in the side by river system; out of these very large part of it, *i.e.*, 22,900 kg/ day is diverted or released from the Calcutta Metropolis to the Hugli river directly which passes in the Sundarbans estuaries (Yadav, 1987). Enough evidences are also available that since 1939, different kinds of pollutants along with large volume of city sewage, different types of industrial effluent and waste water from the Calcutta Municipal Corporation have been indiscriminately discharged in the river Sakha Bidyadhari of Sundarbans (Kulti Gang), *via* (i) 33.8 km. Long Dry Weather Flow Channel (D.W.F. Canal), (ii) 35.40 km. long Storm Weather Flow Channel (S.W.F. Canal) and (iii) Town Headcut Pamarbazar Canal of 8.04 km long (David, 1959). David (1959) also had stated that the Sakha Bidyadhari or Kulti river no longer served as a seed source for culture fisheries, since 1939. On an average rate of 16.20 core gallon/day sewage is discharged to the river Sakha Bidyadhari/Kulti Gang, *via* Kulti Lock Gate at Ghushighata (Naskar, 1999).

The mangrove plants have high percentage tannin in their barks and leaves; mangrove plants can chelate some of these pollutants and its detrimental effect may be reduced naturally (Naskar and Mandal, 1999). The Coast Guard Act, 1978 provides the measure for protection of coastal/ marine environment and over all control of marine environment; but implementation of such rules and strict measure through mass awareness is must.

Photo Plate 128. Sewage discharge point at Ghusighata, Kulti

Protection *vs.* Loss of Bio-diversity

For being the important *'World Heritage Site'* and natural abode of vast group of the threatened mangrove species and mangrove dwelling faunal assemblages, the Sundarbans mangal was included within the important conservation programmes by the Local and Central Government. For these purposes, several National and International aided projects and conservation programmes were undertaken, since the year 1973. The 'Sundarbans Tiger Project' was initiated since 1973 in an area of 2585.10 km^2 of the S.E. part of the Indian Sundarbans mangrove forests. Besides these, 'Bhagabatpur Crocodile Project' was initiated, since 1976 and from the same year three 'Wildlife Sanctuaries' at Sajnakhali, Lothian Islands and Holiday Islands were declared. The 1330 km^2 core area of the Sundarbans Tiger Reserve was also declared as the 'National Park', since 1984 and on the same year the Sundarbans mangrove forest was considered as the *'World Heritage Site'* by the IUCN. The last but most important 'Sundarbans Man and Biosphere Project' in the total area of 9630 km^2 was declared since 1989 and recently it was proposed for declaring as *'Ramsar site'*.

In spite of all these effort, still the mangrove forest wood and timbers are exploited enormously and the killing of the wild animals, *viz.*, tiger, deer, wild boars and birds cannot be checked, due to inadequate protection measures and difficult access in these riverine mangrove forest zones of the Sundarbans. Besides all these protection acts and conservation measures, mangrove forest lands are still cleared for excavating the shrimp farms, here and there; these need strict measures from the local Government, *viz.*, the Gram Panchayet. All these need large scale awareness generation among the rural people by the Government and the NGOs, as well.

Taking these under prior importance, a 'Mangrove Ecological Museum', at CICFR Station, CGO Complex, Salt Lake, Kolkata -700 064 and a 'Mangrove Ecological Park' at Jharkhali, Sundarbans were set up during 1998 for awareness generation among the local people, students, administrators and the planners of the Sundarbans. In these 'Mangrove Ecological Museum' and the 'Mangrove Ecological Park', the important mangrove plants and the animal species were collected, photographed, sketched, described and displayed for ready reference and clear knowledge about the Sundarbans mangrove biota and ecosystem. In this effort, the Department of Environment, Government of West Bengal, Sundarbans Development Board, Fishery Department, W. B. and the Calcutta Wildlife Society - a NGO Body support the National Fellow, ICAR for developing awareness and highlighting the Sundarbans and its important mangrove ecosystems.

It is now well recognised fact that the conservation of the Biodiversity in the Sundarbans mangal is most urgent for the posterity of the conservation of threatened flora and fauna and keeping the existence of the Calcutta Metropolitan city from the devastating natural calamities in the long run. The effect of Sea - level rise is also an alarming question in Sundarbans.

Photo Plate 129. Aerial view of the forest and river, Sundarbans

Major Problems in Indian Sundarbans

Following major problems were identified during field studies in the Indian Sundarbans; these are:

1. Large scale reclamation of forest land, deforestation of mangroves for human settlements as rapid growing needs for lands, fuel wood, fish and natural commodities;
2. Management problems: Problems of *Protection vs. loss of Bio-diversity* in the Sundarbans is acute;

a) Self engagement of vast rural people and the large scale netting out of shrimp seeds *vs.* exploitation and damage of the other fish, prawn and crab juveniles along with damage on river dykes and mangrove forest areas have created lot of problems and degradation of the aquatic resources;

b) Large scale operation of fine mesh nylon nets and exploitation of estuarine fish juveniles indiscriminately, create problems towards loss of aquatic species diversity;

c) Siltation on the river bed is alarming, which also cause frequent flood and over flow of the estuarine rivers and frequent saline water ingress in the human settlement areas and on the crop fields;

d) Transport facilities is not developed, only the country boats are common means of transport;

e) Uncontrolled population growth and very marginal socio-economic status of the rural people of Sundarbans is the main cause for all these aforesaid problems;

f) Agriculture is not much developed in these areas due to lack of irrigation facilities, though these zones are very much potential for having adequate rain-fall;

g) As such, pressure on the natural mangrove forest is enormous and protection of forest resources is not manageable.

h) Last but most common problem is that the Sundarbans is the frequent and severe cyclone prone area.

In an estimate it is seen that during the last four centuries more than ninety severe cyclones, sea surges, earth quakes were partially or fully destroyed the coastal zones of Sundarbans. Among these, most severe cyclone/ storm was during the year 1737; during that time the wind speed was 250 km / hour along with tidal thrust or surges at the height 13.0 m from MSL and severe earth quake fully destroyed the human habitats and mangrove forests.

Furthermore, during 29th November, 1988 cyclone/ wind speed was also 250 km/ hour. Fosberg (1971) wrote, *'the situation of the Ganga delta, at the head of the funnel shaped Bay of Bengal possesses perhaps the most serious threat from the surges, driven by storm waves....'.*

A huge surge in the Bay of Bengal in the year 1688 took a toll of more than 60 thousand in the Sagar Island of Sundarbans alone; again in the year 1707, the cyclone and storm devastated the entire Sundarbans. In the year, 1833, hurricanes and cyclones forced to cease all the development work in the Sundarbans; the cyclones in 1884 took 4137 lives from the Sagar Island alone. During the year 1867, cyclones damaged the entire Sundarbans from Sagar Island to the Pabna district of the Bangladesh. The earthquake of 1895 also affected a lot in the Sundarbans. In the year 1942, there were a terrible devastating hurricane over the Sundarbans, for which several houses, cottages, cattles and lives were collapsed. During the recent time, most disastrous one in the history of this region is probably of the 13th November, 1970, which took a toll of several thousand from the Sundarbans. Again devastation occurred during 1973; on 10th and 11th December of 1981 and 29th November,1988. Several sign of human settlements are still noticed inside the Sundarbans mangrove forests, which may have been destroyed, time to time, due to these severe natural calamities. During the recent time, on the last week of October, 1999 the *'super cyclone'* and tidal thrust devastated the entire Orissa coast at Paradwip, Jagatsinghapur and Erasama and part of the Midnapore district. In this context, mention may be made that these cyclones and the other natural calamities are inevitable and this mangrove forest cannot stop all these natural phenomenon, but dense mangroves can protect as buffer and minimise this natural calamities and the severe effect may be checked considerably, in this deltaic region. So prior to undertaking any developmental activities in this coastal region, due considerations should require to be undertaken for proper planning and strict conservation/ management practices for this important but threatened ecosystem.

Photo Plate 130. Barraging of flowing river at Jharkhali

Large Scale Deforestation of Forest Land

During the last two centuries, *i.e.*, since the latter half of the 18th Century, more than 5000 km^2 mangrove forest area in the Indian Sundarbans alone were cleared and reclaimed for the human settlement. Now, only 4267 km^2 mangroves and mangrove surrounding water area is present out of 9630 km^2 area of the Sundarbans Biosphere Reserve. Till date, several hectares mangrove zones at Jharkhali (Tridiv Nagar), Henry's Island at Bakkhali, Jamboo Dwip and other areas are on the verge of reclamation and clearing, which is dangerous and alarming for the existence of this vulnerable ecosystem.

In spite of large number of projects and protection measures, the mangrove forests in Indian Sundarbans are on the verge of extinction and degradation, very rapidly. This is due to rapid growing human population and for their large scale demand for fuel wood, timber and their livelihood on these natural resources. These rural poor people are mostly landless day labourers and most of them are engaged for destruction and exploitation of these natural forest and estuarine resources. The business-men and the money lenders forced them to be involved in these destruction activities; these moneyed men and selfish business-classes engaged the poor people for harvesting and exploiting mangrove timbers, fish, shrimp and other natural resources and these poor villagers/bonded labourers rather forced to do these destruction activities taking every risk of lives. Here, protection measures and management practices are too inadequate to tackle these bio-diversity conservation activities, which is mostly controlled by the antisocial and selfish people.

Large Scale Netting out of Shrimp Seeds vs. Damage of the Others Fish Juveniles

Lakhs of rural people in Indian Sundarbans are now self engaged for netting out the shrimp (*Penaeus monodon*) juveniles from the vast estuarine rivers and canals. By this netting operation, large scale exploitation of other fish and prawn seeds or juveniles are being killed and damaged. During this netting operation, the river dykes are also eroding very fast and damage of the mangrove plantations and mangrove forest flora become inevitable. All these activities are now noted very much detrimental to the existing ecosystem; but without assigning them the alternative livelihood, these poor rural people cannot be diverted from this shrimp seed collection activities. Besides these, fishing of gravid fish, shrimps, prawns, crabs during breeding period is not checked in spite of existing strict rules and existence of laws and regulations. These uncontrolled practices damage enormously on the biota of estuarine environment in Sundarbans; these need protection and control through awareness programmes.

Siltation on River Bed Cause Frequent Flood

Siltation on the riverbed is also alarming, which cause frequent flood and over flow of the estuarine rivers and frequently saline water enter or ingress in the human settlement areas and on the crop fields. This alarming situation happen every year, here and there in the Indian Sundarbans and the poor rural people suffer a lot by losing their expected crop, domestic cattles and residential houses, too. This river bed siltation is also caused due to rapid rate of deforestation activities in the upstream river bank forests, mangrove forest zones, barraging or closing the upstream of the flowing rivers and canals. Besides all these, the tilting effect of the Ganga (Hugli) water flow due to neo-tectonic movement of the Ganga river from western part in West Bengal towards the eastern part in Bangladesh also caused serious problem in Indian Sundarbans, this is due to less supply of sufficient or adequate upstream freshwater throughout the year and that affects badly on the natural regeneration of mangroves and mangrove habitat fauna.

Transport Facilities Not Much Developed

Manual operating country boats, mechanized country boats (*Bhat Bhati*) and occasionally motor launches are the main vehicles for water transport from island to island. Along with less developed transport facilities the timely marketing of fish, vegetable, rice and other local products become problematic. Health scheme, education facilities, entertainment opportunity, cultural, aesthetic scope and other modern opportunities in Sundarbans are less developed.

Uncontrolled Population Growth

During 1951 census, the population of the Indian Sundarbans was only about 12 lakhs and during 1991 census this rural population in Sundarbans became about 32 lakhs; furthermore the population of Sundarbans is estimated 42 lakh during 2001. But, the agricultural land and production from agricultural fields and brackishwater fisheries almost remain static or less changed. As such, the simple question is how this ever increasing population will survive in this problem stricken land masses, where agriculture is not much developed and uncertain for lack of adequate irrigation facilities. Obviously, pressure on the natural forest ecosystem and natural resources of the Sundarbans mangals will increase, day by day. Then there will be no alternatives than to control the rapid population growth and to combat the destruction and damage of the mangrove forest resources.

Agriculture Not Much Developed

The mangrove reclaimed tidal delta was initially developed from the silted up river flat land, surrounded by number of criss-cross tidal rivers, canals and creeks; through these tidal rivers and canals flow highly saline sea water. As such, there was no freshwater river or any such reservoir, where the rain water or the upstream freshwater can be stored for utilisation during the dry summer months. Besides these, the underground non-saline freshwater remain in much deeper zones, which was not possible to pump out for irrigation in these agricultural fields.Electricity was also not available in these remote areas of different deltaic islands in Sundarbans. Furthermore, adequate measures were not taken to develop the substitute for all those difficulties and to overcome these problems. As such, agriculture in Sundarbans is still dependent on monsoon precipitation, which is also concentrated during 4–5 months, *i.e.*, during middle of June to the middle of October. During rest of the period of the year, rain is irregular and based on these routine agriculture cannot be practiced. Agricultural lands in Sundarbans is most fertile, but occasionally saline and it is worse affected due to lack of adequate irrigation facilities.

Enormous Pressure on the Natural Forest and Inevitable Destruction

Sundarbans rural areas have heavy population pressure and less scope for year long fulltime engagement in the agriculture fields; due to absence of any vital industry, the opportunity for employment for the people of the Sundarbans are scarce or very negligible. As such, these rural poor people have to enter this mangrove forest for searching of fuel wood, timber, honey and wax, fish, prawns and crabs, taking all risk of lives. The large scale interaction and dependence of at least 10-12 lakhs rural people in these Sundarbans mangrove forests and estuarine water bodies for extracting the natural forest and estuarine resources these turned the mangal ecosystem poor to poorer, every day. These mangrove natural resources are very sharply declining here. It is felt outmost important from every corner that the conservation and protection of this fragile ecosystem is very essential and most urgent for the posterity of the human welfare and also for the well being of the people of these rural areas.

Photo Plate 131. Forest Guard in STR

Effect of Developmental Activities Without Environmental Impact Assessment

Every developmental activities for the interest of human society is felt very urgent and important, but care should require to be undertaken that this development must not affect the ecosystem, particularly if that ecosystem is fragile, vulnerable and important in respect of having the unique habitats of threatened or endangered flora and fauna. Sundarbans mangrove ecosystem in West Bengal is such a vulnerable ecosystem and day to day it is degrading. As such, the Environment Department, Government of India and Govt. of West Bengal included this mangrove forest zone of Sundarbans within the Coastal Regulatory Zone-I (CRZ-I); furthermore, within this CRZ-I all sort of developmental activities are declared strictly banned. Protection rules are framed, but the protection practices and strict conservation measures are still badly felt in several places. In most of the cases, the interest and necessities of the common people and the question of development of this remote areas are highlighted and also ignored tactfully the damage of the ecosystem without little thought of its consequences. Though it is very late, still time is there for undertaking strict conservation measures in a holistic manner; if any developmental work is hampered for that conservation activities in Sundarbans areas that is much more desirable for the long run interest and sustainable resource management purposes.

Renewal of Natural Wealth and the Prospects

It is much debatable question that nothing is sustainable in this Universe; that is fully correct, but the exploitation of the natural resources should not be such that within very limited period the natural resources will be exploited without maintaining its basic stock for future. Nature has great power of renewal of its resources, but all these activities follow some process and require some time for renewal. In this connection, it is most urgent to ascertain the basic stock, its renewal abilities; taking this in active consideration, extraction or harvest and exploitation activities may be undertaken for any natural resources. The prior consideration and well thought exploitation measures must help for long term benefit and sustenance of the ecosystem for future. The harvest of forest wood and timbers from the slow growing trees of Sundarbans mangals and the large scale collection of *Penaeus monodon* juveniles and destruction of other fish and prawn seeds in Sundarbans estuarine systems must not give the scope of conservation of these forest ecosystem for long. All these need to be considered very sympathetically without taking any bias opinion and optimistic arguments. Development is must for every society, but that should not be desirable to destroy and destruct the natural ecosystem and resources.

Present Scenario of the Sundarbans Mangals and Few Comments

Last but not the least, few very pertaining points, fact and figures are cited here for stating the ill effect and bad consequences in the Indian part of Sundarbans. In most of the cases the results were not achieved and we have very bad experience. First of all, the example of construction of Piyali Dam at Kella near Kultali; the second example is the renovation of World Bank aided Shrimp Farm Project on the Matla river bed at Canning; the third example is the conversion of several hectors existing mangrove forest land at Herobhanga/Jharkhali area in Sundarbans for excavation of fisheries and human habitations; fourth example is the fisheries development at Henry's Islands near Bakkhali; fifth example is the habitat establishment at Jamboodwip by the fishing folks of Bangladesh; sixth example is the incomplete guard wall development project near Kulpi on the Hugli river mouth; seventh example is the setting up of the gasifire wood burning electricity project at Gosaba and eighth is the blockage of the vast number of canals, creeks and even the river flow in the Sundarbans areas and several others.

These entire development projects may not be well thought of and not properly planned, which damaged considerably the mangrove ecosystem in the Sundarbans.

In most of the cases no good results as proposed, were achieved from all such projects, rather these damaged the river flow, silted up the riverbed and destructed the forest areas. All such projects were framed out and implemented for limited interest and owing to biased attitude of some section of people, having little knowledge and/or interest on conservation of this mangrove forest ecosystem.

Though it is very late, still time is there to give the importance of conservation of the Sundarbans mangrove ecosystem, which have ample natural resources and future prospects. The rapid growing local people are saved from the severe natural calamities due to existence of the dense mangrove forest in the coastal areas of south Bengal. It is out of imagination, if the Sundarbans mangrove forest is lost or degraded in future, the river bed will further rise up and tidal saline water will enter in the densely populated villages and agricultural field; then where this vast population will go and settle ! Will Kolkata not be worst affected ?

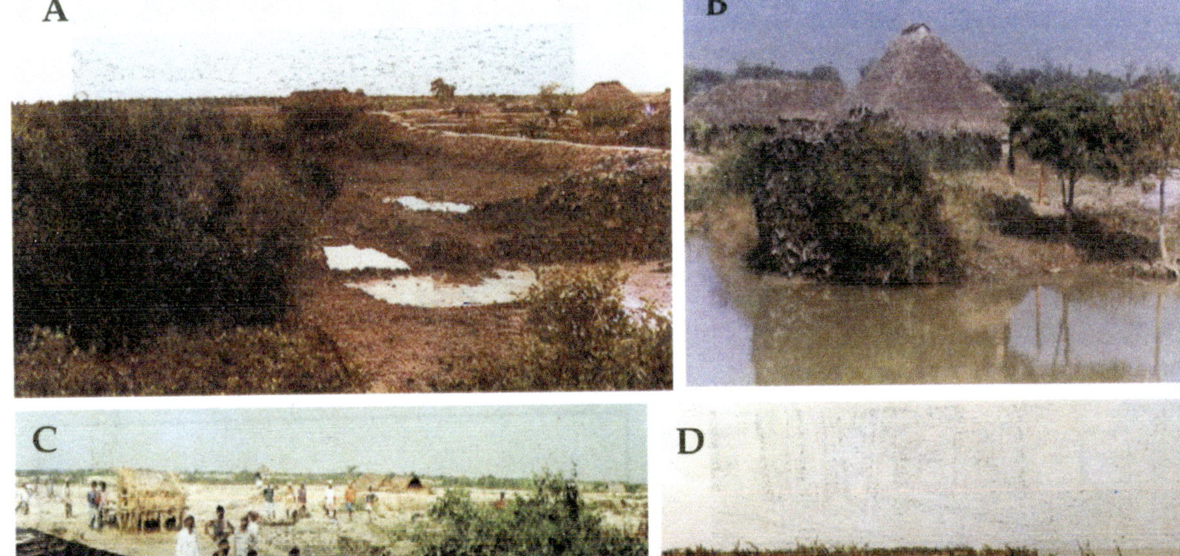

Photo Plate 132. Present scenario of the Sundarbans mangals. A, B & C.. Clearing of mangrove forest area for Brackishwater Fisheries Development at Jharkhali, Sundarbans. D. Degradation of coastal forest areas of the Sundarbans

A. Forest guards engaged in survey & studies

B. Foerst guards engaged in forest survey work.

C. Poacher's ceased boats at STR

D. Exploitation of Mangrove wood

E. Forest Boat in Sundarbans Tiger Reserve

F. Researchers engaged in forest survey

G. Ceased wood in Haldi forest office, STR.

H. Barraging of flowing river at Jharkhali, Sundarbans

Photo Plate 133. Different Research Activities and Investigatyions in The Sundarbans Mangals

9

BIBLIOGRAPHY OF THE MANGROVES AND MANGROVE ECOSYSTEM OF INDIA

Acharji, M. N. & A. K.Mukherjee (1964). Reports on a collection of snakes from Lower Bengal (Reptilia Ophidia). *J. Zool. Soc. Ind.* **16** (1 & 2) : 78 - 81.

Acharya, G. & D. Kamal (1994). Fisheries. pp. 101-114. **In** : Hussain, Z. and Acharya, G. (eds.). *Mangrove of the Sundarbans* Volume two : Bangladesh. IUCN, Bangkok.

Acharya, T. (1985). Captive breeding at Bhagabatpur Crocodile Farm, Sundarbans (West Bengal), India. The case of wild caught female (*Crocodylus porosus* Schneider). *J. Bombay nat. Hist. Soc.*, **82** (2):417-419.

Achuthankutty, C.T. & S. R. Sreekumar Nair (1980). Mangrove Swamps as Fry Source. *Mahasagar Bull of NIO*, **13** (3) : 269 - 276.

Achuthankutty, C. T. & S.R.Sreekumar. Nair (1982). Penaeid prawn population and fry resource a saline grass. *Aeluropus lagopoides* L. *Bot. Mar.*, **24**(7) : 361-364.

Achuthankutty, C. T. & S. R. Sreekumaran Nair (1983). Larval migration and growth of marine prawns in the estuarine habitat. *Mahasagar Bull. of the Nat. Inst. Of Ocean.* **16** : 243 - 250

Achuthankutty, C. T. (1988). Nursery life of the marine prawn, *Metapenaeus dobsoni* (Miers) in the Mandovi estuary, along Goa coast, India. *Indian J. Mar. Sci.*, **17**(4) : 313-316.

Adhikari, M., M. K. Majumdar, S. K. Bandyopadhyay & D. K. Bandyopadhyay (1987). Physico chemical properties of coastal saline soils of Sundarbans, West Bengal. II. Mineralogical make up and Trace Element Status. *J. Indian. Coast. agric. Res.* **5**(1) : 49 - 53.

Adhikari, M., P. P. Chakraborty, D. Ghosh & K. Dey (1987). Physico-chemical properties of coastal Saline Soils of Sundarban, West Bengal. Seasonal effects on physicho-chemical properties & Microbial population *J. Indian. Soc coastal agric. Res*, **5**(1) : 43 - 48.

Adhikari, M., S. K. Bandyopadhyay & M. K. Mazumder (1991). Seasonal effect on soil properties at selected sites and general soil conservation measures for Sundarban, West Bengal *J. Indian. Soc. Coastal agric. Res*, **9** (1&2) : 33 - 36.

Agharkar, S. P. & A.K. Ghosh (1931). The composition of Bengal Flora. *Proc. Indian Science Cong.*, 278.

Agharkar, S. P. (1923). The present position of our knowledge of the aquatic flora of India. *J. Indian Bot. Soc.* **3** : 252 - 260.

Ahmad, A. & Alam, J. M. (1978). The Ganga basin its sub-surfaces their affinity sedimentological and tectonic implications. *Himalayas Geol.* **8**(1). 583-607.

Ahmad, N. (1984). Some aspects of economic resources of Sundarbans mangrove forest of Bangladesh. **In** : *Proc. Asian Symp Mangr. Environ. Res. Management University Malaya*, Kuala Lumpur, Malaysia, pp. 644-651.

Ahmad, S. (1975). Sea snake *Praescutata viperina* (Schmidt) from the coast of Bengal. *Newsl. zool. Surv. India*, **1** (4) : 66-67.

Aksornkoae, S. (1975). Structure regeneration and productivity of mangroves in Thailand. Ph.D. Jhesis, Michigan State University, 109 pp.

Aksornkoae, S. (1980). Distribution growth and survival of seedling of mangrove forest in Thailand. **In** : *Sympo. Forest Regeneration in Southest Asia*, Biotrop. Special Publ. **13** :23-27.

Aksornkoae, S. (1982). Productivity and energy relationships of mangrove populations of *Rhizophora apiculata* in Thailand. *Biotrop Spl. Pub.* **17** : 25-32.

Aksornkoae, S. (1985). Conservation of Mangroves. *The Mangroves : Proc. Nat. Symp. Biol. Util. Cons. Mangroves.* Nov. 1985 : 99-104. (Ed. L. J. Bhosale, Shivaji University, Kolapur).

Aksornkoae, S. (1985). Traditional uses of the mangroves in Thailand. **In**: (eds.) C. D. Field & A. J. Darnall, *Mangrove Ecosystem of Asia and the Pacific, Status, Exploitation and Management*, 104-113.

Aksornkoae, S. (2000). Sustainable use and conservation of mangrove forest resources with emphasis on policy and management practices in Thailand. The United Nations University, Tokyo. 63-76 pp.

Alam, M. K. (1990). Tour report on Sundarbans trip during May. Forest Botany Division, Bangladesh Forest Research Institute, Chittagong 2 pp.

Alam, M. K., N. A. Siddiqi & S. Das (1985). Fodder trees of Bangladesh. UNDP/FAO Project, BGD/79/017, Dhaka. 176 pp.

Ali, A. (1992). Storm surges in Bangladesh. Proceedings of the Workshop on Coastal Zone Management in Bangladesh, (December 27-31). Bangladesh National Commission for UNESCO, Dhaka. 519-528 pp.

Ali, M. L. (1998). Development of fish resource and management technique (in Bangla). Fisheries Resource Development. Fisheries Department, Dhaka. 1-5 pp.

Ali, S. S. (1998). Sundarbans : its resources and ecosystem. Integrated Management of the Ganges flood plains and Sundarbans ecosystem. Khulna University, Khulna 38-49 pp.

Alim, A. (1984). Soil and Plant Interaction in Mangrove Ecosystem. UNDP/UNESCO Training Seminar on Geology, Sedimentology, Erosion and Accretion in Mangrove Areas, Jan. 14 - Feb. 5, Dhaka. 11 pp.

Amarasinghe, M. D. (1988). Socio economic status of human communities of selected mangrove areas on the west coast of Sri Lanka. Mangrove Ecosystems Occasional Papers. UNDP/UNESCO. Mangrove project Asian and Pacific. No. 3. 19 pp.

Ambasi, S. K., H. S. Sen & N.K., Tyagi (1996). Improved crop calendar for maximizing productivity of the land through rainwater management in Sundarbans. *J. of Indian Soc. Coastal Agri. Research* **14** (1&2) : 115-123.

Anand, Swarup & B. K. Bandyopadhyay (1992). Nutrient management in irrigated soil with physical constraints. **In** : *Advances in Research on sustainable irrigation in saline environment.* (N.K. Tyagi, S.K. Mishra, P.S. Minhas & N.T. Singh (eds.). CSSRI, Karnal, India, pp. 137 - 151.

Anderson, J. (1872). On a supposed new monkey from the Sundarbans. *Proc. zoo. Soc. Lond.*, 529-533.

Annandale, N. (1907). The fauna of brackish ponds at Port Canning, Lower Bengal. 4. Hydrozoa. *Rec. Indian Mus.*, **1** : 139-144.

Annandale, N. (1907). The fauna of brackish ponds at Port Canning, Lower Bengal. 6. Observations on Polyzoa, with further notes on the ponds. *Rec. Indian Mus.*, **1** : 197-205.

Annandale, N. (1907). The fauna of the brackish ponds at Port Canning, Lower Bengal. 2. An isolated race of *Metridium schillerianum* (Stoliczka). *Rec. Indian Mus.*, **1**: 47-74.

Annandale, N. (1907). The fauna of the brackishwater ponds at Port Canning, Lower Bengal. I. Introduction and preliminary account of the Fauna. *Rec. Indian Mus.*, **1** : 35 - 43.

Annandale, N. (1908). The fauna of the brackish ponda at Port Canning, Lower Bengal. Further observations on the Polyzoa, with the description of a new genus of *Entoprocta. Rec. Indian Mus.*, **2** : 11-19.

Annandale, N. (1924). Report on bivalve mollusc injuring brick-work in th Calcutta Docks. *Rep. comm. Deter. Structures, Inst. Civil Engin.* (London), **4** : 59-60.

Anthony, M.K.D., V.N. Panakalarao & Sarkaranarayana (1974). Studies on the monsoonal flow of the Mandovi Estuary. *Mahasagar Bull. Natn. Inst. Oceanogr.* **7** :69 - 72.

APHA (1989). Standard Methods for the Examination of Water and Waste Water. 16th Edn., A. J. Taras, A. E. Greenberg, R. D. Hoak & M. C. Rand (Eds.). 874p. APHA, Washington DC.

Ara, R., S. R. Merry & N. A. Siddiqi (1997). Silviculture and Uses of Twelve Medicinal Plants of Bangladesh. *Bulletin 7.* Minor Forest Products Series. Bangladesh Forest Research Institute, Chittagong. 35pp.

Armstrong, D. A., D. Chippendale, A. W. Knight & J. E. Colt (1978). Interaction of ionised and unionised ammonia on short term survival and growth of the Malaysian Prawn, *Macrobrachium rosenbergii. Biology Bulletin* **154** (1) : 15-31.

Ascoli, F. D. (1921). A revue history of Sundarban from 1870 to 1920. Calcutta, The Bengal Secretariate Book Depot. 172 F. 467.2 P. I. vi, ii, 159p. fold map. 25 cm.

Aziz, A. & A. K. M. N. Islam (1979). Marine dinoflagellates from the Bay of Bengal, Bangladesh. *J. Bangladesh Acad. Sci.* **3**(1-2) : 41 - 49.

Aziz, A. & A. K. M. Nurul Islam (1986). Lagoon algae of St. Martin's Island, Bangladesh. *Dhaka University Stud.* E, **1**(1) : 45-52.

Backer, C.A. & C.G.G.J. van Steenis (1951). Sonneratiaceae, Flora Malesiana I, **4** : 280-289.

Badarudeen, A., D. Padmalal & K. Sajan (1998). Organic carbon and total phosphorus in the sediments of some selected mangrove ecosystems of Kerala, Southwest coast of India. *J. Geol. Soc. India*, **51**(5) : 679-684.

Badarudeen, A., K. T. Damodaran, K. Sajan & D. Padmalal (1996). Texture and geochemistry of the sediments of a tropical mangrove ecosystem, southwest coast of India. *Environ. Geol.*, **27**(3) : 167-169.

Bagchi, K. (1956). The Baratala-Matla influence: A preliminary study of Deltaic problems. **In** :Geographical review of India. V.P. 30-33+xviii.

Bagchi, K. & K. N. Mukherjee (1978). Diagnostic Survey of Deltaic West Bengal. Calcutta, pp. 46-75.

Bagchi, K. (1944). The Ganges Delta. Calcutta University Press, Calcutta. pp. 157.

Bagchi, S., A. Matilal, A. K.Shaw & B. B. Mukherjee (1988). Lipids and waxes in leaves of some mangrove plants of Sundarbans, India. *Indian J. Mar. Sci.* **17**(2) : 150-152.

Bahuguna, Anjali, S. Nayak, K. N. Prasad, S. A. Kadri & C. R. Trivedi (1990). Mapping of Long term Changes in the Tidal Wetland Areas of the Gulf of Kachch. Scientific Note. Space Application Centre, Ahmedabad. IRS-UP/SAC/MCE/SN/29/91, 15 pp.

Bainbridge, R. (1953). Studies on the inter-relationships of zoo - and phyto - plankton. *J. Mar. Biol. Ass.* U. K. **32** : 385 - 443.

Baksha, M. W. & M. R. Islam (1997). Insect pests attacking mangrove nursery and plantations of Keora (*Sonneratia apetala*) along the coastal belt of Bangladesh. *Bangladesh Journal of Forest Science* **26**(2) : 69-75.

Baksha, M. W. & M. R. Islam (1999). Infestation intensity of *Zeuzera confera* Walker (Cossidae : Lepideptera) in *Sonneratia apetala* plantations of Bangladesh. *Bangladesh Journal of Forest Sciences*, **28** (1) : 1-6.

Baksha, M. W. (1996). Beehole borer infestation in coastal plantations in Bangladesh and possible management options. *Wallacena* **77** : 17-20.

Baksha, M. W., M.R. Islam & S.H. Chowdhury (1990). Notes on Keora Beehole borer - a pest on keora in Bangladesh. *Proc. Seventh Natl. Conf. Bangladesh.* Dhaka. 113-117.

Bal, A. R. & S. K. Dutta (1984). Studies on the occurrence and distribution of salt glands and stomata in some mangrove species. *J. Indian Soc. Coastal agric. Res.* **2**(1) : 5-9.

Balachandra, N. (1988). A Comprehensive Account of the Mangrove Vegetation of Andaman and Nicobar Islands. *Indian For.* **114** : 741-751.

Balakrishna, K., B. R. Manjunatha & R. Shankar (1997). A laboratory study of the flocculation of dissolved heavy metals in Kali river during estuarine mixing, west coast of India. *J. Geol. Soc. India*, **50**(6) : 753-758.

Balakrishnan Nair, N., K. Dharmaraj, P. K. Abdul Azis, M. Arunachalam, K. Krishnakumar & N. K. Balasubramanian (1984). Ecology of Indian estuaries. 8. Inorganic nutrients in the Ashtamudi estuary. *Mahasagar*, **17**(1) : 19-32.

Balmforth, E. B. (1985). Observations on sundri top dying descriptive sampling in the Sundarbans Reserved Forest. Draft Working Paper of UNDP/FAO Project BGD/79/017. 32 pp.

Bandopadhyay, A. K. & D. Sarkar (1987). Occurrence of Acid Saline Soils in Coastal area of Sundarbans Area of West Bengal. *J. Ind. Soc. Soil Sci.* **35** : 542 - 544.

Bandopadhyay, A. K. (1966). Forests of Sundarbans. **In** : *Centenary Commemoration Volume, West Bengal Forest* (1964) pp. 166 - 175, Published by the D.F.O. Planning and Statistical Cell, Writers' Building, Calcutta.

Bandopadhyay, A. K. (1984). Soil and Water Characteristics of the mangrove forest of Sundarbans (India). **In** : *Report 2nd Introductory Training Course on Mangrove Ecosystems.* UNDP/UNESCO Regional Project RAS/ 79/ 002.

Bandopadhyay, A. K. (1986). Characterization and Management of Coastal Salt Affected Soils. CSSRI, Annual Report, pp. 184 - 188.

Bandopadhyay, A. K. (1987). A note on effect of reclamation through marginal embankments on coastal eco-system of Sundarbans. *J. Indian Soc. Coastal agric. Res.* **5** (1) : 39 - 41.

Bandopadhyay, A.K. (1986). Soil and Water Characteristics of the Mangrove Forests of Sundarbans (India). *Indian For.* **112** : 58 - 65.

Bandopadhyay, B. K., A. K. Bandopadhyay & H. S. Sen (1987). Physico- Chemical Characteristics of Coastal Saline Soils of India. *J. Indian Soc. Coastal Res.* **5**(1) : 1-14.

Bandopadhyaya, B.K., H.S. Sen & B. Maji (1996). Transformation of Sulphate salts and black colour formation in coastal saline soils of Sundarbans under paddy cultivation. *J. Indian Soc. Coastal agric. Res.* **14** (1 & 2) : 39 - 45.

Bandyopadhyay, A. K. & B. K. Bandyopadhyay (1984). Problem and management of coastal saline soils of West Bengal. *Fert. News* **29** (9) : 18 - 22, 47.

Bandyopadhyay, A. K. & H. S. Sen (1975). Quality of surface and sub-surface waters of Sundarbans with reference to irrigation during dry periods. *Technology* **12** : 151-154.

Bandyopadhyay, A. K. (1984). Soil and water characteristics of the mangrove forest of Sundarban (India). **In**: *Report on 2nd Introductory Training Course on Mangrove Eco-system*, UNDP/UNESCO Regional Project RAS/79/002.

Bandyopadhyay, A.K. (1984). Studies on the effect of ground water table on salinisation of heavy texture soils of Sundarban and to effect on crop growth. *J. Indian Soc. Coastal agric. Res.*, **2** (1) : 26-28.

Bandyopadhyay, A. K. (1985). Management techniques for increased crop production in denuded mangal land with special reference to Sundarbans. *J. Indian Soc. Coastal Agric. Res.*, **3**(1) : 7-14.

Bandyopadhyay, A. K. (1986). Soil and water characteristics of the mangrove of Sundarbans (India). **In** : Report of UNDP/UNESCO Training Course on Mangrove Ecosystems, India. Mangrove project Asian and Pacific. 49-51 pp.

Bandyopadhyay, A. K. (1989). Leaching of acid saline soils of Sundarbans. *Journal of Indian Society of Soil Science* **37** : 416-417.

Bandyopadhyay, A.K. (1986). A note on the effect of liming on soil properties and crop growth in acid saline soils of Sundarbans, *J. Indian. Soc. Coastal agric. Res.*, **4**(1) : 73-75.

Bandyopadhyay, A.K. (1988). Effect of lime, superphosphate and oyster shell and rock phosphate on soil properties and crop growth in coastal acid saline soils. *Journal of Indian Society of Soil Science* **36** : 581-583.

Bandyopadhyay, B. K. & A. K. Bandyopadhyay (1983). Effect of salinity on mineral ization and immobilization of nitrogen in a coastal saline soil of West Bengal. *Indian Agric.*, **27** : 41-50.

Bandyopadhyay, B. K. & B. Maji (1995). Nature of acid soils of Sundarbans delta and sustainability of classifying them as acid sulphate or potential acid sulphate soils. *J. Indian Soc. Soil Sci.* **43** : 251 - 255.

Bandyopadhyay, B. K. & B. Maji (1995). Source of acidity in some coastal acid saline soils of Sundarbans. West Bengal. *Salinity Newsletter*. **6** (2-3) : 6.

Bandyopadhyay, B. K. & S. K. Dutt (1982). Agronomy of rice cultivation in coastal saline soils (Sundarbans) of West Bengal. **In** : Rice in West Bengal III (ed. D.K. Mukherjee), Director of Agriclture, Govt. of West Bengal., pp. 101-108.

Bandyopadhyay, B. K., A. K. Bandyopadhyay & H. S. Sen (1987). Physico-chemical characteristics of coastal saline soils, *Journal of Indian Society of Coastal Agricultural Research* **5** : 1-14.

Bandyopadhyay, B. K., G. C. Sahu & B. Maji (1997). Status, nature and composition of organic matter in coastal areas. *J. Indian Soc.Soil Sci. Bull.* **19** : 58 - 67.

Bandyopadhyay, B. K., H. S. Sen & B. Maji (1991).Studies on the growth and nutrition of crops on coastal saline soil under excess soil moisture condition. *J. Indian Soc. Coastal agric. Res.* **9** : 309 - 316.

Banerjee, L. K. & D. Ghosh (1998). Species Diversity and Distribution of Mangroves in India. **In** : *An Anthology of Indian Mangroves*. Envis Centre, CAS in marine biology, Annamalai University. pp. 20-24.

Banerjee, L. K. & T. A. Rao (1985). Mangals of Mahanadi Delta, Cuttack District, Orissa State, India. Krishnamurthy, V. & A. G.Untawale (eds.), *Proc. Marine Plants Their Biology, Chemistry and Utilization*, Goa, 137-152.

Banerjee, L. K. & T. A. Rao (1985). Mangals of the Mahanadi delta, Cuttack district, Orissa State, India. **In** *Marine Plants*, Eds. Krishna Murthy V. and Untawale, A.G. Seaweeds Res. and Util.Assn., Madras.

Banerjee, L. K. (1984). Vegetation of the Bhitarkanika Sanctuary, Orissa State, *J. Econ. Tax. Botany* **5** : 1065-1079.

Banerjee, L. K. (1985). Comparitive Study on the Mangrove of the Sundarbans and that of the Mahanadi Delta in Eastern India. The Mangroves : *Proc. Nat. Symp. Biol. Util. Cons. Mangroves* : Nov. 1985 : 527-530, Kolapur, India.

Banerjee, L. K. (1987). Ecological Studies on the Mangals of Mahanadi Estuarine Delta, Orissa. *Trop. Ecol.* **82** : 117-125.

Banerjee, L. K., A. R. K. Sastry & M. P. Nayar (1989). Mangroves in India - Identification Manual, Botanical Survey of India, Calcutta.

Banerjee, R. K. (1981). Effect of continued submergence on the soil of saline swamps of lower Sundarbans *Indian J. Agri. Chem.* II(I) : 29-34.

Banerjee, R. K., A. C. Nandi, H. C. Joshi & B. B. Das (1988). Impact of Calcutta Metropolitan effluent on the ecology of Kulti Estuaries, *J. Inland Fish. Soc. India.* **18** (1): 36-40.

Banerjee, S. C., R. K. Banerjee, B. B. Pakrasi, & A. Moitra (1976). Salinity problem of pond soils in the deltaic region of lower Sunderbans. *J. Indian Soc. Soil. Sci.*, **24**(1): 84-85.

Banerjee,L. K. & T. A. Rao (1990). Mangroves of Orissa coast and their ecology. Bishen Singh Mahendra Pal Singh, Dehra Dun.

Banerjee,L. K., A. R. K. Sastry & M. P. Nayar (1989). Mangroves in India - Identification manual. Botanical Survey of India, Calcutta.

Banerji, J. (1954). The mangrove forests of the Andamans. *Wld. For. Congr.* : 425-430.

Banus, M.D. & S.E. Kolehmairen (1975). Floating, rooting, growth of Red mangrove (*Rhizophora mangle* L.) Seedlings : Effect on expansion of mangroves in South-Western Puerto Rico. **In** : G.E. Walsh, S.C.Snedakar & H.J.Teas (eds.).

Proceedings of the International Symposium on Biology and Management of Mangrove. Institute of Food and Agricultural Science, University of Florida, Gainesville, Florida, Vol.I, pp.370-384.

Barth, H. (1982). The Biogeography of Mangroves. Chap. 3. **In** : *Tasks for Vegetation Science*, (eds.) D.N.Sen & K.S.Rajpurohit, Vol. 2. The Hague, Junk.

Basak, A. C. (1991). Collar rot of Keora (*Sonneratia apetala*) seedlings in the coastal afforestation nurseries of Bangladesh and its control. *Bangladesh Journal of Forest Science* **20** (1&2) : 60-61.

Basak, R. K. & Guha Bakshi, D. N. (1977): Floristic studies on the Lower Ganga Plain of the State of West Bengal: Present Status and Future Strategies. *Bull.bot.Surv.India*, **19**(1-4):42-55.

Basak, R. K. (1976): The Bibliography on the Flora and Vegetation of Bengal with an introductory note. *Bull. Bot. Surv. India*. **15** (22 - 38).

Basak, U. C., A. B. Das & P. Das (1996). Chlorophylls, Carotenoids, Proteins and secondary metabolites in leaves of 14 species of mangroves. *Bull. mar. Sci.*, **58**(3) : 654-659.

Basu, K. (1965). Cultivation of mangrove seedlings away from their natural habitat, *Cur. Sci.* **34**: 438 - 49.

Basu, N. C. & B. B. Pakrasi (1979). Brackishwater fish and prawn seed potentialities of Bakkhali area in lower Sunderbans, West Bengal. *J. Inland Fish Soc. India*, **11**(1) : 40-48.

Begum, S., M. A. Rahman & S. K. Khisa (1999). Checklist and host Index of Parasitic Algae, Bacteria, Fungi and Mistletoes on Forest Trees and Timber in Bangladesh. Bangladesh Forest Research Institute, Chittagong. 60 pp.

Benakappa, S., M. P. M. Reddy & V. Hariharan (1979). Distribution of Plankton in the Arabian Sea between Mukka and Kaup, South Kanara. *Mysore J. Agri. Sci* **13** : 454 -463.

Bharucha, F. R. & B. S. Navalkar (1942). Studies in the ecology of mangroves : The Chloride-content of sea water, soil solution and the leaf cell sap of mangroves. *J. Univ. Bombay*.**10** : 98-106.

Bharucha, F. R. & B. S. Navalkar (1942). Studies in the ecology of mangroves : The Chloride-content of sea water, soil solution and the leaf cell sap of mangroves. *J. Univ. Bombay*.**10** : 98-106.

Bharucha, F. R. & B. S. Navalkar (1942). Studies in the ecology of mangroves : The Chloride-content of sea water, soil solution and the leaf cell sap of mangroves. *J. Univ. Bombay*.**10** : 98-106.

Bharucha, F. R. & V. S. Shirke (1947). A study of important catabolic changes in the seedlings of *Avicennia officinalis* L. *J. Univ. Bombay*. **15** : 1-14.

Bhat, S. R. & A. G.Untawale (1987). *Bruguiera cylindrica* (L.) BI. (Rhizophoraceae) -a new locality record from the West Coast of India. *Curr. Sci.*, 56(11): 604-606.

Bhattacharya, A. & S. Choudhury (1985). The role of plants and moisture in dune formation: A study from the tidal island deposits of the deltaic Sundarbans facing bay of Bengal, in *'The Mangroves'*, (Ed. L.J. Bhosale), Shivaji University Press, Kolhapur, pp. 170-174.

Bhattacharya, S. (1972). Soils of Sundarbans. *Proc. Interdisciplinary Symp. Bhagirathi-Hoogly Basin*, Calcutta, pp. 101-105.

Bhattasali, N. K. (1941). Antiquity of the Lower Ganges and its Courses. *Sci. & Cult*, **7**:233-239.

Bhosale, L. J. & B. A. Karadge (1975). Photosynthetic efficiency and productivity of mangroves of western India. *Bull. Dept. Mar. Sci. Univ.* Cochin, **7**(1) : 205-211.

Bhosale, L. J. & L. S. Shinde (1983). Photosynthetic products and enzymes in a mangrove, *Aegiceras corniculatum* (L.) Blanco and a halophyte, *Sesuvium portulacastrum* L. *Photosynthetica*, **17**(1) : 59-63.

Bhosale, L. J. (1974). Physiology of salt tolerance of plants. Ph. D. Thesis, Shivaji Univ., Kolhapur.

Bhosale, L. J. (1985). Free amino acids in mangroves - significance. **In** : *The Mangroves : Proc. Nat. Symp. Biol. Util. Mangroves.* pp. 89 - 98.

Bhosale, L. J. (1985). The Mangroves. *Proc. Nat. Symp. on Biol. Util. & Conserv. of Mangroves.* (ed.) Shivaji University, Kolhapur.

Bhosale, L.J. (1978). Ecophysiological studies on the mangrove from western coast of India. U. G. C. Project Report, Shivaji University Publication, Kolhapur, 81 p.

Bhosale, L. J. (1989). Mangroves of Maharashtra : An overview, **In** : *Wetlands, Mangroves and Biosphere Reserves, Govt. of India, MOE&F*, 187-196.

Bhosale, L. J. (1998). Resource Sustainability and Mangrove Ecosystem. **In** : *An Anthology of Indian Mangroves.* Envis Centre, CAS in Marine Biology, Annamalai University. Pp. 37-38.

Bhosale, N. B., M. A. Rokade & M. D. Zingde (1985). Short term variations in particulate matter in Mahi river estuary. *Mahasagar*. **18**(4) : 449-455.

Bhosale, N. B., M. A. Rokade & M. D. Zingde (1985). Total dissolved carbohydrate in Mahi river estuary. *Mahasagar*, **18**(4) : 507-509.

Bhosale, N. B., V. K. Dhargalkar, S. G. P. Matondkar & Sayeeda Bukhari (1976). Biochemical composition of mangroves leaves from Goa. *Ind. J. Marine Science* **5**(2) : 239-241.

Bhunia, A. & A. Choudhury (1982). Studies on primary production and the nutrients of the estuarine water around Sagar Island. *Indian J. Mar. Sci.* **11** : 87 - 89.

Bhunia, A. & A. Choudhury (1985). Estimation of gross primary production of the river Saptamukhi, Sundarbans, West Bengal. **In**: *The Mangroves*,(Ed. L. J. Bhosale), *Proc. National Seminar on Biology, Utilization and Conservation of Mangroves*, Shivaji University Press, Kolhapur, pp. 175 - 179.

Bhunia, A. B. (1979). Ecology of the tidal creeks and mudflats of Sagar Island (Sundarbnans) India, Ph. D. Thesis, Calcutta University, Calcutta.

Biswas, K. (1926). Flora of Salt lakes, Calcutta. *J. Dept. Sci. Univ. Calcutta.* **8** : 1 - 47.

Biswas, K.P. (1934). A comparative study of Indian species of *Avicennia. Notes. Royal Bot. Gard.* **18** : 159-166.

Blasco, F. (1975). The Mangroves of India. Institute Francais de Pondichery. *Trans. Sci. Tech. Vol.* **14** : 1-175 pp.

Blasco, F. (1977). **In** V.J.Chapman (ed.) : *Ecosystem of the world*. Vol I. Wet Coastal ecosystems. Elsevier Science Publishing Co. New York, pp. 241-260.

Blasco, F. (1977). Outlines of ecology, botany and forestry of the mangals of the Indian subcontinent. Chap. 12. In V.J. Chapman (ed.). *Wet Coastal Ecosystems.* Vol. I, New York : Ed. V.J. Chapman, Elesvier Sci. Publ. Co. Australian Oxford, New York, pp.241-260.

Blasco, F. (1982). An inventory of the Gangetic mangrove from Landset satellite data and SPOT stimulations. Conference Leminaire franco-asique 22-26 Nov. 1982.

Blasco, F. (1984). Climate factors and the biology of mangroves plants. In *: The Mangrove Ecosystem : Research Methods* (eds. S.C. Snedekar & U.G. Snedaker) UNESCO, pp. 18-35. Blasco, F. & C. Caratini (1973). Mangrove de Pichavaram (Tamil Nadu), (Indie du sud), Phytogeographic et Polynologie. *Trav. Doc. Cent. Etudes Geigr. Trop.* **8**: 164-185.

Blasco, F., C. Cartini, S. Chanda, & G. Thanikaimoni (1974). Main characteristics of India mangroves; Proc. Int. Symp. Biol. & management of mangroves, pp. 71 - 81, Hawaii.

Blasco, F., R. Kerrest & C. Marius (1985). Consideration of some ecological factors influencing the biology of Indian Mangroves. The Mangroves : *Proc. Nat. Symp. Biol. Util. Cons. Mangroves*, Nov. 1985 : 135-145. L.J.Bhosale (ed.) Kolhapur, Shivaji University.

Blatter, E. (1905). The mangrove of Bombay Presidency and its biology. *J. Bombay nat. Hist. Soc.*, **16**: 644-656.

Blatter, E. (1908). On the flora of Kutch, Part-I. *J. Bombay Nat. Hist. Soc.* **18** (4) : 756-777.

Blatter, E. (1909). On the Flora of Kutch, Part II. *J. Bombay Nat. Hist. Soc.* **19** : 157-176.

Blatter, E. (1928). The flora of Indus Delta. *J. Indian Bot. Soc.* **20** : 22-43.

Blower, J. H. (1985). Sundarbans Forest Inventory Project, Bangladesh : Wildlife Conservation in the Sundarbans. Overseas Development Administration, England. 39 pp.

Bole, P. V. & F. R. Bharucha (1954). Osmotic relations of the leaves of *Avicennia alba* Bl. *J. Univ. Bombay*, **22** : 50-54.

Bole, P. V. & J. M. Pathak (1988). Flora of Saurashtra. B.S.I., Calcutta

Boonruang, P. (1984). The rate of degradation of mangrove leaves. *Rhizophora apiculata* and *Avicennia marina* at Phuket Island, Western Peninsula of Thailand. **In** : Soepadma, E. *et al.* (eds.) Proc. *Asian Symp. Mangr. Env. Res. Mang.*, Univ. Malaya, Kualumpur, Malaysia.

Bose, B. B. (1956). Observations on the hydrology of the Hooghly estuary. *Indian J. Fish*, **3** (1) : 101-118.

Bostrom, T. E. & C. D. Field (1973). Electrical potentialities in the salt gland of *Aegiceras*. In : W.P. Anderson (ed.). *Ion Transport in Plants.* Academic Press, London, pp. 385-392.

Boto, K. G. & J. S. Bunt (1981). Tidal export of particulate organic matter from a northern Australian mangrove system. *Estuar. Coast. Shelf Sci.* **13** : 247-255.

Boto, K. G. & J. S. Bunt (1982). Carbon export from mangroves. **In** *Cycling of carbon, nitrogen, sulphar and phosphorus in terrestrial and aquatic ecosystem* (eds.) J.R. Freney & I.E. Galbally, Springerverlag, pp.105-110.

Boto, K. G., J. S. Bunt & J. T. Wellington (1984). Variations in mangrove forst productivity in northern Australia and Papua New Guinea, *Estuar. Coast Shelf Sci.* **19** : 321-329.

Bowman, H. H. M. (1917). Ecology and physiology of red mangroves. *Proc. Amer. Philo. Soc.* **56** : 589 -672.

Bowman, H. H. M. (1921). Histological variations in *Rhizophora mangle* L., Papers of the Michigan Academy of Science, **22:** 129-134.

Breen, C. M. & I. D. Jones (1969). Observation on the anatomy of foliar nodes of young *Bruguiera gymnorhiza, Journal South African Botany*, **35** : 211-218.

Breteler, F. J. (1969). The Atlantic species of *Rhizophora. Acta. Bot. Neerl.* **18** : 434-444.

Buckley, T. A. (1929). Mangrove bark as a tanning material. *Maly For. Rec.* **7** : 40 pp.

Bunt, J. S. & W. T. Williams (1980). Studies in the analysis of data from Australian tidal forests (Mangrove). I. Vegetational sequences and their graphic representation, *Aust. J. Ecol.* **5** : 385-390.

Bunt, J. S. & W. T. Williams (1981). Vegetational relationship in the mangroves of tropical Australia. *Mar. Ecol. Prog. Ser.* **4** : 349-359.

Bunt, J. S. (1987). Country Report. Australia. In : Mangroves of Asia and the pacific : Status and Management. UNDP/ UNESCO (Ras/79/002), Phillippines. 263-280 pp.

Bunt, J. S., K. G. Boto & G. Boto (1979). A survey method for estimating potential levels of mangrove forest primary production, *Marine Biology*, **52:** 123-128.

Canonizado, J. A. & M. A. Hossain (1998). Integrated Forest Management Plan for the Sundarbans Reserved Forest (Final Draft). Mandala Agricultural Development Corporation and forest Department, Dhaka.

Caratini, C., F. Blasco & G. Thanikaimoni (1973). Relationship between the pollen spectra and the vegetation of a south Indian mangrove . *Pollen et spores*. 15 :281-292.

Carroll, J. & W. S. Moore (1993). Uranium removal during low discharge in the Ganges Brahmaputra mixing zone. Geochim.Cosmochim.Acta, 57(21-22) : 4987-4995.

Carroll, J., K. K. Falkner, E. T. Brown & W. S. Moore (1993). The role of the Ganges Brahmaputra mixing zone in supplying barium and super (236) Ra to the Bay of Bengal. Geochim. Cosmochim. Acta, 57(13) : 2981-2990.

Chaffey, D. R. & J. H. Sandom (1985). Sundarbans Forest Inventory Project, Bangladesh. A Glossary of Vernacular Plant Names and a Field Key to the Trees. Overseas Development Administration England. 23 pp.

Chaffey, D. R., F. R. Miller & J. H. Sandoin (1985). A forest inventing of the Sundarbans, Bangladesh. Main Report. Overseas Development Administration. England. 196 pp.

Chai, P. K. (1980). Mangrove forest of Sarawak. Proc. Workshop Mang. Estu. Veg. Malaysia Univ. Partanian, Malaya.

Chakrabarti, C., S. K.Kundu, P. B.Ghosh & A. Choudhury (1993). A preliminary study on certain trace metals in some plant and animal organisms from mangroves of Sundarbans India. *Mahasagar* **26**(1) : 17-20.

Chakrabarti, K (1986). Plant and animal dynamics in Sundarbans Mangrove - an ecological study. *Mahasagar* **19** (2) : 147-151.

Chakrabarti, K. & A. B. Chaudhuri (1973). Wildlife biology of the Sundarbans Forests- A study of the birds of Sundarbans, *Sci. & cult*. **39** : 8-16.

Chakrabarti, K. & A. B. Chaudhuri (1979). A Decade study on the behaviour pattern of tiger. *Proc. Internatn. Symp. on Tiger*, New Delhi.

Chakrabarti, K. & A. B. Chaudhuri (1979). An ecological reconnaissance of tiger, *Panthera tigris* Linnaeus, habit and conservation of forests and wildlife. *Proc. Intern. Symp. on Tiger*, New Delhi, 22-24 Feb., 1979.

Chakrabarti, K. & A. B. Chaudhuri, (1973). Wildife Biology of the Sundarbans Forests- A study of the habit and habitat of the tigers, *Bull. bot. Soc. Bengal*. **26** (1) : 63-66.

Chakrabarti, K. & A. B. Choudhuri (1972). Wildlife biology of the Sundarbans forests - honey production and behaviour pattern of the honey bee. *Science & Culture* **38** : 28-31.

Chakrabarti, K. (1976). Sundarbans Tiger Project, its purpose and problem in Implementation, *Sci. & Cult*. **42**(3): 131 - 134.

Chakrabarti, K. (1977). A statistical approach to wild animal population analysis. *Cheetal*, **19** (1) : 19-23.

Chakrabarti, K. (1978). Ecology of the Sundarbans tiger with particular reference of range of habitats and adaptibility to changes. *Cheetal*, **20**(2) & (3) 3-15, Dec. 1978/March 1979.

Chakrabarti, K. (1978-79). A preliminary study on the fishery resources of the mangrove swamps of Sundarbans, West Bengal. Indian Fish. Assoc. 8-9: 44-48.

Chakrabarti, K. (1979). A study on statistical ecology of *Crocodylus porosus* in estuarine tract of Sundarbans. Cheetal, 21 (2) : 26-32.

Chakrabarti, K. (1979). Project tiger in Sundarbans. *Proc. Intern. Symp. On Tiger*, Fauna. Chakrabarti, K. (1979). Sundarbans a unique habitat for tiger-a study on its diversity. *Proc. Internatn. symp. on Tiger*, New Delhi, 22-24., 1979.

Chakrabarti, K. (1980). Successful tranquilisation of Sundarbans Tiger. *Tiger Paper*, **7**(4) : 10.

Chakrabarti, K. (1981). A growth study of a crab species *Scylla serrata* (Forskal) in the Sundarbans. *Indian J. Forestry*, **4**(2): 102-106.

Chakrabarti, K. (1982). Dynamics of flora- fauna diversity in the mangroves of Sundarbans in comparison to laterite forests of South Bengal and North Bengal forests-a bio-ecological study. *Indian J. Forestry*, 7(3): 220-232.

Chakrabarti, K. (1982). The Sundarbans Tiger. *J. Bombay. nat. Hist. Coc.* **81**(2) : 459-460.

Chakrabarti, K. (1984). An eco-biometrical study on tiger in estuarine ecosystem of Sundarbans. *Indian forester*, **110**(6) : 540-551.

Chakrabarti, K. (1985). Statistical ecology of Sundarbans tiger. *Tiger paper*, **11**(3). 29-31.

Chakrabarti, K. (1985). The Sundarbans forests where life began. *Sanctuary*. **5**(4): 324.

Chakrabarti, K. (1986). Generic and species diversity of animal, vegetation dynamics of Sundarbans mangrove, south Bengal laterite tracts of West Bengal and no the Bengal Foroeosts- an ecological study. *Indian Forest*, **112**(5): 407-416.

Chakrabarti, K. (1991). **Man, Plant and Animal Interaction**. Calcutta : Darbari Prokashan. 335 and 359.

Chakrabarti, K. (1992). **Man-eating Tiger**. Calcutta : Darbari Prokashan.

Chakrabarti, K.(1983). **Tiger and Man**. *Indus Chronicle*, **5**(15) : 5-15.

Chakrabarti, P. (1995). Evolutionary history of the coastal quaternaries of the Bengal Plain, India. *Proc. Indian natl. Sci. Acad. Phys. Sci.*, **61**(5) : 343-354.

Chakrabarti, P. K. & G. N. Chattopadhyay (1988). Impact of Farakka Barrage on the estuarine ecology of the Hooghly - Matla System. **In** : *Conservation and Management of Inland Capture Fishery Resources*. Eds. A.G. Jhingran & V.V. Sugunan. Bull No. 57. CICFRI - p. 189 - 196.

Chakrabarti, P. K. & K. R. Naskar (1988). Role of Mangrove in Estuarine Fisheries Development. **In**:*Conservation and Resource Management of Inland Capture Fisheries Resource of India. Ed.* A. G. Jhingran & V. V. Sugunan, I.F.S.I., Publication pp.229-233.

Chakrabarti, R. K. & M. L. Bhowmick (1985). Ecology and productivity of a paddy cum fish culture on river Hooghly. *J. Indian Soc. coastal agric. Res.*, **3**(1) : 137-148.

Chakrabarti, R. K. (1971). Migratory winter fishery and dry-fish industry in Hooghly-Matlah estuary (West Bengal). *J. Inland Fish. Soc. India*, **3** : 131-132.

Chakrabarti, R. K. (1986). Pelagic fishery resources off lower Sundarbans, West Bengal and its exploitation by power boats. *Seafd. Export J.*, **18**(5) : 17-20.

Chakrabarti, R. K., D. D. Halder, N. K.Das, S. K.Mandal & M. L.Bhowmik (1986). Growth of *Penaeus monodon* Fabricus under different environmental conditions. *Aquaculture* **51** : 189-194

Chakrabarti, R. K., D. K. De & M. Subrahamanyam (1982). Observations on the availability of Bagda, *Penaeus monodon* (Fabricius) and Chapra, *P. indicus* (H. Milne Edwards) seed in the Hooghly estuary around Narpur and Uluberia, West Bengal. *Sci. & Cult.*, **48**

Chakrabarti, R. K., M. L. Bhowmick & Apurba Ghosh (1987). On a New Method for nursery rearing of *Penaeus monodon* (Fab.) at High Stocking Density. *J. Indian Soc. Coastal agric. Res.* **5**(2) : 427-429.

Chakrabarti, R. K., M. Subrahmanyam & B. B. Pakrasi (1977). A note on the collection and segregation of prawn seed for selective stocking. *J. Inland fish. Soc. India*, **9** : 181-183.

Chakrabarti, R. K., P. Ravichandran, D. D. Halder, S. K. Mandal & D. Sapui, (1985). Physico- chemical characteristics of Kakdwip brackishwater ponds and their influence on the survival, growth and production of *Penaeus monodon* (Fabricius). *Indian. J. Fish.*, **32**(2) : 224-235.

Chakrabarty, K. C. & R. C. Bhattacharjee (1957). An analysis of the wave observations in the Bay of Bengal. *Indian J. Met. Geophys.*, **8**: 313-320.

Chakrabarty, R. & S. Chakraborty (1983). Food habit of large bandicoot- Rat, *Bandicoota indica* (Bechstein) in the fields of West Bengal during rainy season. *Rodent newsl.* **7** : 26.

Chakrabarty, R. D. & M. K. Banerjee (1967). Otter trawling in the Hooghly estuary, West Bengal, *Proc. nat. Acad. Sci. India* (B) **37** (4) : 389-400.

Chakraborti, R. K. & S. R. Das (1999). Chemistry of the brackishwater impoundments in the Indian Sundarbans. **In** : *Sundarbans Mangals*. pp. 150 - 155.

Chakraborty, A. K., N. C. Baral & N. C. Chattopadhyay (1988). Ecological Studies of the Mangrove Flora of Bhagbatpur (Sundarbans).*J. Indian Soc. Coastal agric. Res.* **6** (1) : 1 -4.

Chakraborty, K. (1986). Fish and fishery recources in the mangrove swamps of Sundarbans, West Bengal an indepth study. *Indian Forester*, **112**(6) 538-542.

Chakraborty, S. K. & A. Choudhury (1992). Ecological studies on the zonation of brachyuran crabs in a virgin mangrove island of Sundarbans, *J.mar.biol.Assoc.India*, **34**(1-2): 189-194.

Chakraborty, S. K., A. Choudhury, & M. Deb (1986). Decapod brachyura from Sundarbans mangrove estuarine complex, India. *J. Bengal nat. Hist. Soc.*, N.S. **5**(1) : 55-68.

Chakravarty, N. V. K., K. P. Tripathi & B. Gangwar (1987). A comparative study of coastal climate with special reference to Andamans, India- 1. Temperture and rainfall. *J. Andaman Sci. Assoc.* 3(2): 119-124.

Champion, H. G. (1936). A preliminary survey of the forest types of India and Burma. *Indian For. Rec.*, **1** : 365.

Chan, H. T. & N. Hussain (1985). Propagule dispersal, establishment and survival of *Rhizophora mucronata*. *Malaysia Forester* **48** (4) : 324-329.

Chanda, S. & S.C.Datta (1986). Prospects and problems of a Mangrove Ecosystem in Western Sundarbans (India). *Trans. Bose Res. Inst.* **49**(3) : 47-57.

Chanda, S. (1977). An ecofloristic survey of the mangrove of Sundarbans, West Bengal, India. *Trans. Bose Res. Inst.* **40**(1) : 5-14.

Chantarasri, S. (1994). *Integrated Resource Development of the Sundarbans Resrved Forest. Fisheries Resources Management for the Sundarbans Reserved Forest.* UNDP/ FAO. BGD/ 84/ 056. Khulna, Bangladesh.

Chapman, V. J. & D. J.Chapman (1961). The Algae, Macmillan, London, p.481.

Chapman, V. J. & J. W. Ronatdson (1958). The mangrove and salt marsh flats of the Auckland Isthumus : *New Zealand Dep. Sci. Ind. Res. Bull.* **125** : 1-79.

Chapman, V. J. (1940). The function of the pneumatophores of *Avicennia nitida* Jacq. *Proc. Linn. Soc. Lond.* **158** : 2-6.

Chapman, V. J. (1970). Mangrove phytosociology. *Trop. Ecol.*, **2** : 1 - 19.

Chapman, V. J. (1975). Mangrove biogeography. **In** : Walsh, G.E., S. C. Snedaker & T. J. Teas (eds.) Proceedings of International Symposium on Biology and Management of Mangroves (8-11 October, 1974) Hawaii. Gainesville, University of Florida. 3-22 pp.

Chapman, V. J. (1976). Mangrove Vegetation : I-vii + 1-447. Ed. J.Cramer. A. R. Ganther Verlag Komman ditgesellschaft, FL. 9490, Vaduz.

Chapman, V. J. (ed). (1977). Ecosystems of the World. I. Wet Coastal Ecosystems. Elsevier, Amsterdam. 428pp.

Chapman, V.J. (1960). *Salt marshes and salt deserts of the World.* Leonard Hill Books Ltd., London.

Chapman, V.J. (1962). Respiration studies of mangrove seedlings. *I. Bull. Mar. Sci. Gulf. Caribb.* **12**(1) : 137-167.

Chapman, V.J. (1964). Coastal vegetation, Pergamon Press, 233.

Chapman, V.J. (1970). Sea weeds and their uses. Camelot Press, London.

Chapman, V.J. (1972). Mangroves of the world. Leonard Hill, London.

Chapman, V.J. (1974). In : Ecology of Halophytes. p. 3 -19, (eds.) R. J. Reimold & W. H. Queen, Academic Press, London.

Chapman, V.J. (1975). Mangrove vegetation. Cramer, Lehre, 425 pp.

Chapman, V.J. (1975). The salinity problem in general - its importance and distribution with special reference to natural halophytes. **In** : A. Poljakoff- Mayber and I. Gale (eds.) Plants in saline environments, Berlin, Springer-Verlay.

Chatterjee, D. (1939). Studies on the endemic flora of India and Burma. *J.Asist. Soc. Bengal*, **5**: 19-67.

Chatterjee, D. (1957). The nature and distribution of the mangrove plants. *Proc.Mang. Symp.* Calcutta, pp. 132-135.

Chatterjee, G. C. (1896). The Sundarbans of Bengal, Geographical Journal, London, 8:

Chatterjee, P. K. (1972). Salinity intrusions in the Hooghly estuary. *Proc. Interdisciplinary Symp. Bhagirathi Hooghly Basin*, May 1970, Calcutta, pp. 143-153.

Chatterjee, S., A. Ghosh, S. Misra & A. Choudhury (1986). Changes in triterpenoid compositions of the leaves of three species of mangroves grown under tidal walter stress, in ' The Mangroves' (Ed. L. J. Bhosale), Shivaji University Press, Kolhapur, pp. 312-314.

Chattopadhayay, K. (1983). A preliminary report of a pilot enquiry on the social ecology of Sundarbans. Sòciological Research Unit, ISI, Calcutta. (Mimeo).

Chattopadhyay, G. N. & A. N. Ghosh. (1976). On the behaviour of native phosphorus in water logged saline soils. *J. Inland Fish. Soc. India*, **8** : 121-123.

Chattopadhyay, G. N. & G. C. De (1991). Role of nitrification inhibitors in increasing efficiency of nitrogenous fertilisers in fish ponds. *Proceedings of Nation Synmposium on New Horizons of Freshwater Aquaculture*. CIFA, India. 149-150.

Chattopadhyay, G. N. & H. C. Karmakar (1982). Relationship of water soluble Calcium and Magnesium ion concentration to water salinity of brackishwater ponds of West Bengal. *J. Inland Soc. Agric. Chem.* **15**(1): 73-76.

Chattopadhyay, G. N. & L. N. Mandal (1980). Effect of different water salinity levels and organic matter application on the composition of water soluble and exchangeable bases in a brackishwater fish pond soil. *Hydrobiologia*, **74** (1) : 33-37.

Chattopadhyay, G. N. & L. N. Mandal. (1980). Distribution of different forms of nitrogen in brackishwater fish-pond soils of West Bengal. *Indian J. Fish.*, **27** : 140-144.

Chattopadhyay, G. N. & L. N. Mandal. (1980). Inorganic transformation of applied phosphorus in brackishwater fish pond soil under different salinity levels. *Hydrobiologia*, **71** : 125-130.

Chattopadhyay, G. N. & M. N. Saha. (1980). Distribution of DTPA extractable iron, manganese, copper and zinc in some brackishwater fish pond soils of West Bengal. *J. Inland Fish. Soc. India*, **12** (2) : 56-62.

Chattopadhyay, G. N. (1999). Chemistry of coastal wetlands of India with relation to aquaculture. **In** : *Sundarbans Mangals*. pp. 218 - 232.

Chattopadhyay, G. N., A. Ghosh & P. K. Saha (1983). A note on the possibility of salinisation of soils under paddy-cum-brackishwater aquaculture in some coastal saline soils. *J. Indian Soc. Coastal agric. Res.*, **1** (1) : 43-45.

Chattopadhyay, G. N., A. K. Roy, P. R. Das & R. K. Banerjee. (1985). A note on the determination of available nitrogen in brackishwater fish pond soils. *J. Indian Soc. Coastal agric. Res.*, **3**(1) : 63-66.

Chattopadhyay, G. N., M. Datta Gupta & S. K. Gupta (1986). Studies on characteristics of some mangrove soils of the Sundarbans ecosystem. *J. Indian Soc. Coastal Agric. Res.* **4**(1): 17 - 23.

Chattopadhyay, G.N & R.K. Chakroborty (1986). A comparitive study on the nature and properties of some brackishwater and nearby fresh water fish pond soils. *Proc. Symp. Coastal Aquaculture.* **4**:1110-1114.

Chattopadhyay, G.N. & L. N. Mandal (1980). Inorganic transformation of applied phosphorus in brackishwater fish pond soil under different water salinity levels. *Hydrobiologia*, **17** : 125-130.

Chattopadhyay, G.N. (1998). *Laboratory Manual for Analysis of Fish Pond Soil and Water.* Daya Publishing House. Delhi. p. 132.

Chattopadhyay, G.N., A.K.Roy & P.R. Das (1991). A note on determination of available phosphorus in brackishwater fish pond soils. *J. Indian Soc. Coastal Agric. Res.* **9**(1-2) : 77-79.

Chaubal, P. D. & S. Y. Kotmire (1985). Pollen morphology of some mangroves and their associates. Krishnamurthy, V. & A. G. Untawale (eds.), *Proc. Marine Plants, their Biology, Chemistry and Utilization*, Goa, 311-314.

Chaudhuri A. B. & K. Chakrabarti (1973). Breeding biology of birds of Sundarbans forests. *Cheetal* **15** (2): 7-31.

Chaudhuri, A. B. & A. Choudhury (1994). *Mangrove of the Sundarbans* Volume one : India IUCN, Bangkok. 247 pp.

Chaudhuri, A. B. & H. B. Naithani (1985). A Comprehensive survey of Tropical Mangrove forests of Sundarbans and Andamans. Part 1. International Book Distributors, Dehra Dun, India. 41 pp.

Chaudhuri, A. B. & K. Chakrabarti (1973). Breeding biology of birds of Sundarban forests, *Cheetal*, **15** (2): 7-31.

Chaudhuri, A. B. & K. Chakrabarti, (1974). Wild life biology of Sundarbans forests bio-ecological study on fish and fish researches and biological basis of rational fishey. *Sci. & Cult.* **40** : 93-99.

Chaudhuri, A. B. & K. Chakraborti (1972). Wildife biology of Sundarbans - man eating tigers. *Cheetal* **16** (1) : 52-54.

Chaudhuri, A. B. & K. Chakraborti (1972). Wildife biology of the Sundarbans forests - observations on tigers *Cheetal* (I.U.C.N. Genl. Issue), **15** (1) :11-31

Chaudhuri, A. B. & K. Chakraborti (1973). Wildife biology of the Sundarbans : marine fish. *Sci. & Cult.*, **39** (1) : 8-16.

Chaudhuri, A. B. & K. Chakroborty (1973). Wildlife biology of the Sundarbans forests - a study of the birds of Sundarbans with special reference to the breeding biology of openbilled stork, little cormorant and large egrets. *Science & Culture*. **39**: 8-16.

Chaudhuri, A. B. & K. Chakroborty (1975). Wildlife biology of the Sundarbans Forests. *J. Bot. Soc. Bengal*. **26**(1): 63-66.

Chaudhuri, A. B. (1965). Grasses and grassland types of West Bengal and some aspects of their ecology. *Bull. bot. Soc. Bengal* **19** : 94 - 108.

Chaudhuri, A.B. & K. Chakrabarti (1989). *Sundarbans Mangroves (Ecology and Wildlife).* Jugal Kishore & Co. (Pub. Div.), Dehra Dun - 248 001, India, pp. 1-210.

Chaudhuri, A.B. & K. Chakraborti (1974). Wildlife bilogy of Sundarbans forests- Human casualties from man-eating tigers can be reduced. *Sci. & Cult*, **40** (5) : 207-210.

Chaudhuri, A.B. (1978). Commercial aspects of the Sundarban Division. *The Indian Forester*, Dehra Dun, **64** (10), p. 640.

Chaudhuri, A.B. (1988). Biotic communities in endangered and mangrove environment. *Sci. & Cult.* **54** (6) : 182-187.

Chaudhury, R. N. (1978). An appraisal of the problems and prospects of the Sundarban, West Bengal. *Natn. Geogr. J. India*, **24** (1 & 2), 86-98.

Chengappa, B.S. (1944). Andaman forests and their regeneration. *Ind. Forestor*, **70** : 297-304.

Choudhuri, S. K., N. Mallick & J. Sen (1962). The history of the vegetation of the Lower Gangetic Plain. *Proc. Indian Sci. Congr*. III, pp. 259-260.

Choudhury, A. M., A. M. Quadir & J. Islam (1990). Study of Chakaria Sundarbans Using Remote Sensing Techniques. SPARRSO, Dhaka. 77 pp.

Choudhury, J. H. & M. W. Baksha (1983). Studies on the role of *Chrysocroa* sp. in the initiation of top dying of sundri. *Bano Biggyhan Patrika* **12** : 30-34.

Choudhury, R. N. (1978). An appraisal of the problems and prospects of the Sundarbans, West Bengal. *Nat. Geogr. J. India* **24**(1&2) : 86-98.

Chowdhary, H. J. & J. N. Rai (1980). Microfungi from mangrove swamps of West Bengal, India. 2. Some new records of aquatic fungi. *Nova Hedwigia*, **32**(1) : 237-243.

Chowdhury, M .K. & P. Sanyal, (1985). Some observations on the man- eating behaviour of the Tiger of Sundarbans. *Cheetal*, **26** (3 & 4) : 32-40.

Chowdhury, M. K. & P. Sanyal (1985). Vegetation dynamics of mangrove of Indian Sundarbans - economic importance. *Marine Plants* : Papers presented at the All India Symposium on Marine Plants, their Biology, Chemistry and Utilization, Dona Paula, Goa, Oct.30 - Nov. 1,1983.V. Krishnamurthy & A.G. Untawale (eds.). 325-324.

Chowdhury, M.K. & P. Sanyal (1985). Use of Electroconvulsive shocks to control Tiger predation on human beings in Sundarbans Tiger Reserve. *Tigerpaper* XII/2 : 1 - 5.

Chowdhury, M.K. & P. Sanyal, (1985). Man- eating behaviour of the tigers of Sundarbans, West Bengal. *Eniron. & Ecol.*, **3**(2) : 243-248.

Christensen, B. & S. Anderson Wium (1977). Seasonal growth of mangrove trees in Southern Thailand, I. The Phenology of *Rhizophora apiculata* Bl. *Aquat. Bot*. **3** : 281-286.

Christensen, B. (1978). Biomass and primary production of *Rhizophora apiculata* Bl. in a mangrove in Southern Thailand. *Aquat. Bot*. **4** : 43-52.

Christensen, B. (1983). Mangroves - what are they worth? *Unasylva*, **35** : 2-15.

Cooksey, K. E. (1984). Role of diatoms in the mangrove habitat. **In** : S.C. Snedaker and J. E. Snedaker (eds.) *The Mangrove Ecosystem : Research Methods*. UNESCO.

Cooper, R.E. & S. A. Pasha (1935). The osmotic and sucction pressure fo some species of mangrove vegetation. *Indian J. Bot. Soc*. **14**: 100-120.

Cowan, J. M. (1928): Flora of the Chakaria Sundarbans. *Rec. bot. Surv. India*. **11**: 197-225.

Culshaw, J. C. (1950). Some West Bengal plants. *J. Bombay nat. Hist. Soc*. **49** : 188 - 196.

Dagar, J. C. & A. K. Sharma (1988). Some growth parameters of seedlings *of Rhizophra apiculata* Bl.-a prominent mangrove of the Andamans. *J. Andaman Sci. Assoc*. **4**: 47-50,1988.

Dagar, J. C. & A. K. Sharma (1989). Impact of developmental activities on litterfall and mangal formation in the Andamans. Paper presented in Symposium Development *vis-a-vis* Ecology. INTACH-Forest Dept., Port Blair.

Dagar, J. C. & A. K. Sharma (1989). Multiple viviparity in mangroves. *J. Andaman Sci. Assoc*. **5**(1): 72-73, 1989.

Dagar, J. C. & A. K.Sharma (1991). Litterfall beneath *Rhizophora apiculata* mangrove forests of Andamans, India. *Trop.Ecol.*, **32**(2) : 231-235.

Dagar, J. C. & H. S. Dagar (1986). Mangroves and some coastal plants in ethnobotany of the tribals of Andaman and Nicobar Islands . *J. Andaman Sci. Assoc*. **2**(2): 33-36, 1986.

Dagar, J. C. (1987). Mangrove vegetation, its structure, ecology, management and importance with special reference to Andaman and Nicobar Islands. **In**: *Proc. Symp. Management of Coastal Ecosystems and oceanic Resources of the Andamans* 17-18 July(1987). *Andaman Science Association* (CARI), Port Blair, pp. 8-23, 1987.

Das Gupta, D. K. (1991). Mangrove Vegetation - A Precious Gift of Nature. In: Mangrove of Sundarbans & Elsewhere.(Eds. S. Maity & N. Mukherjee). *Proc. Natl. Symp. Cons. & Manage Mangroves* Dec. 1991: 1-2.

Das, A. K. & N. C. Nandi (1999). Fauna of the Indian Sundarbans mangals and their role in the ecosystem. In : *Sundarbans Mangals* (eds. D. N. Guha Bakshi, P. Sanyal and K. R. Naskar) pp. 417 - 427.

Das, B. M. (1921). Mangrove swamps of the Sundarbans forest division, a valuable source of tanstuffs. *J. Indian Ind. Labour* **1** : 487 - 499.

Das, B.M. (1950). Transtuff resources of the Sundarbans forest Division,*Tanmer*,**4**(7):17- 18.

Das, B.M. (1950). Tanstuff resources of the Sundarbans forest division. *Tanmer*, **4** (8):16-17.

Das, B.M. (1950). Tanstuff resources of the Sundarbans forest division. *Tanmer*, **4** (9) : 12-13.

Das, G. K. & A. Bhattacharya (1991). Studies on mangroves of Sundarbans in tidal flats of the Thakuran river, West Bengal. *Proc. Symp. Farming System for Sustained Productivity in Humid Tropics*, 8 - 10 th December, Port Blair, Andaman & Nicobar Islands, India, pp. 188 - 191.

Das, G. K. & A. Bhattacharya (1991).Physiographic setup and sensitiveness of textural parameters of tidal river related mangrove sediments. *Proc. National Symp. Cons. & Mang.*, B.C.K.V.,W.B. pp. 46 - 52.

Das, G. K. & A. Bhattacharya (1994). A piece meal mechanism of bank erosion following subsidence : A case study from Thakuran river of deltaic Sundarbans, West Bengal. *J. Indian Soc. Coastal; agric. Res.* **12** : 231 - 234.

Das, G. K. & A. Bhattacharya (1995). Morphodynamics of the Thakuran Tidal River of West Bengal, India and its Impact on the Changing Mangrove Floristics, *J. Dept. of Geography*, Utkal University, India pp. 7 - 8.

Das, G. K. & A. K. Bhattacharyya (1991). Pattern of sedimentation with reference to erosional and depositional areas along meander bends in the upper estuarine stretch of the Thakuran river, West Bengal, *J. Indian Soc. Coastal agric. Res.* **9** (1 & 2): 37 - 42.

Das, G.K. & A. Bhattacharya (1991). Sediments baffles by intertidal marsh and swamp vegetation : A study from the distributory channel banks of the Thakuran Tidal River, West Bengal. *J. Indian Soc. Coastal agric. Res.* **9** : 399 - 403.

Das, P. R. (1969). A preliminary note on the toxicity of the plant, *Derris trifoliata* Lour. on Fishes. *J. Bombay Pharm. Manuf.* **7** (4) : 197 - 200.

Das, P. R. (1972). Some observation on the macro-vegetation in and around 'Bheris' of Sundarbans, W. Bengal. *J. Inland Fish. Soc. India* **4** : 210 - 211.

Das, S. & M. Ghose (1993). Morphology of stomata and leaf hairs of some halophytes from Sundarbans, West Bengal. *Phytomorphology*, **43** (1 & 2): 59 - 70.

Das, S. & N. A. Siddiqi (1985). Effect of time and method of storage on the survivality of uprooted keora (*Sonneratia apetala*) seedlings. *Bano Biggyan Patrica.*

Das, S. & N. A. Siddiqi (1985). The Mangrove and Mangrove Forests of Bangladesh. Mangrove Silviculture Division, *Bulletin No. 2*, Bangladesh Forest Research Institute, Chittagong. 142 pp.

Das, S. & N. A. Siddiqi (1985). The Mangroves and Mangrove forests of Bangladesh. UNDP/FAO Profect BGD/79/017, Dhaka. 142pp. (Reprinted in 2000).

Das, S. (1971). The vegetation and the ecology of the coastal belt of Bangladesh.*Forestdale News*, **4** (3) : 36 - 44.

Datta, S. C. & N. C. Majumdar (1966): Flora of Calcutta and vicinity. 20:16-120.

Davis, C. C. (1955). The marine and fresh water plankton. Mich. St. Univ. Press, U.S.A. pp. 1 - 562.

Dev, S. C. (1956). Paleoclimatology and Geophysics of the Ganga Delta, *Geographical Review of India*, **28** : 11 - 18.

Dey, S. C. (1959).Utilization of mangrove bark, *Proc. Mangr. Symp.* 129 - 133.

Dhara, K. P. & M. Chakraborty (1983). Chemical investigations of *Derris trifoliata*. All India Symposium on marine Plants, NIO, Goa.

Dutta, S. K. & A. R. Bal (1987). Mechanism of salt tolerance in some plant species, *J. Indian Soc. Coastal agric. Res.* **5** (1) : 145-150

Dutta, S. K., S. Banerjee & J. Som (1978). Diurnal Changes in Leaf Water Potential of Rice, Barely and Wheat. *Biol. Plantarum.* **20**(6) : 472-474.

Faizuddin, M., M. M. Rahman, M. Shahidullah & A. S. M. Helalsiddiqi (1998). Survival and growth performance of some mangrove species planted in the Sundarbans of Bangladesh. *Bangladesh Journal of Forest Science* **27**(2) : 121-127.

Ferguson, J. & E. H. Pascoe (1973). A manual of geology of India and Burma. New Delhi, Vol, III. pp. (1982).

Fergussen, J. (1863). Recent changes in the delta of the Ganges. *Qr. J. ges. soc.*, **19** : 321-354.

Fergusson, J. (1983). Delta of the Ganges. *Quart. 3. Geol. Soc.*, **59** (1) : 321-354.

Field, C. D. (1985). Salt regulation in Mangroves. In: The Mangroves: *Proc. Nat. Symp. Biol. Util. Cons. Mangroves.* pp. 16-24. (Ed. L.J. Bhosale, Kolapur).

Fox, R. L., R. A. Olsen & H. E. Rhodes (1964). Evaluating the sulphur status of soils by plants and soil tests. *Soil Science Society of America Proceedings* **28** : 243.

Ganeshamurthy, A. N. & A. D. Mongia (1989). Acidity in soils of Andaman and Nicobar Islands. *Journal of Indian Society of Soil Science* **37** : 822-824.

Gangadhara Bhat, H. & K. R. Subrahmanya (1993). Paleoshorelines and coastal processes in Dakshina Kannada, Karnataka, India: A study based on remotely sensed data *Int.J.Remote Sens.*, **14**(17) : 3311-3316.

Ganguli, S. N. & S. M. Sircar (1974). Gibberellins from mangrove plants. *Phytochemistry* (Oxf.) 13(c) :(Engl);

Ganguly, D.K. (1965). Sundarbans. *Indian Forester*, 91 : 661 - 663.

Garg, K. L. (1989). Fungal succession on pneumatophore of mangrove plants. Program of the First International Marine Biotechnology Conference, IMBC 89, 55 pp.

Ghatak, D. B. & S. K. Konar (1994). Field survey on the status of pollution by various industrial effluents in Hooghly estuary ecosystem at Haldia, West Bengal. *Environ. Ecol.* **12**(1) : 128-132.

Ghatak, S. S. & A. Misra (1983). Notes on the occurrence and hermaphroditic nature of the parasitic isopod *Nerocila madrasensis* Ramakrishna and Venkataramaniah from the Hoogly Matla estuary, Sundarbans, India. *Bull. zool. Surv. India*. **5**(1) : 21-25.

Ghosh, A. K. (1941). Submerged forests in Calcutta, *Sci. & Cult.* 6 : 669 - 670.

Ghosh, A. K. (1964). Study of Calcutta Peat and Associated Sediment, *Ind. J. Power & River Valley Dev.* 13 : 14 - 21.

Ghosh, A. N. (1979). Brackishwater fish farming tehniques. *Madras J. fish.*, **8** : 225-229.

Ghosh, A., A. K. Dutta, J. Dutta, S. Misra & A. Choudhury (1985). Occurrence of pentacyclic triterpenses in mangrove plant leaves of Sundarbans forest, in Marine plants (Ed. V. Krishnamurthy) *Seaweed Research and Utilisation Association*, Madras, pp. 263-266.

Ghosh, A., P. K. Chakraborti, K. R. Naskar, G. N. Chattopadhyay, D. Nath & M. L. Bhowmik (1990). Ecology and faunal association of intertidal mangrove habitats in the Hooghly-Matlah estuarine system. *J.Inland Fish.Soc.India*. **22**(1-2), 31-37.

Ghosh, A., P. K. Chakraborty, K. R. Naskar, G. N. Chattopadhyay, D. Nath & M. L. Bhowmik (1987). Ecology & faunal association of intertidal mangrove habitats in the Hooghly - Matla estuarine system . *J. Inland Fish. Soc. India* **22** (1 & 2) : 31 - 37.

Ghosh, A., S. Misra, A.K. Dutta & A. Choudhury (1985). Pentacyclic triterpenoids and sterols from seven species of mangrove. *Phytochemistry*, **24** (8) : 1725-1727.

Ghosh, A., T. K. Jana, B. N. Singh & A. Choudhury (1987). Comparative study of carbon dioxide in virgin and reclaimed mangrove waters of Sundarbans. *Mahasagar*, **20**(3) : 155-161.

Ghosh, B. B. & A. K. Basu (1968). Observations on estuarine pollution of the Hooghly by the effluents from a chemical factory complex at Rishra, West Bengal (India). *Environ. Health.* **10** (3): 204-218.

Ghosh, D. P. (1957). Archaeological discoveries in lower Gangetic valley. *Sci. & Cult.*, **23** : 284-289.

Ghosh, M. & A. Choudhury (1986). Growth of *Acanthamoeba culbertsoni* and *Acanthamoeba rhysodes* (Protozoa: Gymnamoebia) in cultures. *Proc. Ind. Acad. Sci.* (Anim. Sci.) **95**(4); 469-474.

Ghosh, M. & A. Choudhury (1987). Aspects of culture : *Acanthamoeba astronyxis* (Ray and Hayes 1954) from Bay of Bengal coasts, *India. Proc. Indian Acad. Sci. Anim. Sci.* **96**(1) : 63-69.

Ghosh, M. (1972). A large and new prehistoric hovid molar from Chandaketugar, 24-Parganas, West Bengal. *Sci. & Cult.*, **43** :8.

Ghosh, P. B. & A. Choudhury (1989). Some organic components in the sediments of Hooghly estuary. *Environ. Ecol.*, 7(2) : 395-397.

Ghosh, P. B., B. N. Singh, C. Chakrabarty, A. Saha, R. L. Das & A. Choudhury (1990). Mangrove litter production in a tidal creek of Lothian island of Sundarbans, India. *Indian J.mar.Sci.*, **19**(4) : 292-293.

Ghosh, P. B., C. Chakraborty, A. Saha, B. N.Singh & A. Choudhury (1991). Productivity of grass *Porteresia coarctata* in mangrove forest of Sunderbans. *Indian J.mar.Sci.*, **20**(4) : 280-282.

Ghosh, P. B., S. Misra & A. Choudhury (1990). Fatty acids and sterols in sediments of Hooghly estuary, north-east coast of India. *Indian J.mar.Sci.*, **19**(1) : 22-26.

Ghosh, P. B., S. Misra, A. Choudhury & A. Ghosh (1990). Possible sources of aliphatic hydrocarbons in sediments of Hooghly estuary, north-east coast of India. *Indian J.mar.Sci.*, **19**(2) : 145-147.

Ghosh, S. S. & B. S. Neogi (1958). Occurrence of *Carapa* sp. in the so-called Peat Bed near Calcutta, *Curr. Sci.* 27 : 357 - 360.

Ghosh, S., T. K. Jana, B. N. Singh & A. Choudhuri (1985).Comparative study of CO2 system in virgin and reclaimed mangrove waters of Sundarbans during freshet, *Mahasagar*, **20** (3) : 155 -161.

Gopinathan, C. P. & M. S. Rajagopalan (1983). Mangrove resources. *Bull. Cent. mar. Fish. Res. Inst.* **34**: 44-46,.

Griffith. W. (1836). On the family of Rhizophoraceaes, Trans. med. Phys. Soc. Calcutta 8, 1836.

Guha Bakshi, D. N. & K. R. Naskar (1987). Present needs to silvicultural practices in the tidal mangrove forests of the Sundarbans in West Bengal, India. *Recent Res. in Ecol. Env. & Pollution 4 : 105 - 118.* (Ed. Malabika Ray & S. Mandal) Visva Bharati, Santiniketan, pp. 105 - 118.

Guha Bakshi, D. N. & K. R. Naskar (1991). Mangrove Afforestation in The Sundarbans.In :*Plant Science and Man : Problems and Prospects* (*Proceedings of the International Botanical Conference*, Bangladesh Botanical Society, Dhaka, Ed. A.K.M.Nurul Islam, et al.) pp.75 - 83.

Guha Bakshi, D. N. & K. R. Naskar (1991). Succession of the different species of *Sonneratia* L.f. in the Sundarbans mangrove ecosystem, *J. Indian Soc. Coastal agric, Res.* **9** (1/2) : 451 - 454.

Guha Bakshi, D. N. & S. Sen (1969). A botanical exploration of Sagar Island in 24 Parganas, *Bull. bot Soc. Bengal*, 23 : 31 - 34.

Guha Bakshi, D. N., P. Sanyal & Kumudranjan Naskar (eds) (1999). *Sundarbans Mangal.* Naya Prokash, 206 Bidhan Sarani, Calcutta - 700 006. pp. 1- 771+ 8 Photo Plates. (ISBN 81-85421-55-2).

Guha Bakshi, D. N., S. Sen & Jayashri Lahiri (1978). A note on the halophytes in India. *Bull. bot. Soc. Bengal* 30 : 117 - 119.

Guha, B.S. (1979). Studies on social tention among the refugees from East Pakistan. (Dept. of Anthropology, Memoir No. 1).

Guha, M.K. (1991). Coastal climate and Agriculture. *J. Indian Soc. Coastal agric. Res.* **9** (1 & 2): 317-321.

Gunasekaran, S, A. Jayapaul Azariah & P. M. Raju (1992). Distribution of mangrove plants in relation to the chemical characteristics of the soil, Muthupet, Tamil Nadu, India. *Asian Environ.*, **14**(3) : 59-69.

Gupta, A. C. (1957). The Sundarbans, its problems , its possibilities. *Indian Forester* 83 : 481-87.

Gupta, A.C. (1964). Wildlife of lower Bengal with particular reference to the Sundarbans. West Bengal Forests, Centenary commemoration volume, pp. 233-237.

Haldar, S. & K. R. Naskar (1991). Mangroves - the unique but endangered ecosystems in the tropical regions. *J. Indian Soc. Coastal agric, Res.* **9** (1/2) : 447 - 450.

Halder, A. K. & C. J. Thamphi (1983). Soils of Sagar Island, their characteristics and classification, *J. Indian Soc. Coastal agric. Res.*, **1**(2) : 106-111.

Halder, S. & K. R. Naskar (1989). The Metropolis City of Calcutta in the Ganga Basin, its Past, Present and Future. *Monograph* 9, pp.454-456 (Nat. Atl. & Themt. Map. Org., Calcutta): 275-289.

Hassain, M. M. & M. A. Razzaque (1981). A preliminary evaluation of the clay mineralogy of the Sundarbans soils. *Bano Biggyan Patrika* **10** : 21-26.

Hassan, M. M. (2000). Soils and problem soils of Bangladesh. Observer Magazine, Dhaka. 5 May. 23-24 pp.

Heinig, R. L. (1893). Blind root suckers of Sundarbans. *J. Asiat. Soc. Bengal* **62** : 158 - 161.

Heinig, R.L. (1893). Working plan of the Sundarban forests, Khulna and 24 Parganas district. Bengal Secretariate Press, 1 p. 1, 8, ii 25 xxvi p. fold maps 33 mm.

Hendrich, H. (1975). The status of the tiger, *Panthera tigris* in the Sundarbans Mangrove Forest (Bay of Bengal). *Sangilierk. Mitt.* **23** (3) : 161-200.

Hershkovitiz, P. H. (1966). Catalogue of living whales. *Bull. U.S. Natl. Mus.*, (246) : 259.

Hooker, J.D. (1897). Flora of British India, 1-7, Reeve & Co., London.

Hossain, A. T. M. E. (1997). Observation on the fruit and seed germination of *Xylocarpus granatum* Keoning. *Bangladesh Journal of Forest Science* **26**(2) :76-78.

Hossain, M. M. (1994). Concentration of total mercury in shrimps and fishes from the Bay of Bengal, Bangladesh. *Mar.Res.*, **3**(2) : 27-32.

Hunter, W. W. (1875). 24 Parganas and Sundarbans, Trubner, London, 400 p.

Hussain, Z. & G. Acharya (1994) (Eds). Mangroves of the Sundarbans Volume two : Bangladesh IUCN, Bangkok. 257 pp.

Imam, S. A. (1982). The Sundarbans and its future. *Proc. Bangladesh Second National Conference on Forestry*, 1924 pp.

Ingale, L. R. (1984). Physiological studies in halophytes : Studies in *Salvadora persica,* Linn. Ph. D. Thesis, University of Bombay.

Ingole, B. S., L. K. Kumari, Z. A. Ansari & A. H. Parulekar (1994). New record of mangrove clam *Gelonia erosa* (Solander, 1786) from the west coast of India. *J.Bombay nat. Hist.Soc.*, **91**(2) : 338-339.

Islam, M. R. & N. A. Siddiqi (1987). Studies on the seed requirement and method of sowing keora (*Sonneratia apetala* Buch.-Ham.) seeds in the nursery. *Bano Biggyan Patrika* **16** : 73-77.

Islam, M. R., M. A. S. Khan, N. A. Siddiqi & P. Saenger (1990). Optimal planting season for keora (*Sonneratia apetala*). *Bangladesh Journal of forest Science* **19** : 1-9.

Islam, M. R., M. A. S. Khan, N. A. Siddiqi & P. Saenger (1991). Effect of thinning on keora (*Sonneratia apetala*) survival and growth. *Bangladesh Journal of Forest Science* **20** : 8-24.

Islam, M. R., M. A. S. Khan, N. A. Siddiqi & P. Saenger (1991). Effect of thinning on keora (*Sonneratia apetala*) survival and growth. *Bangladesh Journal of Forest Science* **20**: 8-24.

Jackson, M. L. (1958.) Soil chemical Analysis. Prentice Hall Inc., Englewood cliffs. USA.

Jagtap, T. G. (1985). Ecological studies in relation to the mangrove environment along the Goa coast, India.Ph.D.thesis, Shivaji University, Kolhapur, 212pp.

Jagtap, T. G. (1985). Structure and composition of the mangrove forests along Goa coast. *The Mangroves : Proc. Nat. Symp. Biol.Util. Cons.*, pp. 188-195.

Jagtap, T. G. (1985). Studies on littoral flora of Andaman Islands. **In :** *Marine Plants.* Paper presented at the All India Symposium on marine Plants, their Biology, Chemistry and Utilization. Dona Paula -Goa, Oct. 30 – Nov. 1, 1983, Krishnamurthy, V. & A. G. Untawale (eds.). pp. 43-50.

Jagtap, T. G. (1987). Seasonal distribution of organic matter in mangrove environment of Goa. *Indian J.mar.Sci.*, **16**(2) : 103-106.

Jagtap, T. G. (1992). Marine flora of Nicobar group of islands in Andaman Sea. *Indian J. mar. Sci.*, **21**(1) : 56-58.

Jagtap, T. G., V. S. Chavan & A. G. Untawale (1993). Mangrove ecosystem of India. A need for protection. *Ambio* **22**(4) 252-254.

Jahan, M. S., M. A. Mannan & K. N. Manda (1990). Intertidal molluscs of Sundarbans, Bangladesh. *Environ. Ecol.*, **8**(2) : 603-607.

Jamadar, Y. A. & A. Choudhury (1983). Investigation of parasitic ciliates (Protozoa) of some shell-fishes in the coastal waters of Hooghly esuarine system, *Proc. Symp. Coastal Aquaculture,* Cochin, 2, pp. 665-670.

Jamale, D. B. (1975). Physiological studies in saline plants. Ph. D. Thesis. Shivaji Univ., Kolhapur.

Jara, R. S., 1985. Traditional uses of mangroves in the Phillippines. **In** : *Mangrove Ecosystem of Asia and the Pacific : Status, Exploitation and Management* (eds.) C. D. Field and A J.Dartnall, 104-113.

Jayapaul Azariah, V. Selvam & S. Gunasakaran (1992). Impact of past management practices on the present status of the Muthupet mangrove ecosystem. *Hydrobiologia*, **247**: 253-259.

Jhamb, R. B. (1984). Biology of halophytes in Indian desert. Ph. D. Thesis, University of Jodhpur, Jodhpur.

Jhingran, V. G., A. V. Natarajan, S. M. Banerjee & A. David (1988). *Methodology on Reservoir Fisheries Investigations in India*. CIFRI. Bulletin No. 12.

Jones, W. T. (1971). The field identification and distribution of mangroves in eastern Australia. *Queensland Naturalist*, **10** : 35-51.

Joshi, G. V., M. Pimplaskar & L. Bhosale (1972). Physiological studies in germination of mangroves. *Bot. Mar.* **15**: 91-95.

Joshi, G.V. & A. P. Waghmode (1981). Photosynthesis in mangroves. *Indian J.Bot.*, **4**(1) : 15-19.

Kannan, L. (1990). Mangroves - their importance and need for conservation, Biology Education, **7**(2) : 93-103.

Kannan, L., A. Duraisamy & T. T. Ajithkumar (1998). Muthupet mangrove forests sending out an SOS. In: An anthology of Indian Mangroves, ENVIS Publication, CAS In marine biology, Parangipettai, 30-36.

Kannan, L., R. Kannan & S. Ravichandran (1999). How to know the Mangroves? (Annotated checklist of Indian Mangroves). ENVIS Publication, CAS in Marine Biology, Annamalai University, 29pp.

Karim, A. & M. A. Khan (1980). Phytosociological studies of mangrove forest of Chakaria Sundarbans. *Bano Biggyan Patrika* **9** : 17-28.

Karim, M. A. (1988). Environmental factors and the distribution of mangroves in Sundarbans with special reference to *Heritiera fomes* Buch.-Ham. Ph. D. Thesis (unpub. University of Calcutta), 222 pp.

Karkar, M. B. & L. J. Bhosale (1985). Studies on inorganic constitutents of leaf, propagule, soil and flooding water of a mangrove; *K. candel*. The mangroves; Pro. Nat. Symp. Biol. Cons. Mangroves, pp -18-20.

Karmakar, S. M. (1985). Mangroves - A Review. **In :** *The Mangroves*: *Proc. Nat. Symp. Biol. Util. Cons.* Mangroves. pp. 60 - 79.

Kathiresan, K. & L. V. Venkataraman (1991). *Rhizophora* : A walking tree of the sea. *Biol. Edn.*, 8(2) : 147-148.

Kathiresan, K. & M. Pandian (1991). Effect of UV on quality of black tea from *Ceriops decandra. Sci. Cult.*, **57** (3&4) : 93-95.

Kathiresan, K. & M. Xavier Ramesh (1991). Establishment of seedlings of a mangrove. *Indian For.*, **117** (3) : 225.

Kathiresan, K. & P. Moorthy (1993). Influence of different irradiance on growth and photosynthetic characteristics in seedlings of *Rhizophora* species. *Photosynthetica* (Czechoslovakia), **29** (1) : 143-146.

Kathiresan, K. & S. Ravikumar (1993). Two endangered species of mangroves in Pichavaram. *Indian For.*, **119**(9) : 773.

Kathiresan, K. & T. S. Thangam (1990). Effect of phenolics on growth of viviparous seedlings of *Rhizophora apiculata. Geobios*, **17** : 142-143.

Kathiresan, K. & T. S. Thangam, 1987. Light induced effects of latex of *Excoecaria agallocha* L. on salt marsh mosquito *Cules sitiens* L.*J.mar.biol.Assoc.India*, **29**(1&2) : 378-380.

Kathiresan, K. (1989). Multi branching of *Rhizophora* seedlings. *Indian For.*, **115** (1) : 54, 1989.

Kathiresan, K. (1998). Distribution and status of mangroves in India. *Seshaiyana*, **6**(2) : 8-9.

Kathiresan, K., G. A. Ravishankar & L. V. Venkataraman (1997). *In vitro* multiplication of a coastal plant *Sesuvium portulacastrum* L. by axillary bud cultures. **In** : *Recent Advances in Biotechnological Applications of Plant Tissue and Cell Culture.* Ravishankar, G.A. and L.V. Venkataraman (eds.), Oxford & IBH publishing Co. Pvt Ltd., New Delhi, 185-189.

Kathiresan, K., K. S. Bose & A. Sivakumar (1987). Light induced effect of latex of *Excoecaria agallocha* L. on phytoplankton productivity in Pitchavaram mangroves. Nair, N.B. (ed.), Proc. Natl.Sem. Estuarine Management, Trivandrum, 514-515.

Kathiresan, K., P. Moorthy & S. Ravikumar (1995). Studies on root growth in seedlings of a tropical mangrove tree species. *Int. Tree Crops J.*, **8**(2-3) : 183-187.

Kathiresan, K., P. Moorthy & S. Ravikumar (1996). A note on the influence of salinity and pH on rooting of *Rhizophora mucronata* lamk. Seedlings. *Indian For.*, **122**(8) : 763-764.

Kathiresan, K., T. S. Thangam & M. Premanathan (1995). Mangrove halophytes : a potential source of medicines. **In** : *Biology of salt tolerant plants*, M.A.Khan & I.A.Ungar (eds.) University of Karachi, Pakistan, pp. 361-370.

Kathirvel, M. & S. Srinivasagam (1992). Taxonomy of the mud crab, *Scylla serrata* (Forskal), from India. Angell, C.A.(ed.), Report of the Seminar on the Mud Crab Culture and Trade, Bay of Bengal Programme for Fisheries Development, BOBP, Madras, 127-132.

Khan, A. A. (1960). The mangrove forests of Andhra state. *Proc. Silvi Conf.* **9**: 182-186, 1960.

Khan, M. S. (1978). Flora of Bangladesh (Rhizophoraceae). No. 7. Bangladesh Agricultural Research Council, Dhaka. 19 pp.

Khatun, B. M. & M. K. Alam (1987). Taxonomic studies in the genera *Avicennia* Linn. from Bangladesh. *Bangladesh Journal of Botany*, **16** (1) : 39-44.

Kothandaraman, K. R. (1984). Physiological and biochemical studies in the viviparous seedlings of two mangrove species at Pitchavaram, *Rhizophora mucronata* Lamk and *Rhizophora apiculata* Bl. Ph.D.thesis, Annamalai University, India, 82 pp.

Kotmire, S. V. & L. J. Bhosale (1985). A study of mangrove vegetation along Deogad estuary. In *Marine plants* Eds. Krishnamurthy, V. and Untawale, A.G.) Sea weed Res. And Util. Assn., Madras.

Kotmire, S. Y. & L. J. Bhosale (1980). Chemical composition of leaves of *Avicennia officinalis* Linn. and *A. marina* var. *acutissima* stapf and Moldenke. *Indian J.mar Sci.*, **9**(4) : 299-301.

Krishnamoorthy, R.. (1995). Remote sensing of mangrove forests in Tamil Nadu coast, India, Ph.D. Thesis, Anna University, Madras, 202pp.

Krishnamurthy, K. (1985). The changing landscape of the Indian Mangroves. The Mangroves : *Proc. Nat. Symp. Biol. Util. Cons. Mangroves*, L.J. Bhosale (ed)., Kolahpur. 99. 119 - 126.

Krishnamurthy, K. (1998). Mangroves : Future Shock? In: *An Anthology of Indian Mangroves*. Envis Centre, CAS in Marine Biology, Annamalai University. Pp. 64-66.

Krishnamurthy, K. 1993. The Mangroves. ENVIS News Letter, Annamalai University, 1 : 1-3.

Krishnamurthy, K., A. Choudhury & A. G. Untawale (1987). Mangroves in India : Status Report submitted to the Govt. of India, Ministry of Environment and Forests, New Delhi, 150pp.

Krishnamurthy, K., L. Kannan, M. J. Prince Jeyaseelan, R. Palaniappan & M. A. Sultan Ali (1981). A floristic study of halophytes with special reference to mangroves *Bull.bot.Surv.India*, **23** (3&4) : 114-126.

Kulkarni, D. H. (1959). Utilization of mangrove forests in Saurashtra and Kutch. In: Mangrove Symposium, Calcutta, 30-32. (Proceeding. Faridabad, 1959).

Kulkarni, P. K. & L. J. Bhosale (1991). Studies on regeneration in two species of *Rhizophora* under laboratory conditions. *J. Mar. Biol. Assoc. India*, **33**(1-2): 438-442.

Kulkarni, P. K. & L. J. Bhosale (1992). Growth performance of *Rhizophora mucronata* in field and laboratory conditions *J. Mar. Biol. Assoc. India*, **34**(1-2) : 281-284.

Kulkarni, S. & L. J. Bhosale (1985). Studies on excretion of salt in *Avicennia marina* and *Acanthus ilicifolius* under soil culture. Krishnamurthy, V. and A.G.Untawale (eds.), Proc. Marine Plants, Their Biology, Chemistry and Utilization, Goa, 295-298.

Kurz, S. (1874). On two new species of *Heritiera*, *J. Bot.* **12** (n.s.) : 65-66.

Kuthalingam, M. D. K., V. Sundararaj & V. Ramadhas (1982). Utilization of mangrove ecosystem for aquaculture in India. *Atlantica*, **5**(2) : 68.

Lakshmanan, K. K. & R. Jayalakshmi (1985). Floral biology in *Ceriops* (Rhizophoraceae), Krishnamurthy, V. and A.G.Untawale (eds.). Proc. *Marine Plants*. Their Biology. Chemistry and Utilization, Goa, 335-344.

Latif, M. A & R. A. Del Castillo (1991). Rotation ages of keora (*Sonneratia apetala*) in the coastal plantations of Bangladesh. *Bangladesh Journal of Forest Science* **20** : 1-7.

Lawton, J. R., A. Todd & D. K. Naidoo (1981). Preliminary investigations into the structure of the roots of the mangroves *Avicennia marina* and *Bruguiera gymnorrhiza* in relation to ion uptake. *New Phytol.*, (**88**), 713-722.

Lugo, A. E. & S. C. Snedaker (1974). The ecology of mangroves. *Annual Review Ecology and System*. **5** : 39 - 64.

Mac Nae, W. (1966). Mangroves in easten and southern Australia. *Aust. J. Bot.* **14**: 67-104.

Mac Nae, W. (1968). A general account of flora and fauna of mangrove swamps and prospects in the Indo - West pacific region, *Adv. Mar. Biol.*, **6** : 73 - 270.

Mac Nae, W. (1974). Mangrove forest and Fisheries. FAO/UNDP Indian Ocean Programme (Indian Ocean Fishery Commission) IOCF/DEV/74/34, Rome, 1974.

Macintosh, D. J. (1984). Ecology and productivity of Malaysian mangrove crab population. pp. 354 - 377. **In** : E. Soepadma, A. N. Rao & D. J. Macintosh (eds.) *Proc. asian Symp. Mangrove Environment Res. Manage.*, Univ. of Malaya, Kualalumpur, Malaysia.

Macintosh, D.J. (1986). Mangrove animal community. (Report of the 2nd Introductory Training course on Mangrove Ecosystem, Goa, India, 1984,UNDP/UNESCO Regional Project RAS/79/802.

Mahapatra, A. K. & M. K. Mahapatra (1990). An eco-floristic survey of some islands in Sundarbans of West Bengal. *Env. & Ecol.* **8** (14) : 171 - 174.

Mahapatra, A.K. (1978). A brief survey of some unrecorded less known and threatened plant species of Sundarbans of West Bengal, *Bull bot. Soc. Bengal* **32** : 54 - 58.

Majumdar, J. K. (1952). Rivers in the Bengal delta : River problems in West Bengal and their solution. *J. Asiat. Soc., Ser.* **3, 18** (2) : 103-121.

Majumdar, R. B. (1956). Studies on the grasses of 24 Parganas, West Bengal. *Bull. bot. Soc. Benagl* **10** : 1 - 114.

Majumdar, R. B. (1962). Weed flora of the district 24 Parganas. *Indian Agric.* **6** : 89 - 213.

Mall, L. P., S. K. Billore & D. Amritphale (1982). Certain ecological observations on mangroves of the Andaman Islands. *Trop.Ecol.*, **23**(2) : 225-233.

Mall, L. P., S. K. Billore & D. Amritphle (1982). Certain ecological observation on mangroves of the Andaman Islands. *Trop. Ecol.* **23**: 225-233.

Mall, L. P., V. P. Singh & A. Garge (1991). Study of biomass, litter fall, litter decomposition and soil respiration in monogeneric mangrove and mixed mangrove forests of Andaman islands. *Trop.Ecol.*, **32**(1) : 144-152.

Mall, L. P., V. P. Singh, A. Garge & S. M. Pathak (1985). Mangrove forest of Andamans and some aspects of its ecology. **In:** The Mangroves: *Proc. Nat. Symp. Biol. Util. Cons. Mangroves*. pp. 25 - 38.

Mall, L. P., V. P. Singh, A. Garge & S.M. Pathak (1987). Ecological studies of mangrove forest of Ritchie's archipelago in relation to substratum. *Trop. Ecol.* **28**: 182-192.

Mall, L. P., V. P. Singh, A. Garge, S. M. Pathak & M. S. Dandoria (1985). A new approach towards the mangrove foroest flora of Andaman Islands. *Indian For.* **111:** 290-300.

Mandal, A. K. & N. C. Nandi (1989). Fauna of Sundarbans mangrove ecosystem, West Bengal, India, Fauna of Conservation areas, *Zoological Survey of India,* Calcutta, **3** : 1 - 116.

Mandal, A. K. & R. K. Ghosh (1989). *Sundarbans - A socio-biological study*, Bookland, Calcutta, pp. xv + 194.

Mandal, D. & A. Choudhury (1979). Studies on some gastrointestinal protozoa of wild boars (*Sus scrofa*) of Sunderbans forest, West Bengal. *Proc. 2nd. Nat. Kongr. Parasit.*, B.H.U., Varanasi, p. 7.

Mandal, D. & A. Choudhury (1980). Studies on the parasites of wild animals of a mangrove swamp forest, Sunderbans, India. *Proc. Asian. Symp. on Mangrove Environment*, Research and Mangement, Kualalumpur, Malaysia. (Abstract).

Mandal, R. N., Dolanchampa Ghosh, A. K. Ghosh, A.K. Sarkar & K.R. Naskar (1995). On distributional Record of *Scyphiphora hydrophyllacea Gaertn.* and *Atalantia correa* M. Roem. from the inter-tidal mangrove forests of the Indian Sundarbans. *J. Nat. Bot. Soc.* **49** : 71-74..

Mandal, R.N. & K. R. Naskar (1991). Impact of Aquatic Flora for Sustained Productivity in humid Tropics. **In**: *Proc. Symposium Farming for Sustained Productivity in Humid Tropics*, Port Blair, pp. 207 - 209.

Mandal. R. N. & K. R. Naskar (2000). Morpho - anatomical studies of *Ipomea pes - caprae* Sweet in relation to its halophytic and sand dune adaptation. *J. Interacad* **4**(3): 356 - 359.

Martosubroto, P. & N. Naamin (1977). Relationship between tidal forests (mangroves) and commercial shrimp production in Indonesia. *Mar. Resources Indonesia* **18**: 81-86, 1977.

Masters, J. M. (1840). Calcutta Flora, containing a synopsis of plants indigenous to or cultivated in the vicinity of Calcutta. *Trans. Agri. Hort.Soc. India* **7** : 39 :85.

Matilal, S., B. B. Mukherjee & N. Chatterjee (1986). Studies on soil and vegetation of Mangrove forests of Sundarbans, *Indian J. Mar. Sci.* **15** (3) : 181 - 184.

Megeri, S. & J. H. Davis (1981). Global status of mangrove ecosystem. Paper No. 3 Gland Switzerland.

Misra, J. N. (1960). The ecology, distribution and seasonal succession of the littoral algae on the west coast of India. *Proc. Symp. Alg. ICAR* : 187 - 203.

Misra, K. C. (1989). *Manual of Plant Ecology.* 3rd Edition. Oxford & IBH Publishing Co. Pvt. Ltd. pp. 491.

Misra, R. & G. S. Puri (1954). Indian Manual of plant Ecology. The English Book Depot, Poona and Dehradun, 1954.

Misra, R. (1968). Ecology Workbook. Oxford and IBH , Calcutta.

Misra, S. & A. Choudhury (1985). Changes of leaf surface waxes in the adaptation of flooded plants of Sundarban mangrove forest, Krishnamurthy, V. and A.G.Untawale (eds.). *Marine Plants.* Proc.Marine Plants. Their Biology. Chemistry and Utilization Goa, 267-272.

Misra, S., A .Choudhury, A. Ghosh & J. .Dutta (1984). The role of hydrophobic substances in leaves in adaptation of plants to periodic submersion by tidal water in a mangrove ecosystem. *J.Ecol.*, **72**(2) : 621-625.

Misra, S., A. Choudhury & A. Ghosh (1987). Fatty acid decomposition of detritus and of some detritivoroux benthic animals of Sunderbans mud flat, *Int. J. Aquatic Org.* **1**, 7-14.

Misra, S., A. Choudhury, A. Ghosh & J. Dutta (1983). Enzymatic reactions on thin-layer chromatographic plates. IV. Lipolysis of wax esters and separation of products on a single plate. *J. Chromatography*, **280** : 313-320.

Misra, S., A. Choudhury, A. Ghosh & J. Dutta (1984). The role of hydrophobic substances in leaves in adaptation of plants to periodic submersion by tidal water in a mangrove ecosystem. *J. Ecol.* **72** : 621-625.

Misra, S., A. Choudhury, S. Chattopadhyay & A. Ghosh (1987). Sterols and terpenoids from the seagrass *Porteresia coarctata* grwon in two different mangrove habitats of Sunderbans. *J. Mar. Biol. Assoc.India,* **29**(1-2) : 365-366.

Misra, S., A. Choudhury, S. Chattopadhyay & A. Ghosh (1988). Lipid composition of *Porteresia coarctata* from two different mangrove habitats in India, *Phytochemistry,* **27** (2) : 361 - 364.

Misra, S., A. Ghosh & A. Choudhury (1983). Fatty acids of the mud skipper *Boleophthalmus bodderti. J. Sc. Food Agric.*, **34** : 1413-1418.

Misra, S., A. Ghosh & A. Choudhury (1986). Changes of leaf hydrpcarhons and wax esters of three species of mangroves grown under tidal water stress, in *'The Mangroves'*, (Ed. I. J. Bhosale), Shivaji University Press, Kolhapur, pp. 544-547.

Misra, S., A. Ghosh & A. Choudhury (1986). Lipids and fatty acids of the gastropod mollusc, *Cerithidea obutsa, Food Chemistry*, **22**, 305-312.

Misra, S., A. Ghosh, A. K. Dutta, J. Dutta & A. Choudhury (1983). Penta cyclic triterpene profiles of some submerged and normal plant leaves of Andaman mangrove forest. Marine Plants. Paper presented at the All India Symposium on *Marine Plants, their biology, chemistry and utilization*, Dona Paula, Goa, Oct,30 - Nov. 1,1983 (eds.) V. Krishnamurthy & A.G. Untawale, No. 10, Goa.

Misra, S., A. K. Dutta, A. Ghosh & A. Choudhury (1984). Sterols and Fatty acids from three species of mangrove. *Phytochemistry.* **23**, 2823-2827.

Misra, S., A. K. Dutta, A. Ghosh & A. Choudhury (1985). Oxidation of oleanolic acid of *Avicennia officinalis* leaves to olanonic acid in the natural environment of Sunderban mangrove Ecosystem. *J. Chem. Ecology*, **11**(3), 339-343.

Misra, S., A. K. Dutta, J. Dutta, A. Ghosh & A. Choudhury (1983). Enzymatic reactions on TLC plates, IV Lipolysis of wax esters and separation of products on a single plate. *J. Chromatography.* **280**, 313-320.

Misra, S., A. K. Dutta, J. Dutta, A. Ghosh & A. Choudhury (1984). Lipid contaminants in commercial lipases, Lipids, **19**, 302-303.

Misra, S., J. Dutta, A. Ghosh & A. Choudhury (1985). Microbial decomposition of leaf litter of *Avicennia officinalis* in a mangrove forest biome: biochemical changes during decomposition, in *Marine Plants* (Ed. V. Krishnamurthy) Seaweed Research and Utilisation Association, Madras, pp. 257-262.

Misra, S., J. Dutta, A. Ghosh & A. Choudhury (1985). Role of hydrophobic components of leaves in the adaptation of plants to periodic submersion by tidal water in mangrove ecosystem, *J. Ecology*, **72**, 621-625.

Misra, S., P. K. Pal, A.Ghosh & A. Choudhury (1986). Effect on the leaf lipids of three species of mangrove of periodic submergence in tidal water, *Phytochemistry*, **25,** 1083-1087.

Misra, S.,A. Ghosh, A.K.Dutta, J. Dutta and A. Choudhury. 1985. Microbial oxidation of pentacyclic triterpene acids of *Avicennia officinalis* leaves in the mangrove detrital system of Sundarbans. Krishnamurthy, V. and A.G. Untawale (eds.), Proc. Marine Plants, their Biology, Chemistry and Utilization, Goa, 257-262.

Misra,S., A. Ghosh, A.K. Dutta, J. Dutta & A. Choudhury (1993). Microbial oxidation of pencyclic triterpenes adds of *Avicennia officinalis* leaves in the mangrove detrital system of Sundarbans. Marine Plants, Paper presented at the All India Symposium on *Marine Plants, their biology, chemistry and utilization*, Dona Paula, Goa, Oct,30 - Nov. 1,1983 (eds. V. Krishnamurthy & A.G. Untawale, No. 10, Goa.

Misra. S., A. Choudhury, K. Dutta & A. Ghosh (1984). Sterols and fatty acids from three species of mangrove. *Phytochemistry*, **23**(12) : 2823-2827.

Mitra, R. L. & L. K. Banerjee (1979). The genus *Bruguiera* Lamk. (Rhizophoraceae) in India. *Bull. bot. Srv. India*, **21**(1-4): 142-150.

Mitra, S. N. & K. Mitra (1980). Pollen analysis of honey from Sundarbans (W. B.) *Geophytology* **10**:137-139.

Montgomery, S. Y. (1995). *Spell of the Tiger*. The man eaters of Sundarbans. A Peter Davison Book, Houghton Mifflin Company, Boston, New York, pp. 1 - 230.

Mookherjee, H.K., D.N. Ganguly & J.C. Mazumdar (1949). On the food and feeding babits of the leopard pomphret, *Scatophagus argus* (Pallas) and possibility of its culture near the estuaries of Bengal. *Sci. & cult.*, **15** : 76-77.

Morgan, J. P. & W. G. McIntire (1959). Quaternary Geology of the Bengal Basin, East Pakistan and Burma. *Bull. geol. Soc. Am.* **70** : 319-342.

Morgan, J. P. (1970). Depositional Processes and products in the deltaic environment: **In** deltaic sedimentation: S.E.P.M. SP. Pub. 15, P. 31-47.

Mukherjee, A. B. (1978). Design and construction of sluice for shrimp farms in India. *Proc. 1st. nant. Symp. Shrimp Farming*, Bombay, 16-18 August, 1978 : pp. 118-119.

Mukherjee, A. K. & K. K. Tiwari (1984). Mangrove ecosystem changes under induced stress : The case history of Sundarban, West Bengal, India. *Proc. Asian symp. on Mangrove Environment, Research and Management*, Kualalumpur, Malaysia .pp. 633 - 643

Mukherjee, A. K. & L. K. Banerjee (1968). Three new plants records in West Bengal. *J. Bombay nat. Hist. Soc.* **65** : 268 -269.

Mukherjee, A. K. & S. Gupta (1965). Habits of the Rhesus macaque, *Macaca mulatta* (Zimmerman) in the Sundarbans, 24-Parganas, West Bengal. *J. Bombay nat. Hist. Soc.* **62** : 145-146.

Mukherjee, A. K. (1959). Pakhirala : Sajnakhali- an introduction to a bird sanctuary in the Sundarbans. *J. Bengal nat. Hist. Soc.*, **30** : 161-165.

Mukherjee, A. K. (1960). The genus *Bruguiera* in the Sundarbans. *Bull. bot. Surv. India.* **2**(1-2): 173-174.

Mukherjee, A. K. (1962). Some observation fo the nesting behaviour among vultures, eagles and kites in the Sundarbans, 24-parganas District, West Bengal. *Proc. First All India Congr. zool. Soc. India (Jabalpur)* : (1959) :247-248.

Mukherjee, A. K. (1969). Food habits of water birds of the Sundarbans, 24 Parganas District, West Bengal, India. *J. Bombay nat. Hist. Soc.* **66** (2) : 345 - 360.

Mukherjee, A. K. (1970). As the correct name of mangrove whistler *Pachycephala cinerea* of *Pachycephla grisola* (Blyth). *J. Bombay nat. Hist. Soc.*, **67** (1): 112-113.

Mukherjee, A. K. (1970). Wildlife in the Lower West Bengal. *Indian Museum Bull.* **5** (2): 28 - 34.

Mukherjee, A. K. (1971). Experience of a naturalist in the Sundarbans. *Indian Tourist*, **2** (1) : 33-35.

Mukherjee, A. K. (1971). Food habits of water birds of the Sundarbans, 24-,Parganas District, West Bengal, India-Herons and Bitterns. *J. Bombay nat. hist. Soc.*, **68** (1) : 37-64.

Mukherjee, A. K. (1972). Food-habits of the water birds of the Sundarbans, 24-Parganas District, West Bengal, India. III. Egrets. *J. Bombay nat. Hist. Soc.*, **68** (3) : 691-716.

Mukherjee, A. K. (1974). Food habits of water-birds of the Sundarbans, 24-Parganas District, West Bental, India-IV. Stork, Teal, Moorhen and coot. *J. Bombay. nat. Hist. Soc.*, **71** (2) ;188-200.

Mukherjee, A. K. (1975). The Sunderbans of India and its biota. *J. Bombay nat. Hist. Soc.*, **72** (1) : 1-20.

Mukherjee, A. K. (1975). Food habits of water -birds of the Sundarbans, 24-Parganas District, West Bengal, India- V. Lapwing, Sandpiper, Stint, Tern, Kingfisher. *J. Bombay nat. Hist. Soc.* **72**(2) : 422-447.

Mukherjee, A. K. (1975). Mangrove Wealth of Indian Sundarbans, utilization and conservation. *J. Eco. & Tax. Botany*, **5** : 227 - 230.

Mukherjee, A. K. (1976). Food habits of water-birds of the Sundarbans, 24-Parganas District, West Bengal. India-VI. *J. Bombay nat. Hist. Soc.*, **73**(3) 482-486.

Mukherjee, A. K. (1979). Ecological studies on the Sagar Island and its enviroms in the Hooghly estuary, West Bengal, India. Ph. D. thesis, University of Calcutta, p. 168.

Mukherjee, A. K. (1981). Wildlife in the Sundarban, West Bengal. *Proc. Wildlife Workshop,* Dehra Dun. 1979. (Zool. Sur. India). pp. 123-127.

Mukherjee, A. K. (1982). Endangered animals of India. Zoological Survey of India, Calcutta, pp. X+122.

Mukherjee, A. K. (1984). Mangrove wealth of Indian Sundarbans-utilization and conservation. *J. Economic & Taxonomic Botany.* **5**(1) : 227-230.

Mukherjee, A. K. 1975. The Sundarban of India and its biota. *Journal of Bombay Natural History Society* **72** (1) : 1-20.

Mukherjee, B. B. & J. Mukherjee (1978). Mangroves of Sundarbans, India. *Phytomorphology* **28**(2):177-192.

Mukherjee, B. B. (1969). *Vegetational history of Southern part of West Bengal with an introduction to the pollen flora of Sundarbans.* Ph. D. Thesis, Calcutta University.

Mukherjee, B. B. (1972). Pollen analysis of a few quartenary deposits of lower Bengal Basin. *Proc. Sem. Palaeopalynol. India,* Calcutta, pp. 177-192.

Mukherjee, B. B. (1980). Mangrove forest types and their distribution in Sundarbans, India Proc. Asian Symp. on Mangrove Environment, Kualalumpur. pp. 75-81.

Mukherjee, D. (1987). A note on coastal region of West Bengal. *J. Indian Soc. Coastal agric. Res.* **5** (9) : 25 - 30.

Mukherjee, J. & S. Chanda (1973). Biosynthesis of *Avicennia* L. in relation to taxonomy. *Geophytology,* 3 : 85 - 88.

Mukherjee, K. N. (1959). Some aspects of Historical geography in Bengal. *J. City College Com.* Calcutta, pp. 115-118.

Mukherjee, K. N. (1960). Nature and problems of neo-reclamation in the Sundarban. *Geog. Review of India.* Calcutta, **31** (4) : 1-8.

Mukerjee, S. K. (1960). The Genus-*Bruguiera* in the Sundarbans, *Bull. bot. Surv. India,* **2**(1-2): 173 -174.

Mukerjee, S. K. (1966). *Bruguiera hainesii* Rogers from the Sundarbans, a new record. *Bull, bot. Surv. India,* **8** (3/4) : 357.

Mukhopadhayay, K. (1983). Preservation of natural environment-A case study in Sandeshkhali, P. S. (West Bengal). *Proc. Int. Seminer on Envo Maps and Atlas,* Calcutta, Dec. 1-3, 1983, pp. 134-135.

Mukhopadhyay, A. (1968). Fishermen of the Sundarbans. *Frontier,* August 24, **1** (20) : 11-12.

Mukhopadhyay, A. (1980). Some observations on the biology of the Openbill stork, *Anastomus oscitans* (Boddaert), in Southern Bengal. *J. Bombay nat. Hist. Soc.,* **77**(1) ; 133-137.

Mukhopadhyay, K., Jayjayanti Choudhuri & S. Maity (1991). In vitro response of four different mangrove species for callus induction. **In**: *Proceedings of National Seminar on conservation and Management of Mangrove Ecosystem with Special Reference to Sundarbans.* Decmber 6 - 8, 1991. pp. 280-283.

Mukhopadhyay, M. K. & P. U. Verghese (1979). Observations on the larvae *of Lates calcarifer* (Bloch) from Hooghly estuary with a note on their collection. *J. Inland Fish. Soc. India,* **10** : 138-141.

Mukhopadhyay, M.K. & H.C. Karmakar (1984). Effect of salinity on food intake, growth and conversion efficiency in juveniles of *Lates calcarifer* (Bloch). *J. Inland Fish. Soc. India,* **13** (1): 8-14.

Mukhopadhyay, M.K. and B.N. Saigal (1986). Distribution of *Coscinodiscus granii* and its possible use as bio-indicator for demarcating salinity regime in the Hooghly estuary. *J. Inland Fish. Soc. India,* 18(1) : 24-28.

Mukhopadhyay, M.K., B.B. Ghosh & M. M. Bagchi (1987). Preliminary observation on the impact of some industrial wastes on prawn fisheries of Hooghly estuary. *J. Indian Soc. Coastal agric. Res.* **5** (1) : 293-296.

Mukundan, U. (1981). Physiological studies in marine plants : Studies in *Acanthus ilicifolius,* Linn. Ph. D. Thesis, University of Bombay.

Mulik, N. G. & L. J. Bhosale (1989). Flowering phenology of the mangroves from the west coasts of Maharashtra. *J. Bombay nat. Hist. Soc.,* **86**(3) : 355-359.

Muniyandi, K. (1985). Studies on mangroves of Pitchavaram (Southeast coast of India), Ph.D.thesis, Annamalai University, 215pp.

Naik,U. G. & B. Neelakantan (1989). Seasonal abundance of phytoplankton in the inshore waters of Karwar. Comp.Physical Ecol. 14 : 219-226.

Nair, N. C. & P. Daniel (1986). The floristic diversity of the Western ghats and its conservation : A review. Proceedings Indian Academy of Science (Anim.Sci.Pl. Sci.) Suppl. November : 103-125.

Nakamura, T. (2000). Some ecological problems on the *Sonneratia* and *Avicennia* pneumatophores. The United Nations University, Tokyo. 117-128 pp.

Nalampoon, A. (1979). Harvesting and silvicultural techniques in the mangrove forest of Thailand. *Biotrop* No. 10. 81-83 pp.

Nandi, N. C. & A. Misra (1987). Bibliography of the Indian Sundarbans with special reference to fauna, *Records of the Zoological Survey of India*, Occasional Paper No. 97, pp. 1 -135.

Nandi, N. C. & S. K. Pramanik (1994). Crab and crab fisheries of Sundarbans, Hindusthan Publishing Corporation (India), Delhi, pp. - 1 -192.

Nandi, N. C., A. K. Das & N. C. Sarkar (1993). Protozoa fauna of Sundarbans mangrove ecosystems, *Rec. Zool. Surv. of India*, **93** (1-2) : 83 - 101.

Nandi, N.C. & S. Ghatak (1985). Crabs of commercial importance from coastal West Bengal. *J. Indian Soc. Coastal agric. Res.*,**3**(2): 133-135.

Nandi, N.C. & S.K. Pramanik (1986). A note on crab fishery and landing of *Scylla serrata* (Forskal) from Budhakhali, Sundarbans, West Bengal. *J. India Soc. Coastal agric Res.*, **4**(2): 151-153.

Nandi, N.C. & S.K. Raut (1985). A note on the occurrence of a bopyrid isopod parasite in the prawn *Palaemon (Exopalaemon) styliferus*. *J. Indian Soc. Coastal agric. Res.*, **3**(1) : 67-68.

Nandi, N.C. & S.K. Raut (1986). A note on a tail abnormality of the spotted house Gecko (*Hemidactylus brooki*). *Hamadryad*, **11** (1) : 22.

Nandi, N.C. (1971). Studies on myxoporidan parasites from some estuarine fishes of West Bengal. M. Sc. dissertation. Calcutta University.

Nandi, N.C. (1984). A note on the swarms of the jelly-fish affecting fishing in the estuarine waters of Sundarban, West Bengal. *J. Indian Soc. Coastal agric. Res.*, **2**(2): 103.

Nandi, N.C. (1984). Faunal resources of Sunderbans its perspective and predicaments. *Environ. & Ecol.*, **2**:99-102.

Nandi, N.C., A.K. Mandal & A. Choudhury (1984). Blood parasites of some birds from West Bengal, India. *Bull. zool. Surv. India*, **5**(2 & 3): 45-51.

Nandi, N.C., S.R. Das & J.G. Chatterjee (1983). On the occurrence of the penacid prawn *Penaeus japonicus* Bate in Hooghly estuary. *Indian J. fish.*, **30**(2) : 343-344.

Nandi, S. & A. Choudhury (1983). Quantitative studies of the benthic macro-fauna of Sagar Island, intertidal Sundarbans, India. Mahasagar - *Bull. Nat. Inst. Oceangr.* **16** (3) : 409 - 414.

Nandi, S. & A. Choudhury (1983). A quantitative study of benthic macrofauna in the littoral waters of Hooghly estuary, *Mahasagar, Bulletin of the National Institute of oceanography* **16**(3): pp. 409-414.

Nandy, A. C., M. M. Bagchi & S.K. Majumdar (1983). Ecological changes in the Hooghly estuary due to water release from Farakka Barrage. *Mahasagar-Bull. Nat, Inst. Oceanogr.*, **16** (2) : 209-220.

Naskar, K.R. & R. Mandal (1999). **Ecology and biodiversity of Indian mangroves**. Vol. I & II. Daya Publishing House, New Delhi, 361-775.

Naskar, K. R. & D. N. Guha Bakshi (1983a). A brief review on some less familiar plants of the Sundarbans. *J. Econ. Tax. Botany*. **4** (3) : 699 - 711.

Naskar, K. R. & D. N. Guha Bakshi (1983b). An eco-taxonomical studies of the typical halophytic flora of Sundarbans in the District 24-Parganas, West Bengal with special reference to their socio-economic impact.. *J. Econ. Tax. Botany*. **4** (3) : 713 - 746.

Naskar, K. R. & D. N. Guha Bakshi (1981). Some of the important medicinal plants of Apocynaceae from West Bengal. *Bull. Bot. Soc. Bengal*. **35** : 7 -14.

Naskar, K. R. & D. N. Guha Bakshi (1983). An eco - taxonomical studies of the tropical halophytic flora of the Sundarbans in the District 24 Parganas, West Bengal with special reference to their socio-economic impact, *J. Econ. Tax. Botany*, **4**(3): 713 - 746.

Naskar, K. R. & D. N. Guha Bakshi (1986). On the verge of extinction of some important Mangrove Species from the Sundarbans Delta in West Bengal . *J. Econ. Tax. Botany* **8** (2): 431-437.

Naskar, K. R. & D. N. Guha Bakshi (1987). Different Phyto-Ecological Zones in the Twenty Four Parganas District of West Bengal with special reference to its land utilization patterns. *J. Indian Soc. Coastal agric.Res.* **5** (1) : 183-187.

Naskar, K. R. & D. N. Guha Bakshi (1987). *Mangrove Swamps of the Sundarbans - An Ecological Perspective*, Naya Prokash, 206, Bidhan Sarani, Calcutta - 700 006. pp. 1- 264.

Naskar, K. R. & D. N. Guha Bakshi (1989). Distribution of Mangroves in India and Abroad and their Roles in the Estuarine Ecosystems. *Monograph* **9** : 22-23(National Atlas & Thematic Mapping Organisation, Calcutta).

Naskar, K. R. & D. N. Guha Bakshi (1993). The Biological Spectrum of Floral Elements of the 24 Parganas Dist. in W. Bengal. *Bangladesh J. Bot.* **22**(1) : 15-20.

Naskar, K. R. & D.N. Guha Bakshi (1982). Sundarbans - the world famous mangrove forest of the district 24-Parganas in West Bengal India. *J.Econ. tax. Bot.*, **3** : 883-918.

Naskar, K. R. & D.N. Guha Bakshi (1983). An Eco-taxonomic studies of the typical halophytic flora of Sundarbans in the district of 24-Parganas, West Bengal with special reference to their socio-economic impact. *J. Econ. Tax. Botany*, **4** (3) : 713-746.

Naskar, K. R. & N. M. Chakraborti (1984). Studies on the economic fauna from the Sundarbans delta in West Bengal. *J. Indian Soc. Coastal agric. Res.*, **2** (1) : 17-25.

Naskar, K. R. & N. M. Chakraborty (1987). Studies of the family Lemnaceae from the Ganga delta. *Bull. Bot. Soc. Bengal.* **41** : 5 - 12.

Naskar, K. R. & S. C. Santra (1985). A note on *Enteromorpha tubulosda* in brackish mixed sewage fed fisheries from the Sunderbans in West Bengal . *J. Indian Soc. Coastal agric Res.* **5** (2) : 471 - 472.

Naskar, K. R. & S. C. Santra (1986). Studies on *Enteromorpha tubulosa* in brackish water mixed sewage fed fisheries from Sundarbans . *Sci. & Cult.* **52** : 210.

Naskar, K. R. & S. K. Saha (1988). Utilization of Domestic Wastes for Integrated Farming in the Tropics. *J. Aqua. Trop.* **3** (1) : 37-41.

Naskar, K. R. & Santanu Biswas (1988). Hydrocharitaceae from the lower Ganga delta and its role in aquaculture and agriculture. *Environ. & Ecology*, **6** (3) : 620 - 626.

Naskar, K. R. (1981). *Floristic Studies of the District 24 Parganas in West Bengal (India) with Special Reference to the Mangrove Vegetation of the Sundarbans*, Ph. D. Thesis, University of Calcutta, 3 Vols. 1800 pp.

Naskar, K. R. (1983). Halophytes and their unique adaptation on the Sundarbans mangrove swamps in India. *J. Indian Soc. Coastal Agric. Res.*, **1** (2) : 91-105.

Naskar, K. R. (1985). Population pressure on the tidal mangrove forest of the Sundarbans - Its cause and problems. *Monograph - 4*, pp. 159 - 161. National Atlas and Thematic Mapping Organisation, Calcutta, India.

Naskar, K. R. (1985). A short history and the present trends of brackish water fish culture in paddy fields of the Kulti-Minakhan areas of the Sunderbans in West Bengal. *J. Indian Soc. Coastal agric Res.* **3** (2) : 115 - 126.

Naskar, K. R. (1986). Economic Importance of the dominant Mangrove Family- Rhizophoraceae from the Sundarbans Delta of 24 Parganas District, W.B., *J. Econ. Tax. Bot.* **8**(2): 359-372.

Naskar, K. R. (1986). Role of *Avicennia* L. plantation on the Brackishwater fisheries with special reference to their taxonomy and ecological notes in the tidal mangrove forests of Sundarbans. *J. Ind. Soc. Coastal agric. Res.* **4**(2): 139-142.

Naskar, K. R. (1986). Economic importance of the dominant mangrove family - Rhizophoraceae from the Sunderbans delta of Twenty Four Parganas District, W.B. *J. Econ. Tax. Botany*. **8** (2) : 359-372.

Naskar, K. R. (1987). Mangrove Ecology of the Sundarbans Delta in West Bengal and its role on the brackishwater fisheries, *J. Aqua. Organisms* **1**(1): 1-9.

Naskar, K. R. (1987). Mass Production of Duckweeds for carp culture. *Proc. USAID Training Programme on ' Fresh Water Aquaculture'* , CIFA, Bhubaneswar, pp. 56-65.

Naskar, K. R. (1987). Recent trends of aquatic weeds management through proper utilizations. **In**: *Plant Science Research.* Department of Botany, Visva-Bharati, Santiniketan. India. pp.188-193.

Naskar, K. R. (1988). Economic potentialities of the tidal mangrove forests of Sundarbans in India, *J. Indian Soc. Coastal agric. Research* **6** (2) : 149-157.

Naskar, K. R. (1990). *Aquatic & Semi-Aquatic Plants of Lower Ganga Delta,* Daya Publishing House, Delhi-110 035, pp. 1-402.

Naskar, K. R. (1990). Status of the Sunderbans Mangrove Forests of India and Its Ecological Stresses. International Symposium on Ecology of the Mangroves and Related Ecosystems. Mombasa, Kenya. (24th - 30th September 1990).

Naskar, K. R. (1991). Mangrove Ecosystems in the Indian Sub-continent with special reference to the Sundarbans. W.B. *J. Indian Soc. Coastal agric. Res. 9* (1/2) 455 -462.

Naskar, K. R. (1995). *Biosphere Reserve of The Sundarbans- Its Aims & Objectives.* **In**: Souvenir West Bengal Government Fisheries Association.

Naskar, K. R. (1996). Role of Aquatic Plants in Sewage Treatment , *vis-a-vis*, its application. **In :** Fish Culture Bulletin, Sewage Fed Division, CIFA, Rahara, pp.

Naskar, K. R. (1996). State of the coastal vegetations in West Bengal with special reference to the Sundarbans Mangals. pp. 72 - 130. **In** : *Integrated Coastal Zone Management : A Manual.* (ed. A. K. Ghosh & P. Sanyal). Department of Environment, W. B. & Department of Ocean Development, Government of India.

Naskar, K. R. (1996). *Status of the Coastal Zone Vegetation in West Bengal with Special Reference to the Sundarbans* .STATUS PAPER. Dept. of Environment, West Bengal. pp. 51-79.

Naskar, K. R. (1996). Status of the mangroves in the Indian Sundarbans in the perspectives of India and the world mangals. **In**: *Background Information, Programme & Abstracts of Papers of 'William Roxburgh Memorial Seminar on the Seminar on the Sundarbans Mangals'*. pp. 19 - 55 + table pages 1 - 29.

Naskar, K. R. (1996). *Sundarike Bhalobasi*, Souvenir, William Roxburgh Memorial Seminar on Sundarbans Mangals, Calcutta.

Naskar, K. R. (1997). Final Reports of Project Entitled - *Potentialities of the Tidal Mangrove Forests of the Sundarbans With Special Reference to Estuarine Fisheries & Forestry*, Ministry of Environment & Forest, New Delhi.

Naskar, K. R. (1999). Status of the mangroves in Indian Sundarbans - In the perspectives of India and world mangals. **In :** *Sundarbans Mangals* (ed. Guha Bakshi, D. N., Sanyal, P & K. R. Naskar). pp 20 - 89.

Naskar, K. R. (2000). General Recommendations from the 'Sir William Roxburgh Memorial Seminar on the Sundarbans Mangals'. **In**: *Sundarbans Mangals*, eds. D. N. Guha Bakshi, P. Sanyal & K. R. Naskar. Naya Prokash, 206, Bidhan Sarani, Kolkata.

Naskar, K. R., A. Banerjee, N. M. Chakraborty & Apurba Ghosh (1986). Yield of *Wolfia arrhiza* (L.) Horkel ex Wimmer from cement cisterns with different sewage concentration and its efficacy as carp feed. *Aquaculture* (Netherland) **51**: 211- 216

Naskar, K. R., D. N. Guha Bakshi & G. G. Maity (1979). On the occurrence of *Moghania macrophylla* (Willd.) O.Kuntze - from the lower Ganga Plain. *Bull. bot. Soc. Bengal* **33**:39-43.

Naskar, K. R., Dolan Champa Ghosh, Neera Sen, R. N. Mandal & A. K. Sarkar (1997). Mangrove Ecology of the Indian Sundarbans : Its Impact on the Rural Economy and Coastal Environment. *J. Interacad.* **1** (1) : 49 - 60.

Naskar, K. R., P. K. Saha, Rita Mazumdar & D. N. Guha Bakshi (1986). Compost Value of Water hyacinth(*Eichhornia crassipes* (Mart.) Solms.) in the field of agriculture and aquaculture. *Bull. bot Soc. Bengal* **40** (1&2) :11-17.

Naskar, K. R., R. K. Das, A. Ghosh & Apurba Ghosh (1991). Productivity Constraints in Estuarine Wetlands Fisheries Caused by Different Aquatic Weeds Infestation - A Case Study. *J. Indian Soc. Coastal agric. Res.* **9**(1/2) : 479-484.

Naskar, K. R., R. N. Mandal, Dolanchampa Sarkar, Neera Sen & A. K. Sarkar (1997). Investigation on seedling development, *vis - a - vis*, plantation of Heritiera fomes. Buch. - Ham. Beyond the intertidal non - saline zones. *J. Interacad.* **1**(3) : 177-182.

Naskar, K. R., S.K. Saha & A. K. Datta (1988). Utilization of Domestic Wastes for Rice Production. *Environ. & Ecol.* **6** (1) : 125-128.

Naskar, K. R., S.K. Saha & P. K. Saha (1985). Compost of water hyacinth for the fields of agriculture and aquaculture. *Environ. & Ecology* **3** (2): 247-253.

Naskar, K.R. & D. N. Guha Bakshi (1989). Stages of Land Formation and Comparative Studies of the Different Microecological zones of Sundarbans (West Bengal), India. *Fragm. Flor. Geobot.* (Poland). **34** : (3-4) : 341 - 354.

Naskar, K.R. & D.N. Guha Bakshi (1982). Sundarbans - the world famous mangrove forests of the District 24 Parganas in West Bengal. *J. Econ. Tax. Botany.* **3** : 883 - 918.

Naskar, K.R. & N.M. Chakraborty (1984). Studies on the economic fauna from the Sunderbans delta in West Bengal. *J. Indian Soc. Coastal agric. Research.* **2** (1) : 17 - 25.

Naskar, K.R. & S.K. Saha (1988). Utilisation of domestic wastes for integrated farming in the tropics. *J. Aqua. Trop.* **3** (1) : 37 - 41.

Naskar, K.R. (1985). Ecological succession of the different mangove species of tidal Sundarbans in India and their biology, utilization and conservation. *Nat. Sump. on Biology, Utilzation and consevation of Mangroves,* 18-20 November, 1985, Abstract No. 19, p.10.

Naskar, K.R. (1987). Mangrove ecology of the Sunderbans delta in West Bengal and its role on brackish water fisheries. *Jour. Aqua.* **1** (1) : 1 -9.

Naskar, K.R. (1987). Riverbed vegetation in the tidal estuary of Sunderbans and its impact on brackishwater fisheries. Symposium on the 'Impact of Current Land Use Pattern & Water Resources Development on Riverine Fisheries', Central Inland Capture Fisheries Research Institute, Barrackpore. (25th - 27th April, 1987).

Naskar, K.R. (1991). Mangrove ecosystems in the Indian sub-continent with special reference to the Sunderbans, W.B. *J. Indian Soc. Coastal agric, Res.* **9** (1/2) : 445 - 462.

Naskar, K.R. (1991). Present Status and Problems of the Mangrove Forests of the Sunderbans in West Bengal. National Seminar on Conservation & Management with Special Reference to Sunderbans (6th - 8th December 1991)

Naskar, K.R. (1991). Natural Resource Conservation in the Tidal Mangrove Forest of Sunderbans.National Symposium on *'Farming Systems for Sustained Productivity in Humid Tropics'.* Andaman Science Association & the State Council of Science & Technology, Andaman & Nicobar Islands. (16th - 17th December 1991).

Naskar, K.R.(2001). Status of Mangrove Ecosystems in Indian Sundarbans. In Proc. *Of the Workshop on Sundarbans,* 3rd June, 2001. Organising by Sundarbans Development Board, Zilla Parisad 24 Parganas (S), Calcutta Wildlife Society.

Naskar, K.R., Dolan Champa Ghosh, Neera Sen, R.N. Mandal & A. K. Sarkar (1997). Mangrove ecology of the Indian Sundarbans : Its impact on the rural economy and coastal environment. *J. Interacad.* **1** (1) : 49 - 60.

Navalkar, B. S. (1951). Succession of the mangrove vegetation of Bombay and Salsette Islands. *J. Bombay Nat. hist. Soc.* **50**: 157-160, 1951.

Navalkar, B. S. (1953). The analytical characters of some marshy vegetation of Bombay and Salsette Islands. *J. Bombay Nat. hist. Soc.* **51**: 636-652.

Navalkar, B. S. (1967). The soil requirements and the comparative list of species of the halophytic plants of Bombay and Salsette Islands with those of other places. *J. Univ. Bombay* **35**: 1-4.

Navalkar, B. S. (1973). Ecology of mangroves. **In**: R. Misra, B. Gopal, K.P. Singh and J.S. Singh (eds.). Progress of Plant Ecology in India. Today and Tomorrow Printers and Publ. New Delhi Vol. I pp. 101-109,1973.

Oldham, R. D. (1893). A Manual of the Geology of India : Stratigraphical and Structural Geology (2nd ed.). pp. 432-444.

Olivia, J. Fernando (1982). Ecological studies in the intertidal region of the Vellar estuary (Porto Novo,S.India). Ph.D.thesis, Annamalai University, India. 140 pp.

Pakrasi, B. B., N. C. Basu & R. K. Banerjee (1977). Role of grey mullets in polyculture in coastal tanks of West Bengal. **In** : Contributions to Estuarine Biology. ed. C.V. Kurian, Cochin, Department of Marine Sciences, University of Cochin: 31-40, *Bull. Dept. Mar. Sci. Univ. Cochin,* **7** (4) : 31-40.

Pakrasi, B. B., P. R. Das & S. C. Thakurta (1964). Culture of brackishwater fishes in impoundments in West Bengal. Indo-Pacific Fish. Coun., 11th Session, Tech-Pap., 19, 1-19.

Pakrasi, B. B., S. C. Banerjee & R. K. Banerjee (1980). Ecology of Ponds in coastal region of Sunderbans. ***Symp. Coast. Aquaculture***, MBAI, Cochin, 12-18 January, 1980. Abstract. No. 20 :12.

Pal, A. (1978). Role of Banks in commercial brackishwater pisciculture. *Proc. Training in brackishwater prawn & fish farming* held at CIFRI, Kakdwip. 5 pp. (Mimeo.).

Pal, A. K. & M. K. Bandyopadhyay (1986). The role of mangroves in deltaic morphology: A study in Prentice and Lothian islands, Sunderbans, West Bengal, in 'The Mangroves', (Ed. L. J. Bhosale). Shivaji University Press, Kolhapur, pp. 218-221.

Pal, M. K., M. K. Guha Roy & S. Pathak (1978). Performance of multiple cropping with jute in irrigated and rainfed areas. *A Review of Agro Animal Sciences and Health.* **3**(11) : 533-536.

Pal, R. N. & A. K. Ghosh (1979). Diseases of fishes and their control. Souvenir : **In** commemoration fo the ICAR Golden Jubilee Year CIFRI, Barrackpore 2 : 139-142.

Pal, R. N. & S. D. Tripathi (1979). Use of terramycin for fish disease in carp and catfish culture in Indian water. *J. Inland Fish. Soc. India,* **10** : 166-168.

Pal, R. N. (1963). Observation on fluctuations in parasitization of the Indian shad, *Hilsa ilisha* (Hamilton) of the Hooghly Estuary, *Indian J. Helminth.,* **15**(2) : 119-126.

Pal, S. G. & S. Modak (1981). Histophysiology of amoebocytes of marine and estuarine bivalves. *Biol. Bull. India,* **3** (2): 110-116.

Pal, U. C. & S. C. Santra (1986). Leaf inhabiting blue - green algal flora of West Bengal. *Indian J. Bot.* **9**(2) : 175 - 176.

Pal, U. C., K. R. Naskar & S. C. Santra (1988). A check list of algal flora on Sundarbans delta of West Bengal, India. *Phykos,* 27 : 48 - 53.

Pandit, P.K. & N. C. Datta (1991). Feeding behaviour and food capture process of young Bhetki, *Lates calcarifer* (Bl.) inhabiting the mangroves of Sundarbans. **In**: Mangroves of Sundarbans and elsewhere : 329 - 334.

Pannier, F. & R. F. Pannier (1975). Physiology of vivipary in *Rhizophora mangle.* **In**: Proc. Intntl. Symp. on Biol. and managt. of mangroves. G.E. Walsh, S.C. Snedaker and H.J. Teas (eds.) Inst. Food and Agrl. Sci. Univ. of Florida, Gainesville florida. Vol. 2pp. 632-639.

Pant, A. V. K., N. B. Dhargalkar, L. Bhosale & A. G. Untawale (1980). Contribution of phytoplankton photosynthesis to a mangrove ecosystem. *Mahasagar- Bull. natn. Inst. Oceanoger.*, **13**: 225-234.

Pargiter, F. E. (1889). The Sundarbans. *Calcutta Review*, **89**: 280.

Patil, R. P. (1961). A note on the vegetation of Sazinakhali Sundarbans. *Indian Forester*, **87** : 481 - 483.

Patil, R.P. (1964). Cultivation of mangrove seedlings in pots at Allahabad, U.P. *Sci. & Cult.* **30** : 43 - 44.

Percival, M. & J. S. Womersley (1975). Floristic and Ecology of the Mangrove Vegetation of Papua New Guinea. *Botany Bulletin No 8*. Department of Forests. Lae, 96 pp.

Pillay, T. V. R. (1958). Land reclamation and fish culture in the deltaic areas of West Bengal, India. *Proc. Fish Cult.*, F.W.S., **20** (3) : 99-103.

Pillay, T. V. R. & B. B. Bose (1957). Observations on the culture of brackishwater fishery in paddy fields in West Bengal (India). *Proc. Indo-Pacif. Fish. Coun.*, **7** (2 & 3) : 187-192.

Prain, D. (1903). Bengal Plants, 2 Vols. (Rep.Ed. 1963), BSI, Calcutta.

Prain, D. (1903). Flora of Sundribuns. *Rec. bot. Surv. India*, **2** : 231 - 370.

Prain, D. (1903). Notes on Sundribuns Plants, *Proc. Asiat. Soc. Bengal*, 72 : 107-108.

Prain, D. (1905). Vegetation of the districts Hoogly, Howrah and 24-Parganas. *Rec.bot. Surv. India*, **3** : 143-329.

Prakasam, V. R. & P. Johnson (1992). Ecology of Quilon canal (T S Canal) with reference to physico-chemical characteristics. *J. Environ.Biol.*, **13**(3) : 221-225.

Pramanik, M. & B. Mazumdar (1983). Comparative studies of stomata in some Halophytes, cultivated rice and rice mutants in relation to salt resistance. *J. India Soc. Coastal agric Res.*, **1** (2) : 78-83.

Prasad, B. (1922). Observations on the luminosity of some animals in the Gangetic delta. *J. Asiat. Soc. Bengal.*, **18**: 585-591.

Prasad, R. R. (1963). Study of primary production and its importance in an integrated fisheries research programme. *Fish Tech.* **1** : 1- 4.

Premanathan, M., K. Chandra & K. Kathiresan (1992). Antiviral activity of marine plants from India. *Antiviral Res.* (Amsterdam), 17 (Suppl.1) : 154.

Premanathan, M., K. Kathiresan & H. Nakashima (1999). Mangrove halophytes : A source of antiviral substances. *South Pacific Study*, **19** (1-2) : 1-9.

Premanathan, M., K. Kathiresan & K. Chandra (1993). Broad spectrum antiviral activity of mangrove plants. *Anti. Res.* (Netherlands), 20S(1) : 235.

Premanathan, M., K. Kathiresan & K. Chandra (1995). Antiviral evaluation of some marine plants against Semliki forest virus. *Int.J.Pharm.*, (Netherlands), **33**(1) : 75-77.

Premanathan, M., K.Kathiresan & A. Sivakumar (1988). Effect of marine plant extracts on *Ceratonereis costae* Fauvel. Sriramulu, V. and PR. Karpagaganapathi (eds.), *Proc. National Symposium on Ecotoxicology*, Annamalai University, India, 1-3.

Premanathan, M., R. Arakaki, H. Izumi, K. Kathiresan, M. Nakano, N. Yamamoto & H. Nakashima (1999). Antiviral properties of a mangrove plant, *Rhizophora apiculata* Blume, against human immunodeficiency virus. *Anti. Res.*, **44** : 113-122.

Preston, H. B. (1911). Description of six new species of shells from Bengal and Madras. *Rec. Indian Mus.*, **6** : 39-42.

Preston, H. B. (1915). Report on a collection of mollusca from the out skirts of Calcutta. *Rec. Indian Mus.*, **11**(6) : 479-482.

Puri, G. S. (1960). Indian Forest Ecology, 2 Vols. Oxford Book and Stationery Cp, New Delhi, India.

Quereshi, T. (2000). The conflicting interests of mangrove resource use in Pakistan. Asia-Pacific Cooperation on Research for conservation of Mangroves. The United Nations University, Tokyo. 31-38 pp.

Qureshi, J. M. (1957). Botanical silvicultural features of mangrove forest of Bombay State. *Proc. Mangrove Symp.* Calcutta, pp. 20-26, 1957.

Qureshi, M. T. (1990). Experimental plantation for rehabilitation of mangrove forests in Pakistan. Mangrove Ecosystems Occasional Paper. UNDP/UNESCO. No. 4. 37 pp.

Rabinowitz, D. (1978). Early growth of mangrove seedlings in Panama, and a hypothesis concerning the relationship of dispersal and zonation. *Journal of Biogeography* **5** : 113-133.

Radhakrishna, K. (1978). Primary productivity of the Bay of Bengal. *Indian J. Mar. Sci.* **7**(1): 58-60.

Ragothaman, G. & Y. R. Reddy (1982). Hydrobiology of Tapi Estuary, Surat. Indian *J.mar.Sci.*, **11**(3) : 256-258.

Rahaman, A. A. (1990). Living resources as indicators of sea level variations. G. V. Rajamanickam (ed.), Sea Level Variation and its impact on coastal environment, 245 pp.

Rahaman, A.A. (1987). Ecosystem studies and management in coastal belt of Cauvery delta in Muthupet, Tamil Nadu. Nair, N.B. (ed.), *Proc. Natl. Sem. Estuarine Management*. Trivandrum, 168-171.

Rahman, M A. (1990). A comprehensive report on sundri (*Heritiera fomes*) trees with particular reference to top dying in the Sundarbans. Proceedings of the seminar on top dying of sundri (*Heritiera fomes*) trees (August 11, 1988). Bangladesh Agricultural Research Council, Dhaka. 12-63 pp.

Rahman, M. A. (1998). Disease and wood decay of tree species with particular reference to top dying of sundri and the magnitude of its damage in the Sundarbans in Bangladesh. Integrated Management of Ganges Flood Plains and Sundarbans Ecosystem. Khulna University. 50-76 pp.

Rahman, M. A., S. K. Khisha & A. C. Basak (1983). Top dying of sundri in the Sundarbans. *Bano Biggyan Patrika* **12** (1&2) : 69-71.

Rajagopalan, V. R. (1952). Ecological adaptation of the mangrove vegetation of Pichavaram. *J. Annamalai Univ. Annamalainagar*, India.

Rajaguru, A., G. Shantha & R. Natarajan (1988). Mangrove waters as nursery grounds for juveniles of marine flat fishes. *CMFRI Spec.Publ.*, **40** : 62.

Rajendran, N. & K. Kathiresan (1996). Effect of effluents from a shrimp pond on shoot biomass of mangrove seedlings. *Aqua. Res.* (Netherlands) **27** : 745-747.

Rajendran, N. & K. Kathiresan (1998). Mangrove vegetation trap technique for improving fishery resources in coastal waters. *Curr. Sci.* **75**(5) : 429.

Rajendran, N. & K. Kathiresan (1999). Do decomposing leaves of mangrove attract fishes? *Curr. Sci.*, **77**(7) : 972-976.

Rajyalakshmi, T. (1979). A coastal zone management programme for India-need for a multidisciplinary perspective. *Seafd. Export J.*, **11** (7): 19-27.

Ramachandran, P., V. Nair & C. P. Gopinathan (1983). Primary production in coastal waters. *Bull. Cent. Mar. Fish. Res. Inst.* **34:** 20-32.

Ramachandran, S. & R. Krishnamoorthy (1998). Application of Remote Sensing Techniques for Mangrove Studies. In : *An Anthology of Indian Mangroves*. Envis Centre, CAS in Marine Biology, Annamalai University. pp. 39-47.

Ramdial, B. S. (1991). Role and importance of mangrove forests. Deshmukh, S.V. and R.Mahalingam (eds.) A Global Network of Mangrove Genetic Resource Centres, Project Formulation Workshop, Madras, India, 75-84.

Rao, T. A. & A. R. K. Sastry (1972). An ecological approach towards classification of coastal vegetation in India. I. Strand Vegetation. *Indian Forester*, **98** : 594 - 607.

Rao, T. A. & A. R. K. Sastry (1974). An Ecological approach towards classification of coastal vegetation of India and Estuarine border vegetation. *Indian Forester*, **100** : 438-452.

Rao, T. A. & A. R. K. Sastry (1976). An outline of the coastal vegetation of India. *Bull. bot Surv. India*, **16** : 101 - 115.

Rao, T. A. & K. R. Aggarwal (1963). An ecological account of the vegetation of Rameshwaram island. *Bull. Bot. Surv. India*. **6** : 201-323.

Rao, T. A. & K. R. Aggrawal (1966). Ecological studies of Saurashtra coast and neighbouring islands. *Bull. Bot. Surv. India*. **8**: 60-67.

Rao, T. A. & S. Chakraborti (1987). Distributional resume of coastal floristics elements in the Andaman and Nicobar islands. *Current. Sci.* **56** : 1045-1051.

Rao, T. A. & Sastry, A. R. K. (1973). Studies on the flora and vegetation of coastal Andhra Pradesh, India, *Bull. Bot. Surv. India*. 15: 92-107.

Rao, T. A. & Sastry, A. R. K. (1974). An outline of the coastal vegetation of India. *Bull. Bot. Surv. India*. **16** : 101-115.

Rao, T. A., A. K. Mukherjee & L. K. Banerjee (1973). *Acrostichum aureum* L truely a mangrove fern. *Curr. Sci.* **42** (15) : 546-547.

Rao, T. A., P. V. Suresh & A. N. Shereiff (1987). Little known attributes of some mangrove plants of the Karnataka coast, *Curr. Sci.* **56** : 725.

Rao, T. A., P. V. Suresh & A. N. Sherieff (1986). Multiple viviparity in a few taxa of mangroves. *Curr. Sci.* **55**: 295-261.

Rao, T. A., S. Chakraborti & R. K. Premananth (1982). A typical vivipraro us condition in *Burguiera cylindirca* (L) Bl. (Rhizophoraceae). *Bull. Bot. surv. India*. **24**, pp. 183-184.

Rao, T. S. S. & S. R. K. Nair (1984). Trophic relationships in mangrove ecosystems, Report on the 2nd Introductory Training Course (Goa), UNDP/UNESCO Regional Mangrove Project, 87 - 91.

Rao, U. M. (1974). Addition to the algal flora of the Gulf of Mannar and Palk Bay from Mandapam area - II. *Phykos* **13**(2) : 56–59.

Ratha, D. S. & B. K. Sahu (1993). Seasonal variation of geochemical data in sediment samples from two estuaries in western India. *Indian J.Environ.Health,* **35**(4) : 294-300.

Ray, Animesh (1991). Mangrove Ecosystem in India. Mangrove of Sundarbans & Elsewhere.(Eds. S. Maity & N. Mukherjee).*Proc. Natl. Symp. Cons. & Manage Mangroves* Dec. 1991: 6 - 26.

Reddy, T. K. K., A. Rajasekhar, B. Jayasunderamma & R. Ramamurthi (1991). Studies on marine bioactive substances from the Bay of Bengal : Bioactive substances from the latex of the mangrove plant *Excoecaria agallocha* L : Antimicrobial activity and degradation. Thompson, M.F., R.Sarojini and R.Nagabhushanam (eds.), Bioactive compounds from marine organisms with emphasis on the Indian Ocean. An Indo - United States Symposium, Goa, 75-78.

Yadav, V. K. & A. Choudhury (1985). Remote sensing applications to the study of mangrove ecosystem of Sundarbans, West Bengal. **In**: *The Mangroves,*(Ed. L. J. Bhosale), *Proc. National Seminar on Biology, Utilization and Conservation of Mangroves*, Shivaji University Press, Kolhapur.

Roy, H. K. (1955). Plankton ecology of the river Hooghly at Palta, West Bengal. *Ecology,* **36**(2) : 167-175.

Roy, R. & N. C. Sarkar (1985). On the occurrence of intestinal coccidea of wild Boar, *Sus scrofa* at Sundarbans forests, West Bengal. *J. Beng. nat. Hist. Soc.* NS, **4**(1): 29-33.

Roy, S. D. (1997). Study of litterfall and its decomposition in a mangrove stand, South Andaman. *J.Andaman Sci.Assoc.,* **13**(1-2) : 119-121.

Saenger, P. (1979). Morphological and reproductive adaptations of Australian mangroves. Proceedings of the Australian National Mangrove Workshop (April 18-20), Cape Ferguson. Australian Institute of Marine Science. 153-191 pp.

Saenger, P., E.J. Hegere & J.D.S. Davie (1983). *Global status of mangrove ecosystem.* IUCN and Natural Resources. Switzerland.

Sahni, K. C. (1953). Botanical exploration in the Great Nicobar. *Indian For.* **79**: 3-16, 1953.

Sahni, K. C. (1957). Mangrove forests in the Andaman and Nicobar Islands. Paper in Mangrove Symp. 16-19 Oct. (1957). Calcutta, India, 1957.

Sahoo, A. K., K. D. Sah & S. K. Gupta (1986). Studies on nutrient status of some mangrove mud of the Sundarbans, The Mangroves. Proc. Symp. Biol. Util. Cons. Mangroves: Nov. (1985). pp. 375-377.

Sanyal, P. & A. K. Bal (1985). Some observations on abnormal adaptations of mangroves in Indian Sundarbans, Nat. Symp. on Biology, Utilization and conservation of Mangroves, Kolhapur 18-20 November, 1985, Abstract No. 25, p.14.

Sanyal, P. & A. R. Bal (1986). Some observations on abnormal adaptations of mangrove in Indian Sundarbans, *J. Indian Soc. Coastal agric. Res.* **4** (1) : 9 - 16.

Sanyal, P. (1981). Sundarbans Tiger Reserve - An Overview. *Cheetal* (1) : 5 - 8.

Sanyal, P. (1983). Mangrove tiger land-the Sundarbans of India. *Tigerpaper,* **10** (3): 1-4.

Sanyal, P. (1983). Saving the sea turtle. In: A. Dasgupta ed. *The Statesman*, Calcutta. 4th July 1983.

Sanyal, P. (1984). 'Bird - Watchers' Paradise in Sundarbans. School Science, NCERT, New Delhi, XXII/3 : 29 - 30.

Sanyal, P. (1984). Saving the Ridley Turtle in Sundarbans India. *Hamadryad,* Madras Snake Park Trust, **9**(3) :21-23.

Sanyal, P. (1987). Managing the Man-eaters in the Sundarbans Tiger Reserve of India - A case study. Tiger of the World, (Ed.) R. Tilson, Noyes Publication, U.S.A. : 427 - 434.

Sanyal, P. (1987). Sundarbans - The Mangrove Tigerland of India. Vivekananda Kendra Patrika, pp. 129 - 136.

Sanyal, P. (1991). Mangal Disposition in Indian Sundarbans. Mangrove of Sundarbans & Elsewhere.(Eds. S. Maity & N. Mukherjee).*Proc. Natl. Symp. Cons. & Manage Mangroves* Dec. : pp. 36-40.

Sanyal, P., L. K. Banerjee & M. K. Chowdhury (1984). Dancing mangals of Indian Sundarbans. *J. Indian Soc. coastal agric. Res.* **2** (1) : 10 - 16.

Sanyal, P., R. N. Mandal, Dolanchampa Sarkar & K. R. Naskar (1998). Studies on the mangrove patch at Subarnarekha river mouth of Orissa State. *J. Interacad.* **2**(2) : 140-149.

Sanyal, R. B. (1892). A hand book of the management of animals in captivity in lower Bengal, xiix+351 pp. Bengal Secretariat Press, Calcutta.

Sathe, S. S. & L. J. Bhosale (1991). Physiology of *Avicennia* : A review. Agate, A.D., S.D., Bonde and K.P.N. Kumaran (eds.), *Proc. Symp. Significance of Mangroves,* Pune.

Schlich, W. (1875). Remarks on the Sundarbuns. *Indian For.* **1** : 6 -11.

Schlich, W. (1880). The Sundarbans (islets of the Ganges delta) Les Sundarbans (ilots dev delta dn Gange). *Rev. Eanx. For.,* **19**: 451-455.

Scholander, P.F. (1968). How mangroves desalinate water. *Physical Plant* **21**: 251-261.

Schwamborn, R. and V.S. Paul (1996). Mangroves Forgotten Forests. *Naturl. Res Develop.*, 43/44 : 13-36.

Sekar, T., Karunakaran, M.S. & Joseph, S.J. (1989). Preliminary studies on the regeneration of *Rhizophora* species in Pitchavaram Mangroves, Tamil Nadu. Indian Forester **115**(12) : 851-859.

Selvam, V., R. Mohan, R. Ramasubramanian & J. Azariah (1991). Plant communities and soil properties of three mangrove stands of Madras coast. *Indian J.mar.Sci.*, **1**(20): 67-69.

Selvaraj, A. M., G. Santhana Kumar & Augustine Xavier (1995). Biomass production through mangrove coastal social forestry. Proc *Nat. Symp. Mahatma Gandhi Institute of Integrated Rural Energy.* Planning and Development, Bakoli, New Delhi.

Sen, D. N. & K.S. Rajpurohit (1982). *Contributions to the Ecology of Halophytes.* Dr. W. Junk Publishers, The Netherlands, 272 p.

Sen, D. N. (1985). Biology and utilization of halophytes of Indian desert. In: The Mangroves: *Proc. Nat. Symp. Biol. Util. Cons. Mangroves.* pp. 39 - 48. (Ed. L.J. Bhosale, Kolhapur).

Sen, Neera & K. R. Naskar (2001). Occurrence of a Brown Algae in the Indian Sundarbans. *Seshaiyana-ENVIS Newsletter on Estuaries, Mangroves, Coral Reefs and Lagoons.* **9**(1):1-2.

Sen, Neera, Bratati Bagchi, S. Baidya & K. R. Naskar (2001). Food and feeding habits of the estuarine fish *Scatophagus argus* (L.) Day from the Sundarbans Mangals of West Bengal. *J. Interacad.* **5**(1): 87 - 90.

Sen, Neera, Dolanchampa Ghosh & K. R. Naskar (1997). Socio-Economic Status of the Rural Population in the Indian Sundarbans and Their Role in The Conservation of Mangroves. p. 267-272. **In** : K. K. Vass and M. Sinha (eds). Changing Perspectives of Inland Fisheries. Proceedings of the National Seminar, March 16 - 17, 1997, Inland Fisheries Society of India, Barrackpore.

Sen, Neera, Dolanchampa Sarkar, A. K. Sarkar & K. R. Naskar (1997). A note on the algal flora of Sundarbans. *J. Interacad.* **1**(3) : 177 - 182.

Sen, Neera, K. R. Naskar, D. K. Bhattacharya, B. K. De (2000). Fatty acid composition of the lipids of different algae from estuarine habitat of the Indian Sundarbans. *J. Oil Tech. Asso. India* **32**(3) : 120 - 123.

Sen, Ratna, Ankhi Sau, K. R. Naskar & A. Bhattacharya (1980). Studies on nodulation of some useful Leguminous plants. *Bull. Bot. Surv, India.* **22** (1-4) : 166-172.

Sen, T.K. (1975). Fishes and Fisheries aspects of the Sundarbans, West Bengal, *Sci. & Cult.*, **41** (1) : 31-34.

Sengupta, A. & S. Chaudhuri (1991). Ecology of heterotrophic dinitrogen fixation in the rhizosphere of mangrove plant community at the Ganges river estuary in India. *Oecologia*, **87**(4) : 560-564.

Sengupta, A. & S. Chaudhuri (1991). Ecology of heterotrophic dinitrogen fixation in the rhizosphere of mangrove plant community at the Ganges river estuary in India. *Oecologia*, **87**(4) : 560-564.

Sengupta, A. & S. Choudhuri (1990). Halotolerant *Rhizobium* strains from mangrove swamps of the Ganges River Delta. *Indian J. Microbiol.*, **30**(4) : 483-484.

Sengupta, S. (1966). Geological and Geophysical studies in western part of Bengal basin India. *Bull. Amer. Assoc. of Petrologeologists.* **5** : 1001-1017.

Senthilnathan, S. & T. Balasubramanian (1998). Heavy metal concentration in oyster *Crassostrea madrasensis* (Bivalvia / Anosomyaria) from the Uppanar, Vellar and Kaduviar estuaries of southeast coast of India. *Indian J.mar.Sci.*, **27**(2) : 211-216.

Shafi, M. (1982). Artificial regeneration of keora in Bangladesh. In: *Proceedings of the Bangladesh Second National Conference on Forestry*, Dhaka. 136-139 pp.

Shahidullah, M. & N. A. Siddiqi (1994). Notes on the feasibility of reforestation in the Chakaria Sundarbans mangroves. *Bangladesh Journal of Forest Science* **23** (1) : 65-66.

Shailaja, M.S. and A.Pant (1986). Relationship between particulate and extracellular carbon compounds of phytoplankton photosynthesis in a tropical estuary. *Mar..Ecol Prog.Ser.*, **32**(2-3) : 161-167.

Shanmukhappa, H. & K. Neelankantan (1989). Concentration of humic acids in mangrove habitat of Karwar, west coast of India. *Indian J. Mar. Sci.*, **18**(4) : 284-285.

Sharma, M. & H. S. Garg (1996). Irodoid glycosides from *Avicennia officinalis. Indian J.Chem.B Org.Med.*, **35**(5) : 459-462.

Shetty, H. P. C., S. B. Saha & B. B. Ghosh (1961). Observations on the distribution and fluctuations of the plankton in the Hooghly - Matla estuarine systems with notes on their relation to commercial fish landings. *Indian J. Fish.*, **8**(2) : 326 - 363.

Shetye, S.R. and A.D. Gouveia (1992). On the role of geometry of cross section in generating flood dominance in shallow estuaries. *Estuar. Coast.Shelf Sci.,* 35(2) : 113-126.

Shibu, S., S. D. Ritakumari & N. B. Nair (1990). Environmental inventory and the distribution of inorganic nutrients in a tropical estuary of the south west coast of India. *J.Indian fish.Assoc.,* **20**: 59-68.

Siddiqi, N. A. & J. H. Choudhury (1987). Man eating behaviour of tiger (*Panthera tigris* Linn.) in the Sundarbans - 28 years record analysis. *Tiger Paper* **14**(3) : 26-32.

Siddiqi, N. A. & Husain, K.Z. (1994). The impact of deer on the natural regeneration in the Sundarbans Mangrove Forest. *Bangladesh Journal of Zoology* **22** (2) : 223-234.

Siddiqi, N. A. & M. A. S. Khan (2000). Plantation of *Phoenix paludosa* a mangrove palm in the coastal areas of Bangladesh. *Journal of Asiatic Society of Bangladesh,* Science **26**(2) : 259-264.

Siddiqi, N. A. & M. R. Islam (1988). Studies on fruit size, seed production and viability of seeds of keora (*Sonneratia apetala*). *Bano Biggyan Patrika* **17** : 15-19.

Siddiqi, N. A. & M.A.H. Shahjalal (1997). Feasibility of underplanting in the mangrove plantation along the eastern part of Bangladesh. *Bangladesh Journal of Forest Science* **26**(1) : 76-78.

Siddiqi, N. A. & S. Das (1988). Effect of time and method of storage on survival of uprooted keora (*Sonneratia apetala* Buch. -Ham.) seedlings. *Indian Journal of Forestry* **11**(4) : 326-330.

Siddiqi, N. A. (1986). Impact of forest management practices in Bangladesh on wildlife and the environment. *Tiger Paper* **13**(3) : 29-31.

Siddiqi, N. A. (1986). Preliminary trial of mangrove and mainland species in the Sundarban highlands. *Bano Biggyan Patrika* **15** : 25-30.

Siddiqi, N. A. (1987). Observation on initial spacing in a keora (*Sonneratia apetala*) plantation along the coastal belt of Bangladesh. *Malaysian Forester* **50**(2) : 204-216.

Siddiqi, N. A. (1988). Growth, natural thinning and wood production in a keora (*Sonneratia apetala*) stand. *Bano Biggyan Patrika* **17** : 15-19.

Siddiqi, N. A. (1989). Seedling regeneration in relation to time and degree of salinity in the Sundarbans Mangrove Forest. *Bano Biggyan Patrika* **18** : 9-17.

Siddiqi, N. A. (1994). The importance of mangroves to the people in the coastal areas of Bangladesh. *Proc. 7th Pacific Science Inter Congress,* Mangrove Session (Okinawa, Japan, 1-2 July, 1993), International Society for Mangrove Ecosystems. 5-10pp.

Siddiqi, N. A. (1995). Role of crabs on the natural regeneration of mangroves in the Sundarbans Forest of Bangladesh. *Australian Journal of Ecology* **20** : 340-343.

Siddiqi, N. A. (1995). Site suitability for raising *Nypa fruticans* in the Sundarbans mangroves. *Journal of Tropical Forest Science* **7**(3) : 405-411.

Siddiqi, N. A. (1996). A review on the impact of animals on regeneration in the mangroves with particular reference to Bangladesh. *Tiger Paper* **23**(1) : 3-10.

Siddiqi, N. A. (1996). A review on the regeneration problem of mangroves with particular reference to Bangladesh. *Mangroves Newsletter* (Japan) No. 16. 2-8pp.

Siddiqi, N. A. (1998). Enrichment planting in the mangrove of Sundarbans - a review. *Bangladesh Journal of Forest Science* **27**(1) : 103-113.

Siddiqi, N. A. (1999). Effect of abiotic and biotic factors on the seedling recruitment of *Heritiera fomes* Buch.-Ham. In the mangrove of Sundarbans, Bangladesh. *Journal of Trcpical Forest Science* **11**(4) : 690-703.

Siddiqi, N. A. (1999). Raising plantations of nipa palm on new accretions along the coastline of Bangladesh. *Journal of Non-Timber forest Products* (1/2) : 52-56.

Siddiqi, N.A. (2001). Mangrove Forestry in Bangladesh. Institute of Forestry & Environmental Science, University of Chittagong, Bangladesh. 201 pp.

Siddiqi, N. A., A. K. F. Hoque & M. S. Alam (1993). The performance of some non mangrove species in the coastal areas of Bangladesh. *Bangladesh Journal. of Forest. Science* **22** : 71-72.

Siddiqi, N. A., M. J. Alam & M.A. Habib (1995). Natural regeneration of mangroves in the coastal areas of Bangladesh with particular reference to Noakhali. *Bangladesh Journal of Forest Science* **24**: 62-99.

Siddiqi, N. A., M. R. Islam & M. A. S. Khan (1989). Effect of salinity on germination success in keora (*Sonneratia apetala* Buch Ham) seeds. *Bano Biggyan Patrika* **18** : 57-62.

Siddiqi, N. A., M. R. Islam, & M. Shahidullah (1993). Mangrove Nurseries in Bangladesh. Mangrove Ecosystems Occasional Papers. International Society for Mangrove Ecosystems, Japan. No.1 : 14pp.

Siddiqi, N. A., M. Shahidullah & A.K.F. Hoque (1994). Present status of the Chakaria Sundarbans the oldest mangrove forest of the subcontinent. *Bangladesh Journal of Forest Science* **23**(1) : 26-34.

Siddiqi, N. A., M. Shahidullah & M. A. H. Shahjalal (1991). Studies on the size viability and germination of seeds of Sundri (*Heritiera fomes* Buch. Ham.) *Indian J. Forestry*, **14** (2) : 119-124.

Siddiqi, N.A., Shahidullah, M. & Shahjalal, M.A.H. (1991). Studies on seed, germination success and raised seedlings of *Nypa fructicans. Indian Forester* **111**(7) : 553-559.

Siddiqi, N.A., Shahidullah, M. & Shahjalal, M.A.H. 1991. Studies on the size, viability and germination of seeds of sundri (*Heritiera fomes* Buch. -Ham.). *Indian Journal of Forestry* **14**(2) : 119-124.

Sidhu, S. K. (1963). Studies on the mangroves of India. I. East Godavari Region. *Indian Forester*, **89**: 336-351.

Sidhu, S. S. (1963). Studies on the mangroves of India. I. East Godavari Region, Reprinted, *Indian Forester*, Vol. **89**. 337-51.

Sidhu, S. S. (1963). Studies on the mangroves of India. *Proc. Indian Academy of Sciences*, **33**(8): 129-136.

Sil, H., K.Ray & A.Kaviraj (1995). Nutritive values of leaf and leaf litters of two mangroves on the Sundarbans for prawn and fish culture. *Natl. Acad.Sci.Lett.*, **18**(1-2) : 43-46.

Silas, E. G., M. Rajagopalan & S. S. Dan (1983). Marine turtle conservation and management : A survey of the situation in West Bengal 1981/82 and 1982/83. *Mar. Fish. Infor. Surv.* T 7 E ser., No. **50** : 24-32.

Singh, U. N., V.K., Srivastava & Sengar, R.B.S. (1982). A comparative study of the rate of dry matter production in forest tree seedlings (*Diospyros montana* Roxb. and *Buchanania latifolia* Roxb.) *Indian Journal of Forestry* **5**(1) : 8-14.

Singh, V.P., L.P., Mall, A., Garge and S.M., Pathak (1990). Human impact assessment on mangrove forests of Andaman Islands. *Indian Forester* **116** (2) : 131-139.

Sivabalan, A., U.G.Naik, N.Kusuma & B. Neelakantan (1991). Mangrove Genetic Resource Centres, Project Formulation Workshop, Madras, India, 27-33.

Smith, W. B. (1859). The Gangetic delta. *Calcutta Review*, **32** (65) Stobart, Tom, 1975. Tiger trail. London, Davis-Poynter, 199p. maps. 21.5 cm.

Smyth, Ralph, (1857). Statistical and geographical report of the 24 Parganas district. Calcutta, John Gray, ii, 120 p. map.

Snedekar, S. C. & J. G. Snedekar (eds.) (1984). *The mangrove ecosystem. Research Methods.* Paris - France UNESCO. 265 pp. 8 volumes.

Soegiarto, A. (2000). Research and conservation of mangrove ecosystem in Indonesia. The United Nations University, Tokyo. 77-86 pp.

Southwell, T. (1920). Notes on estuarine fisheries in the Sundarbans. *Bull. Dep. Fish. Bengal.* **15** :1.

Srinivasan, K. S. (1946). Ecology and seasonal succession of the marine algae at Mahabalipuram (Seven Pagodas) near Madras. *J. Indian bot. Soc., Iyengar Commem. vol.* : 267 - 78.

Stebbing, T. R. R. (1907). The fauna of brackish ponds at Port Canning, Lower Bengal. Part. V. Definition of a new genus of Amphipoda and description of the typical species. *Rec. Indian Mus.*, **1** : 159-162.

Stephenson, J. (1908). The fauna of brackish ponds at Port Canning, Lower Bengal, part VIII. Preliminary description of an Oligochaete worm of uncertain position. *Rec. Indian Mus.*, **2** : 39-42.

Stephenson, J. (1910). Matla bengalensis : a correction. *Rec. Indian Mus.*, **5** : 82.

Subramanian, A.N. & V.K.Venugopalan (1983). Phosphorus and iron distribution in two mangrove species in relation to environment. *Mahasagar*, **16**(2) : 183-191.

Subramanyam, C., K.B.Rao, C.V. Rao & B.V. Rao (1992). Chemical examination of the mangrove species *Suaeda maritima* and *S. monaica. Acta Ciencia Indica.*, **18** (1) : 7-8.

Subramanyam, K. (1977). History of botanical exploration in India. *Bull. Bot. Surv. India* **19**(1 - 4) : 1 -4.

Sumitra, Vijayaraghavan, V.Ramadhas, L.Krishna Kumari & J.P.Royan (1980). Biochemical changes and energy content of the mangrove, *Rhizophora mucronata* leaves during decomposition. *Indian J.mar.Sci.*, **9**(2) : 120-123

Sumitra, Vijayaraghavan & Sayeeda Wafar. Further studies in using mangroves foliage as a prawn feed. *Mahasagar Bull.Nat.Inst.Oceangr.*, **16**(3) : 309-316.

Sun, Z. X. (1980). Observation of the regeneration plants from somatic tissue of hybrid rice. *Acta Phytophysoil, Sin.* **6**: 243-249.

Sundararaj, V. (1978). Suitability of a mangrove biotope for brackishwater aquaculture. *Seafood Export J.*, **10**(12) : 23-27.

Tapaswi, M. P. (1982). Estuarine research in India. A bibliographic study. Doss-2. *Nat. Inst. Oceanogr., Goa, India.*

Tapaswi, M. P. (1983). Bibliography on Bay of Bengal. Doss-7 *Nat. India. Inst. of Oceanogr., Goa, India.*

Tattersall, W. M. (1914). Further records of Indian Brackishwater Mysidae with description of a new genus and species. *Rec. Indian mus.*, **10** : 75-80.

Teas, H. J. (1985). Nutrient element cycling in *Rhizophora* Mangrove ecosystem. In: The Mangroves: *Proc. Nat. Symp. Util. Cons. Mangroves*. pp. 7 - 15.

Thirumalai Raj, K. (1959). Mangrove Forests of Thanjavore Division, *Proc. Mangr. Symposium*, Faridabad, India, pp. 18-19.

Thorpe, C. F. (1905). The submarine canyons of the Ganges and Indus. *Geol. J.*, **26** : 568-669.

Thotharthri, K. (1960). Botanical exploration in Car Nicobar and Nancouri islands, *Bull. bot. Surv. India*, **2** :341 - 346.

Thotharthri, K. (1960). Studies on the Flora of Andaman Islands, *Bull bot. Surv. India*, **2** : 357 - 373.

Thotharthri, K. (1962). Contribution to the Flora of the Andaman & Nicobar Islands, *Bull bot. Surv. India*, **4** : 281 - 286.

Thotharthri, K. (1981). Studies on the mangroves of Peninsular India *vs.* the Andaman & Nicobar Islands. *Bull bot. Surv. India*, **23** : 151 - 154.

Tissot, C. (1987). Recent evolution of mangrove vegetation in the Cauvery Delta : A palynological study. *J.mar.biol.Assoc.India*, **29** : 16-22.

Tomlinson, P.B. (1986). *The Botany of Mangroves*. Cambridge University Press, Cambridge, London. New York, pp. 1-414.

Tuan, L. D. (1998). Sustainable management of the mangrove ecosystem in Can Gio, Ho Chi Minh City. Proceedings of the CRES/Mac Athur Foundation of coastal biodiversity in Vietnam Halong City (24-25 Dec. 1997). pp. 27-39.

Untawale, A. G. & A. H. Parulekar (1976). Some observation on the ecology of an estuarine mangrove of Goa, *Mahasagar Bull. Nat. Inst. Oceanography*, **9** (1 & 2) : 57 - 62.

Untawale, A. G. & S. Wafar (1982). Annotated Bibliography of Indian Mangroves. *National Institute of Oceanography,* ***Goa, India.***

Untawale, A. G. & S. Wafar (1986). Distribution of Mangroves along the Gulf of Kuchchh. **In** *Marine Archaeology of Indian Ocean Countries*, pp. 135 - 138. (1985).

Untawale, A. G. & T. G. Jagtap (1991). Floristic composition of the Deltaic Regions of India *Memoirs Geol. Sci. India*, **22** : 243 -263.

Untawale, A. G. & T. J. Jagtap (1985). Marine macrophytes of Minicoy (Lakshadwip) Coral atoll of the Arabian Sea, *Aqua. Bot.* **19** : 97 - 103.

Untawale, A. G. (1978). Country Report on India. *Proc. Reg. Seminar on Human Uses of the Mangr. Environ. and Management Implication UNESCO*,Dacca, Bangladesh.

Untawale, A. G. (1980). Present status of mangroves along the West coast of India. *Proc. Asian Symp. Mangrove Env. Res. Management*. University of Malaya, Kualalumpur, pp.1-26.

Untawale, A. G. (1985). India (Mangrove ecosystem). *Mangroves of Asia and The Pacific : Status and Management, Technical Report of the UNDP/UNESCO Research and Training Pilot Prog. on Mang.Ecosyst. In Asia and the Pacific*, Ras 79/ 002:51 -87.

Untawale, A. G. (1985). Mangroves in India : Present Status and multiple use practices, *UNDP/UNESCO Regional Mangrove Project*, 67 p.

Untawale, A. G. (1985). Status of Mangrove Research in India, The Mangroves : *Proc. Nat. Symp. Biol. Utl. Cons. Mangrove* (Ed. L. J. Bhosale), Kolahpur, pp . 127 - 134.

Untawale, A. G. (1986). How to Grow Mangroves, October, 1986, pp. 1-9.

Untawale, A. G. (1987). Conservation of Indian Mangroves - A National Perspectives . In S. Z. Qasim (ed.). *Contribution in Marine Science*, pp. 85 - 104.

Untawale, A.G., *et al.* (1982). Application of remote sensing techniques to study the distribution of mangroves along the estuaries of Goa. In. Ecol. & Management. *Proc. 1 st Inter. Wetlands Conference*, New Delhi, pp. 51 - 57.

Untawale, A. G., N. B. Bhosale, V. K. Dhargalkar, S. G. P. Matondkar & S. Bukhari (1977). Biochemical changes in Mangrove foilage during growth and decomposition. *Indian J. Mar. Sci.* **6** : 104 - 106.

Untawale, A. G., N. B. Bhosale, V. K. Dhargalkar, S. G. P. Matondkar & S. Bukhari (1978). Seasonal variation in major metabolites of mangrove foliage, *Mahasagar*, Vol. **11**, No. 1 & 2 : 105-110.

Untawale, A. G., S. N. Dwivedi & S. Y. S. Singbal (1973). Ecology of Mangrove in Mandovi and Zuari Estuaries and the inter-connecting Cumbarjua Canal of Goa. *Indian J. Marine Sci.* **2** : 47 - 53.

Untawale, A. G., S. Wafar & T. G. Jagtap (1986). Status of mangroves along the countries bordering the Arabian Sea. **In** B.N. Desai (ed.) *Oceanography of the Indian Ocean* pp. 239 - 245.(1985).

Untawale, A. G., V.K.Dhargalkar & S.A. Bukhari (1976). Bibliography on Indian Mangroves. National Institute of Oceanography (C.S.I.R.), 1 fasc. 38 p. Dona Paula, Goa, India.

Untawale, A.G. (1985). Status of mangrove research in India. *The mangroves* : Proc.Nat.Symp.Biol.till cons. Mangroves, pp. 127-134.

Untawale, A.G. (1998). Management of Mangroves for Energy Needs. In : *An Anthology of Indian Mangroves.* Envis Centre, CAS in Marine Biology, Annamalai University. pp. 52-56.

Untawale, A.G., Jagtap, T.G. & Wafar, S. (1989). Status report on mangrove research in Goa. In Wetlands, Mangroves and Biosphere Reserves, pp. 155-183. Proceedings Indo-US workshop, New Delhi.

Venkatesan, T. (1966). The Mangroves of Madras State. *Indian For.* **92** : 27 - 34.

Venkateshwarlu, V. (1944). The estuarine flora of the Godavary, *J. Bombay Nat. Hist. Soc.* **44** : 431 - 433.

Venkateswarlu, J. & R.S.P. Rao (1964). The wood anatomy and taxonomic position of Sonneratiaceae. *Current Science,* **33** : 6-9.

Vijayaraghavan, S. & S. Wafar (1983). Further studies in using mangrove foliage as a prawn feed. *Mahasagar,* **16**(3) : 309-316.

Voight, J. O. (1845). Hortus Suberbanus Calcuttensis, Calcutta, India.

Wafar, S. & A. G. Untawale (1991). Role of Mangrove Swamps in Brackishwater Fish Culture. **In** *: Aquaculture Productivity* (eds.) V.R.P. Sinha & H.C. Srivastava, Oxford & IBM Publishing Co. Pvt. Ltd. pp. 235 - 239.

Wafar, S. (1985). Observation of the foliage production in some mangrove sp. of Goa. *Marine Plants, Papers presented at the All India Symposium on Marine plants, their biology, chemistry and utilization,* Donapaula, Goa, Oct. 30-Nov. 1,1983, V. Krishnamurthy & A. G. Untawale (eds.) 1985:245 - 252.

Wafar, S., A. G.Untawale & M.Wafar (1997). Litter fall and energy flux in a mangrove ecosystem. *Estuar. Coast.Shelf Sci.,* **44**(1) : 111-124.

Wagale, B. G. (1982). Geomorphology of the Goa Coast. *Proc. Indian Acad. Sci.* (Earth, Planet Sci.) **91** : 105 - 117.

Waheed Khan, M. A. (1959). Ecological studies on the mangrove forests in India. *Proc. Mangr. Symp., Calcutta,* pp. 97-109.

Walsh, G.E. (1974). Mangrove : A review. Pp. 51-174 In : Reimold, R. and Queen, W. (eds.). Ecology of Halophytes. Academic Press. New York.

Walter, H. (1961). The adaptations of plant Saline Soils. Arid Zones, Proc. Teheran Symp. 14: 129-134.

Ward Kingdon, F. (1943). A note of the journey from Chittagong to N. Demagiri, Chittagong Hills Gard. *Chron.* 176 - 177.

Watson, J.G. (1928). Mangrove Forests of the Malay Peninsula. Malay Forest Records No. 6. Singapore Fraser & Neave. 275 pp.

Zamski, E. (1979). The mode of secondary growth and the three dimensional structure of the phloem in *Avicennia. Bot. Coat.* **140** : 67-76.

Zingde, M. D., P. V. Narvekar, P. Sharma & M. M.Sabnis (1986). Environmental studies of the Ambika and associated river estuaries. *Mar.Pollut.Bull.,* **17**(6) : 267-274.

10

INDEX TO THE MANGROVES AND MANGROVE ECOSYSTEM

INDEX TO THE SCIENTIFIC NAMES, LOCAL NAMES, COMMON NAMES OF THE FLORA AND FAUNA AND OTHER SCIENTIFIC NAMES/COMMON NAMES, DEFINITIONS AND TERMS

(Number/numbers against each name or term indicate page numbers where mentioned in the book)

(Scientific names of the flora and fauna are in italics)

Acanthaceae 12, 44, 121
Acanthus ebracteatus Vahl. 12, 121, 124
Acanthus ilicifolius L. 12, 44, 121, 122
Acanthus volubilis Wall. 12, 44, 121, 123
Acrostichum L. 170
Acrostichum aureum **L.** 15, 170
Acrostichum danaeifoltum Langsd & Fisch. 15, 170
Acrostichum speciosum Willd. 15, 170
Acrstichum aureum L. 170
Aegialitidaceae 12, 44, 113
Aegialitis annulata R. Brown 12, 112
Aegialitis R. Br. 113
Aegialitis rotundifolia Roxb. 12, 44, 113, 114
Aegiceraceae 112
Aegiceras floridum Roem. & Schult. 12, 111
Aegiceras Gaertn 111
Aegiceras corniculatum (L.) Blanco 12, 44, 111, 112
Aeluropus lagopoides (L.) Trin. Ex. Thw. 13, 138
Aerial root systems 6
Aerial roots 3
Aerial stilt roots 3
Aerva lanata (L.) Juss. Ex Schult. 14
Aglaia cucullata (Roxb.) Pellegrin 12, 44, 104, 108
Aglaia Lour. 104
Aglaieae 104
Agriculture not much developed 176
Aizoaceae 14, 151
Ajmalmari forest block outside STR 34
Allophyllus cobbe (L.) Bl. 14
Amaryllidaceae 14, 45, 148
Ambarvel 127
Amoora cucullata Roxb. 108
Amphibian habitat dwelling fauna 46
Amphileena latifolia (Mill.) A. Gentry 13
Amur 108
Anacardiaceae 15
Andaman & Nicobar Island of the Bay of Bengal 17
Andaman & Nicobar Islands 7, 18, 19
Anemopaegma chrysoleucuma (IIBK) Sand 13
Angola 10
Animal species reported from Sundarbans 43
Anthropogenic pressure 17
Antomul 130
Apis dorsata 48
Apocynaceae 13
Aquatic food chain 65
Araceae 14, 45, 150
Arbesi block 34
Ardisia w. 111
Ardisia elliptica Thunder 12
Ardisia Sw. 111
Areas under Rivers, Canals, Roads, Railways 30
Arecaceae 12, 44, 100
Asclepiadaceae 13, 44, 126
Asteraceae 15
Atalantia Corr 137
Atalantia correa M. Roem. 13, 45, 137
Atolls Mangroves of Arabian Sea 17
Australia 10
Avicenna schaueriana Stapf. & Leechman ex Mold 89
Avicennia africana Palisot de Beauv. 11, 89
Avicennia bicolor Standley 11, 89

Avicennia intermedia Griff. 92
Avicennia tomentosa Wild 91
Avicennia alba Blume 11, 44, 88, 89, 91
Avicennia balanophora Stapf. & Mold. 11, 89
Avicennia eucalyptifolia (Zipp. Ex Miq.) Moldenke 89
Avicennia germinanas (L.) Stearn 11, 89
Avicennia lanata Ridley 11, 89
Avicennia maraina (Forsk.) var. *alba* Bakh. 91
Avicennia marina (Forsk.) Vierh. 1, 44, 88, 89, 92
Avicennia marina var. *acutissima* Stapf. 11, 89
Avicennia mindanaense Elmer 92
Avicennia officinalis L. 11, 44, 88, 89, 90
Avicenniaceae 11, 44, 88

Back mangroves 9
Bagmara block 34
Bakul Kankra 82, 83
Balasore coast 26
Ban Jhampi 142
Ban Jnui 142
Ban Lebu 137
Bangladesh 10
Baole lata 128
Bara Manda 167
Baratang, Andaman 24
Barringtonia acutanguls (L.) Gaertn. 14, 45, 152
Barringtonia racemosa Roxb. 14, 45, 152
Barringtoniaceae 14, 45, 152
Batidaceae 15
Batis maritime L. 15
Batraj 142
Benthic 66
Bhara 79
Bhola 134, 135
Bignoniaceae 13, 45, 131
Bilati Bakul 146
Bilati Mehendi 147
Black mangrove 99
Blinding tree 110
Blocks in South 24-Parganas 30, 31
Blocks in North 24 - Parganas 30, 31
Bones/jaw of Rhinoceros 54
Boodleopsis sundarbanensis 9
Boraginaceae 14, 45, 153
Brackishwater mixed *Heritiera* sp. Forest 40
Brazil 10
Broom like stilt roots 4, 5
Brownlowia argentata Kurz. 13, 136
Brownlowia tersa (L.) Kosterm. 13, 44, 136
Brownlowia lanceolata (L.) Kosterm 13, 44, 136
Bruguiera Lamk 73, 74
Bruguiera conjugata Merr. 80
Bruguiera cylindrica (L.) Blume 11, 44, 82
Bruguiera decandra Griff. 84
Bruguiera eriopetala Wight & Arnold 81
Bruguiera exarislata Ding Hou 11
Bruguiera gymnorhiza (L.) Lamk. 11, 44, 80, 89
Bruguiera hainesii Rogers 11
Bruguiera madagascariensis DC. 98
Bruguiera malabarica F. Vill. 81
Bruguiera parviflora W. & A. ex Griff. 11, 44, 83
Bruguiera rheedii Blume 80
Bruguiera sexangula (Lour.) Poir 11, 44, 81
Buffer areas within STR 30
Buffer Zone 34
Bulbosstylis spp. 165
Buttress root 4, 5

Cactus 168
Caesalpinia bonduc (L.) Roxb. 14, 45, 159, 160
Caesalpinia crista L. 14, 45, 159
Caesalpinia L. 154, 159
Caesalpiniaceae 14, 45, 154, 159
Calamus erionaceus (Recc.) Dransfield 12, 103
Calamus L. 100
Calophyllum inophyllum L. 14, 163
Camaroon 10
Candle mangrove 86
Cannon Ball Mangrove 106
Canopy structures 5
Carallis brachiata (Lour.) Merr. 73
Carex spp. 165
Cassine viburnifolia (Juss.) Hou 15
Cassytha filiformis L. 15
Catenella sp. 9
Cauveri 17
Cauvery delta, Pichavaram Muthupet & Chhatram 17
Caw Phal 129
Cedar Mangrove 107
Celastraceae 15
Ceperus L. 165
Cerbera L. 125

Cerbera floribunda Schum. 13, 125
Cerbera lactaria Hamilt. Ex DC. 125
Cerbera manghas L. 13, 125
Cerbera odollam Gaertner 13, 125
Ceriops Arnold 73, 74, 75
Ceriops boiviniaa Tulasne 85
Ceriops candalleana Arn. 85
Ceriops decandra (Griff.) Ding Hou 11, 44, 84
Ceriops roxburghiana Arn. 84
Ceriops tagal (Perr.) Robinsonl 11, 44, 85
Ceriops timoriensis Domin 85
Cervis axis 138
Chak keora 97
Chamta block 34
Chandkhali block 34
Chandra Dweep 29
Char pata jal 65
Chenchco 165
Chenopodiaceae 14, 161
Chhagal Knuri 164
Chhotohardi block 34
Chulia kanta 155
Chulkati forest block outside STR 34
Clearing of mangrove area 16
Clerodendron inerme (L.) Gaertn. 13, 45, 141
Clerodendron maerocarpun Kurz. 141, 142
Clerodendron neriifolium var. *macrocarpa* 13, 45, 141
Club mangrove 114
Clusiaceae 14, 163
Coast cotton tree 135
Cocos nucifera 102
Columbia 10
Combretaceae 11, 98
Common and higher group of fauna 55
Compartmental model of the food chains 68
Conocarpus erectus L. 12
Convolvulaceae 15, 164
Convolvulus pes-caprae L. 164
Core Area 34
Core area within STR 30
Crabs 66
Crinum defixum Ker - Gawler 14, 148, 149
Crinum L. 148
Crocodile 48, 66
Crocodiles porosus 48
Crustacean 66
Cryptocoryne ciliate (Roxb.) Fischy 14, 150
Crypto-viviparous germinated seedlings 8
Crypto-viviparous germinated seeds 1, 8
Cuscuta reflexa Roxb. 15
Cuscutaceae 15
Cynometra L. 154
Cynometra ramiflora L. 14, 44, 154
Cyperaceae 15, 165
Cyperus rotundus L. 165
Cyperus spp. 165

Dabur 125
Dalbergia Linn. f. 154
Dalbergia spinosa Roxb. 14, 44, 155
Deer 66
Deforested mangrove swamp 40
Dendrophthoe falcate (Linn.f.) Ettingsh 15, 166, 167
Dendrophthoe mart 166
Dense mangrove forest 40
Dense mangrove forest floor dwelling fauna 46
Derris indica (Lamk.) Bennet 14, 44, 155, 158
Derris scandens Benth 14, 44, 155, 157
Derris Lour 154, 155
Derris trifoliate Lour 14, 44, 155, 156
Dhani Ghash 139
Dhulibhasani forest block outside STR 34
Dhundul 106
Distinguished Indian Mangrove Researchers 25
Dodonaea viscosa (L.) Jack. 14, 147
Dolichandrone spathacea (Linn.f.) Schum. 13, 131
Dominant Indian mangrove families 71, 72
Duabunga Ilam 93
Dudhi lata 127

East coast 17
East coast of Indian territory 17
Ebenaceae 15
Ecology & Biodiversity of Indian Mangroves 18
Ecuador 10
Effect of developmental activities 176
Eggs of marine turtle 54
Eleocharis spp. 165
Estuarine and river water dwelling fauna 46
Estuarine or deltaic mangroves 17
Euphorbiaceae 12, 109
Euterpe spp. 12

Excoecaria agallosha L. 12, 44, 109, 110, 166
Excoecaria dallachyana (Ball.) Benth. 12, 110
Excoecaria indica (Willd.) Muell. 12, 110
Excoecaria L. 109
Exploitation of mangrove wood 16
Extensive stilt roots of *R. micrometer* 34
Extinct Mammalian species in Indian Sundarbans 64

Fabaceae 14, 154
Fani Mansa 167, 168
Fimbristylis spp. 165
Finlaysonia obovata Wall. 13, 44, 165
Finlaysonia Wall. 126
Fire fly mangroves 93, 95
Fish cradle 42
Fisherman along with their poor fish catch 68
Fishing cats 47
Fizi 10
Flacourtiaceae 15
Flora of the Indian mangals 70
Forest land within STR 34
Freshwater mangrove 152
Fringing waterways 41
Fuirena sp. 165

Gabon 10
Gada bani 151
Galedupa indica Lamk. 158
Gambia 10
Ganga and the Brahmaputra 29
Ganga delta/Sundarbans in W.B. 17, 117
Ganga River 117
Ganga-Brahmaputra delta 42
Garjan 78, 79
Geoa or Gncoa 110
Geographical areas 30
Gire Sak 161, 162
Glochidion J.R. & G. Forst 109
Glochidion littorale Blume 12
Gluta velutina Blume 15
Goa 17
Goa coast/Mandavi, Zuary, etc. 17
Goasaba block 34
Godavari 17
Gol-pata 101
Gona block 34
Goria 86
Gorshingiah 131
Gramincae 138
Green algae 9
Green ridley 47
Grey Mangrove 92
Grey mangrove family 88
Gujarat coast/Gulf of Cambay & Kuchchh 17
Gujarat State 17
Guttiferae 163

H. fomes is now rare in Indian Sundarbans 40
Halophytes 1
Harinbari block 34
Heliotropium curassavicum L. 14, 153
Heliotropium L. 153
Helophilous 1
Hental or bokra 103
Heritiera Aiton 115
Heritiera fomes Buch.-Ham. 12, 44, 115, 116
Heritiera globosa Kostermanas 12
Heritiera kanikensis Banerjee & Majumder 115
Heritiera littoralis Dryand 12, 115, 117
Herobhanga forest block outside STR 34
Hibiscus L. 132
Hibiscus tiliaceus L. 13, 132, 134
Hibiscus tortuosus Roxb. 13, 132, 135
Higher group of fauna 51
Highest Bio-diversity 43
Highest productive zone 1
Hijal 152
Hippomane L. 109
Hippomane mancinnella L. 12, 109
Holiday Island Wildlife Sanctuary 30
Holly mangrove 122, 124
Hong Kong 10
Horgoja 122, 124
Horkoch Kanta 122
Hoya parasitica Wall. 13, 44, 130
Hoya R. Brown 126
Hudo fern 170
Hugli 17
Hugli-Padma-Meghna 42
HWNTL 38
Hydrophyllax Linn.f. 118
Hydrophyllax maritime Linn.f. 44, 118, 120

Important roles played by tiger 47
India 10
Indian Beach 158
Indian mangals 17
Indian Sundarbans 29
Indian Tulip tree 132
Indonesia 10, 42
Intertidal mangrove plants 66
Intsia bijuga (Colebr.) Kunt. 14
Inundated with all the high tides 39
Inundation - Above the tidal reach delta 39
Inundation - Inundation during abnormal tidal rise 39
Inundation Sub-tidal shore facing or river flats 39
Inundation classes 39
Inundation classes - Sea board or Estuarine river 39
Inundation - Supra-tidal, inundated during spring tide 39
Ipomoea pes-caprae (L.) R. Brown 15, 164
Ipomoea tuba (Sch.) G. Don 164
Irian java 42
Island Mangroves 17
Iviry coast 10

Jat Baine 90
Jat Ban 90
Jat ban, jat bani or jat baine 90
Jat Bani 90
Jat Goran 85
Java 42
Jele Goran 85
Jele Goran or Jhampti Goran 85
Jhila block 34

Kakra 80
Kal Baine
Kal Bani
Kalimantan 42
Kalo Ban
Kaloban, kal bani or kal baine 91
Kampuches 10
Kandela candle (L.) Drucee 11, 73, 86
Kandela Wight & Arnold 73
Kandela Wight & Arnold 73, 86
Kandelia rheedii W. & A. 86
Kankinada Bay of Andhra Pradesh 18, 19
Kankra 80
Karanja 158
Karnataka 17
Karnataka coast/Coondapur, etc. 17
Kath Champa 163
Kenya 10
Keora 95
Kerala coast/Cochin estuary 17
Kerali 150
Keya 169
Keystone species 47
Khalsi 112
Khamu 78
Khatuajhuri block 34
Konkan Railway Bridge construction at Zuari 27
Kripa or Kripal 99
Krishna 17

Lakshadweep Atols 1, 228
Lagancularis Gaertn.f. 98
Laguncularia racemosa (L.) Gaertn.f. 11, 98
Lakshadweep 17
Lal Jhau 143
Large Leafed Orange Mangrove 80
Large scale deforestation of forest land 188
Large scale exploitation of shrimp seed 69
Large scale netting out of shrimp seeds vs. damage 188
Lata Gurbegun 145
Lata Ilorgoja 123
Lata Sundari 136
Latmi 108
Lauraceae 15
Lecythidaceae 152
Leguminosac 154
Lesser cats 66
Liberia 10
Litter traps for bio-mass estimation 66
Long fruited stilted mangrove 78
Looking Glass mangrove 115, 116, 117
Loranthaceae 15, 166
Loranthus cochinchinesis Lour. 166
Lothian forest block outside STR 34, 35
Lothian Island Wildlife Sanctuary 30
Lumnilzera littorea (Jack.) Voigt 11, 98
Lumnitzera racemosa Willd. 11, 44, 98, 99
Lumnitzera Wild. 98
Lumnizera rosea (Gaud.) Presl. 11, 98
Lythraceae 93

Macrosolen (Bl.) Reichenb. 166
Macrosolen cochinchinensis (Lour.) van Teig. 15, 166, 167
Mahanadi delta 17, 19
Mahanadi delta/Bitarkanika of Orissa coast 17
Maharashtra/Ratnagiri 17
Maharastra 17
Major elements of mangroves 9
Major problems in Indian Sundarbans 187
Malagasy Republic 10
Malvaceae 13, 132
Manda 166
Mandovi 17
Mangal 9
Mangrove - Rhizophoraceae 73
Mangrove apple 96, 97
Mangrove Areas in New World 10
Mangrove Areas in Old World 10
Mangrove associates 9, 109
Mangrove Fern 170
Mangrove Lili 149
Mangrove lime 137
Mangrove litter collection 66
Mangrove palm 101
Mangrove palm or water coconut 101
Mangrove thickets at Sipighat, Port Blair 23
Mangrove wood 16
Mangrove wood at Ghushighata, Sundarbans 28
Mangrove zones and succession 38
Mangroves 9
Mangroves of Bitarkanika of Mahanadi delta 18, 19, 26
Mangroves of Orissa coast 19
Mangroves of Ratnagiri 18, 22
Mangroves of Andaman Islands, Port Blair 24
Mangroves of Bitarkanika, Orissa 26
Mangroves of Cochin estuary 18, 21
Mangroves of Coondapur & Malpe Bay 17, 18, 21
Mangroves of Coral reef 18, 22
Mangroves of Coringa 18, 19
Mangroves of Gulf of Cambay 22
Mangroves of Mandovi & Zuary 18, 21, 27
Mangroves of Muthupet, Pichavaram & Chhatram area of the Cauveri Delta 18, 19
Mangroves of Port Blari 23
Mangroves of the Elephanta cave, Bombay 25
Mangroves of the Gulf of Kuchchh 18, 22
Mangroves of the Mahim Creek, Bombay 25
Mangroves on marshy back water 17
Mangroves on the Gulfs 17
Mangroves support coastal and estuarine fisheries 65
Mangroves Swamp of Indian Sundarbans 18
Mangue 41
Manicaria sp. 103
Manicaria sp. 12
Manikara hexabdra (Roxb.) Dub. 14, 146
Marshy mangrove tiger land 42
Mat Goran
Mat Goran or Jat Goran 85
Matla black 34
Mauritia sp. 103
Mayadwip block 34
Mayanmar 10
Mean high water neap 67
Mean high water spring 67
Mean low water neap 67
Mean low water spring 67
Mean tidal level 67
Meliaceae 12, 104
Merope angulata 13, 137
Merope Roem. 137
Metropolitan pollutants 42
Mexico 10
Microbs 66
Milky mangrove 110
Mineralization processes 37
Minicoy Islands of Arabian Sea 18, 23
Minor elements of mangroves 9
MLWN 38
Molluscas 66
Monitor 66
Monkey 66
Mozambique 10
MSL 38
Mucuna gigantean (Willd.) DC 14
Myriostachya wightiana (Nees ex Steud) Hb.K. 13
Myrsinaceae 12, 111
Myrsine 111
Myrsine umbellulatum A. DC. 12

Najadaceae 171
Naked mangrove forest floor dwelling fauna 46
Naked river flat/river slope dwelling fauna 46
Nalai 140

Nalor Dharma 140
Nata 160
Natural calamities, viz. cyclones etc. 36
Natural succession of *Porteresia* sp. 70
Netidhopani block 34
Netting out of shrimp and other fish juveniles 69
New Zealand 10
Nigeria 10
Nipa fruticans Thumberg 101
Noa lata 157
Noalata 157
Nona Durba 140
Nona Hatisnur 153
Nona Jhangi 171
Nona Jhau 143
Nona Palang 161
Nut or Fever nut 160
Nypa fruticans (Thunb.) Wurmb. 12, 44, 100, 101, 102
Nypa Steek 100

Olive ridley 47
Oneosperma Bl. 100
Oncosperma tigilarium (Jack.) Ridl. 12, 103
Opuntia dillenii (Ker. Gawler) Haw. 15, 168
Opuntiaceae 15, 168
Ora 96
Oryza coaretata Roxb. 139
Other fauna of the Indian Sundarbans 54
Other Pacific Islands 10

Pacific oceans 42
Pakisthan 10
Palm swamps dominated by *P. paludosa* 40
Palmae 100
Pan lata 156
Panama 10
Panchamukhani bloack 34
Pandanaceae 15, 169
Pandanus odoratissimus Soland ex Park 15, 169
Panthera tigris tigris 47
Paras 133, 135
Pargachha 130
Passur 107
Pavonia rhizophorae Killip 13
Pavonia spicata Cav. 13
Pelliciera rhizophoreae Triana & Planchon 15
Pellicicraceae 15
Penaeus monodon 65
Peninsular Malaysia 10
Pentatropis capensis Wall. 13, 44, 126
Pentatropis W. & A. 126
Periphytic algae 66
Petalema alternifolia Roxb. 98
Peyara Baine 92
Peyara ban or peyara bani or peyara baine 92
Philippines 10
Phoenix L. 100
Phoenix paludosa Roxb. 12, 44, 100, 103
Phoenix reclinata 12, 103
Phragmites kakra (Retz.) Trib. Ex. Steud 13
Phryganocydia phellosperma (Hemsl.) Gentry. 13
Phytoplankters 66
Pirkhali block 34
Plantation of *Avicennia* spp. 70
Plantonic 66
Pneumatophores 1
Pneumatophores of *A. marina* 3
Pneumatothods 1, 4
Poaceae 13
Pongamia glabra Vent. 158
Pongamia pinnata (L.) Pierre 158
Population Growth in Indian Sundarbans 32
Porteresia coactata (Roxb.) Takeoka 13, 138
Pouteria obovata (R.Br.) Bachni 14
Premna corymbosa (Burm.f.) Rottl. & Wild. 13, 45, 141
Present scenario of the Sundarbans mangals 177
Pressure on the natural forest is enormous 176
Primary consumer 66
Producer 66
Protection vs. loss of Bio-diversity 173
Pseudo-viviparous germination mechanism 1
Pterdaceae 13, 45, 170

Radhanagar beach, Ilavelock 24
Ramsar site 33
Raphia sp. 12, 103
Rare, threatened and endangered mangroves 46
Reclaimed Agricultural Lands in Sundarbans 31
Reclaimed Sundarbans Mangroves 31
Renewal of natural wealth and the prospects 176
Rhabdenia biflora (Jack.) Muell-Arg. 125
Rhesus monkey 66

Rhizophora X. lamarekii Montr. 11
Rhizophora apiculata Blume 11, 44, 79
Rhizophora candelaria DC. 79
Rhizophora caryophylloides Burm.f. 82
Rhizophora conjugate Arn. 79
Rhizohpra corniculatum L. 112
Rhizophora cylindrica L. 82
Rhizophora decandra Roxb. 84
Rhizophora gymnorhiza L. 80
Rhizophora L. 73, 74
Rhizophora mangle L. 11, 87
Rhizophora micrometer Lamk. var. *typical* Sch. 78
Rhizophora mucronata Lamk. 11, 44, 78
Rhizophora racemosa Meyer 11, 87
Rhizophora samoensis (Hochr.) Salv. 11, 87
Rhizophora sexangula Lour. 81
Rhizophora stylosa Griff. 11, 73
Rhizophora tagal Perr. 85
Rhizophora X harrisomii Deechm 87
Rhizophora X lamarckii Mont.
Rhizophora X selala (Salv.) Tom 11, 87
Rhizohpraceae 11, 73
River mangrove 112
Riverbank - row of *Rhizophora* sp. 16
Royal Bengal tiger 47
Rubiaceae 12, 118
Ruppia maritima L. 15, 171
Ruppia rosetelle Koch 171
Ruppiaceae 15, 171
Rustia occidentalis (Benth.) Hemsl. 12
Rutaceae 13

Sabah 10
Sajnakhali Wildlife Sanctuary within STR 30
Salicornia brachiata Roxb. 161
Salt accumulation 2
Salt excludion 2
Salt extrusion 2
Sapindaceae 14
Sapium indicum Willd. 110
Sapotaceae 14
Saptamukhi forest block outside STR 34
Sarcolobus carinatus Wall. 13, 44, 126
Sarcolobus globosus Wall. 13, 44, 127
Sarcolobus R. Brown 126
Satellite map of the Sundarbans 33
Sawara 10
Scirpus L. 165
Scirpus littoralis Sehrad. 15, 165
Scirpus spp. 165
Scolopia macrophylla (Wight & Arnold) Clos 15
Seylla serrala 65
Scyphiphora Gaertn.f. 118
Scyphiphora hydrophyllacea Gaertn.f. 12, 44, 118, 119
Sea date 103
Sea-land interface zones 9
Sea-shore or coastal mangroves 17
Secondary consumer 66
Senegal-Gambia 10
Sesuvium protulacastrum L. 151
Shinger 154
Shingri lata 159
Shingri lata nut or fever nut 160
Siltation on river bed cause frequent flood 188
Singapore 10
Small-leaved orange mangrove 82
Solanaceae 14, 145
Solanum L. 145
Solanum surattense Burm.f. 145
Solamam surallensis Burm 14, 145
Solanum trilobatum L. 14, 145
Sonneratia Linn.f. 93
Sonneratia ovata Backer 11
Sonneratia alba J. Smith 11, 93
Sonneratia apetala Buch. - Ham. 11, 44, 93, 94, 95
Sonneratiaceae 93
Sonneretia griffithii Kurz. 11, 44, 93, 94, 96
Sonneretia cassowaries (L.) Engler 11, 44, 93, 94, 97
Sonneretia ovata Backer 93
South Africa 10
South Vietnam 10
Spring high level 67
Spring low level 67
Sri Lanka 10
Sterculiaceae 12, 115
Stilt-roots 2
Suaeda Forsk. 161
Suaeda maritime Dumort. 14, 161
Suaeda monoica Forsk. Ex. Gmel 14, 161
Suaeda nudiflora Roxb. 14, 161
Succession of *P. coactata* & *S. apetala* 38
Sukha Darsan 149

Sulawesi 42
Sultan Champa 163
Sumudra + ban 29
Sumudra 152
Sumatra 42
Sundarbans mangals act as the sink for pollutants 172
Sundar (beautiful) + ban (forest) 29
Sundarbans Biosphere Reserve 35
Sundarbans mangals - the largest single mangroves 172
Sundarbans mangrove reclaimed areas 30
Sundarbans National Park 35
Sundarbans Reserve Mangrove Forest Outside STR 30
Sundarbans - the marshy tiger-land 47
Sundarbans tiger Reserve 30
Sundari 115, 116, 117
Sundri 29
Sundri tree 115
Supra tidal level 67

Tabebuia palustris Hemsl. 13
Tagri Bani 119
Tamaricaceae 14, 143
Tamarix aphylla (L.) Kurze 14, 143
Taxarix dioica Roxb. 14, 45, 143
Tamarix gallica L. 14, 44, 143
Tamarix L. 143
Tamarix troupii Ilole 14, 143
Tanzania 10
Terminalia catappa L. 12
Thailand 10
Thakuran forest block outside STR 34
Thespesia lampus (Cav.) Dalz. & Gibs. 13, 45, 133
Thespesia populnea (L.) Solander ex Corr. 13, 45, 133
Thespesia populneoides (Roxb.) Kostel. 13, 45, 133
Thespesia Solander ex Correa 132
Threatened Birds species in Sundarbans 64
Threatened Mammalian species in Indian Sundarbans 64
Threatened Reptilian species 64
Threatened, endangered and extinct fauna 64
Tidal amplitude/fluctuation 67
Tidal creek & Mangroves at Vieroli, Bombay 25
Tidal creek at Subarnarekha mouth 26
Tidal inundated mangrove forest 16
Tiger 66
Tiger - deer - wild boar interaction 48
Tiger - honey collector interaction 48
Tiger fisherman interaction 48
Tiger human conflict 50
Tiger in Sundarbans Tiger Reserve 52
Tiger population in Indian Sundarbans 49
Tiger straying incidents in Sundarbans 49
Tiger wood collector interaction 48
Tiliaceae 13
Top consumer in land 66
Top consumer in water 66
Tora 114
Total Sundarbans Mangrove Areas 30
Total World Mangrove Zones 10
Transport facilities is not much developed 175
Tropic level in Indian Sundarbans 66
True mangrove species 5
True-mangroves 1
Tuberostylis axillaris Blakel 5
Tuberostylis rhizophorae Steetz. 15
Tylophora R. Brown 126
Tylophora tenuts Bl. 13, 44

U.S.A. 10
Ulva lactuca 9
Umbrella tree 133
Uncontrolled population growth 175
Undivided Sundarbans of Bangladesh & India 29
UNDP, UNESCO, IUCN & WWF 35, 47
Unique Sundarbans mangals 43

Vegetated river flats and river slopes dwelling fauna 46
Venezuela 10
Verbenaceae 13, 141
Viscum L. 166
Viscum monoicum Roxb. 15, 166
Viscum orientale Willd. 15, 166
Viviparous germinated hypocotyls 7
Viviparous germination 1, 73

Wandoor Bay, Andaman 24
Water areas within STR 34
Water coconut 100, 101
West African Mangrove Areas 10
West Coast 17
White mangrove 90, 91
Wild Animals of Sundarbans 53
Wild boar 66

Wild Rice 139
World Heritage Site 33, 42
World's largest single mangrove chunk 29
Wrinkle mangrove 154

Xylocarpeae 104
Xylocarpus granatum Koening 12, 44, 104, 105, 106
Xylocarpus Koening 104
Xylocarpus mekongensis Pierre 12, 44, 104, 107
Xylocarpus moluccensis (Lamk.) Roem. 12, 104

Yellow mallow Tree 134
Yellow Mangrove 84, 85

Zoobenthos 66
Zooplankters 66
Zuari 10, 17